Lake Functioning

This book explores the interconnections of internal phosphorus loading, cyanobacteria, and climate change and their role in determining water quality in freshwater.

It goes on to discuss the sometimes-elusive process of internal phosphorus loading with its chemical and biological roots. Reviewing recent observations on present and future climate change, the book explores its effects on lake functioning. It concludes with the abatement and prevention of cyanobacteria proliferation, including techniques that deal with internal phosphorus loading. Some key topics covered are:

- In-depth exploration of internal phosphorus loading and its quantification in global lakes with diverse morphometry, hydrology, geochemistry, and degree of eutrophication.
- Determination of climate change effects on physical, chemical, and cyanobacteria-related variables in tables based on more than 100 reviewed journal articles.
- Evidence for the enhancing influences of climate change on internal phosphorus load and cyanobacteria, and of internal load on cyanobacteria.

This book will be of interest to limnologists, environmental and engineering professionals, and natural science students. It will also be an interesting read for government agents, lake managers, and others interested in maintaining lake water quality and understanding algal blooms.

Lake Functioning

Internal Phosphorus Loading, Cyanobacteria, and Climate Change

Gertrud K. Nürnberg

CRC Press is an imprint of the
Taylor & Francis Group, an **informa** business
A BALKEMA BOOK

Designed cover image: Gertrud K. Nürnberg

Cover art: The background displays *Aphanizomenon* as dispersed throughout the water column in hyper-eutrophic Mitchell Lake reservoir, South Dakota, June 30, 2001 (Figure 3.1 #3). The graph insert compares August–September internal phosphorus load in Lake Erie, Central Basin, to mean August–October cyanobacteria biomass at a northern shoreline site for 1985–2012 (Figure 5.3B)

First published 2025
by CRC Press/Balkema
4 Park Square, Milton Park, Abingdon, Oxon, OX14 4RN

and by CRC Press/Balkema
2385 NW Executive Center Drive, Suite 320, Boca Raton FL 33431

CRC Press/Balkema is an imprint of the Taylor & Francis Group, an informa business

British Library Cataloguing-in-Publication Data
A catalogue record for this book is available from the British Library

Library of Congress Cataloging-in-Publication Data
Names: Nürnberg, Gertrud K., author.
Title: Lake functioning : internal phosphorus loading, cyanobacteria, and climate change / Gertrud K. Nürnberg.
Description: Abingdon, Oxon ; Boca Raton, FL : CRC Press/Balkema, 2025. | Includes bibliographical references and index.
Identifiers: LCCN 2024019541 (print) | LCCN 2024019542 (ebook) | ISBN 9781032294407 (hbk) | ISBN 9781032294421 (pbk) | ISBN 9781003301592 (ebk)
Subjects: LCSH: Water—Phosphorus content. | Water chemistry. | Biogeochemical cycles. | Cyanobacterial blooms. | Climatic changes. | Lakes.
Classification: LCC TD427.P56 N87 2025 (print) | LCC TD427.P56 (ebook) | DDC 628.1/6841—dc23/eng/20240723
LC record available at https://lccn.loc.gov/2024019541
LC ebook record available at https://lccn.loc.gov/2024019542

ISBN: 978-1-032-29440-7 (hbk)
ISBN: 978-1-032-29442-1 (pbk)
ISBN: 978-1-003-30159-2 (ebk)

DOI: 10.1201/9781003301592

Typeset in Times New Roman
by Apex CoVantage, LLC

I dedicate my analysis to our lakes, worldwide.

I applaud the efforts to bestow specific legal rights (e.g., environmental personhood) onto many natural lakes, rivers, and selected reservoirs.

These legal resolutions exist for New Zealand's Whanganui River, India's Ganges and Yamuna Rivers, Colombia's Atrato River, Quebec Canada's Magpie River, and the Spanish estuary, Mar Menor. The right of nature *to exist, persist, maintain, and regenerate its vital cycles* is set into the Ecuadorian constitution, and the Vilcabamba River successfully defended that right in 2011. There is a similar *law of the Rights of Mother Earth* as a *dynamic living system* in Bolivia.

Such legal actions would encourage society to protect the environment, acknowledge its impact on our well-being, and combat climate change, which augments internal phosphorus loading and cyanobacteria.

Contents

Foreword

Climate change adaptation in lake management is largely absent around the world as we know from the UNEP World Water Quality Alliance Ecosystem Restoration Workstream, which I co-chair with Professor Kenneth Irvine of IHE-Delft. This leaves many communities facing a future of poor water quality, biodiversity loss, human health risks, and harmful algal blooms. The next generation of lake managers must meet this new challenge of managing lakes through an era of multiple stressors, with climate change and nutrient pollution at the fore.

The science points toward a bleak future for lakes due to our current path of unsustainable nutrient use and climate change. Yet, preventative management is hampered by a lack of evidence on the intricate relationships between multiple pressures and their combined effects on ecosystems, including harmful algal blooms. How can we select and design effective lake management measures if we don't know what is causing the problem? The book *Lake Functioning, Internal Phosphorus Loading, Cyanobacteria, and Climate Change* will be an important resource in addressing this problem.

I consider myself a long-term student of Gertrud K. Nürnberg. I have enjoyed many hours of reading Gertrud's papers and applying the models and methods on internal phosphorus loading as I attempt to decipher the inner workings of polluted lakes. I now pass these methods on to my own students and colleagues. Gertrud's meticulous presentation of processes and methods has become a staple in the field of lake restoration, breaking down complex scientific problems to component parts and explaining them clearly to others. Gertrud's commitment to working at the interface of theoretical and applied limnology shines throughout this new book.

The book is structured around a description of evidence from an impressive array of real-life case studies. Process understanding is central, from catchment loading to redox processes in lake bed sediments, and from hydrodynamic processes to biogeochemical interactions and complex ecological responses. Evidence for contemporary and projected climate change forcing of these processes is presented, providing a detailed blueprint for assessing risks of future stressors across lakes. Critically, Gertrud combines her own experiences of managing lakes across the world with findings from the international scientific community to outline measures for lake management in the new era of multiple stressors. The book includes a large number of extremely helpful illustrations that have been carefully and thoughtfully selected by the author. Notable is the commitment to the process of peer review with contributions from talented experts to broaden the scope.

When Gertrud informed me of her plans for a new book to untangle the effects of phosphorus cycling and climate change on lakes, I instantly knew we were talking about a seminal work in the field of lake restoration research; and, this is exactly what Gertrud has produced. An expert review and synthesis of a vast body of literature and experiences that are now accessible to all.

On behalf of the wider scientific community working on lake restoration – thank you, Gertrud, for your important contribution to our field and for producing this book.

Professor Bryan M. Spears
Limnologist
UK Centre for Ecology and Hydrology, Edinburgh, UK
Professor Spears serves as Director of the Global Environment Facility – United Nations Environment Programme 'uPcycle Lakes' project and works within UKCEH's Freshwater Restoration and Sustainability Group, Edinburgh. He is co-chair of the International Society of Limnology Working Group on Lake Restoration and the United Nations Environment Programme coordinated World Water Quality Alliance (WWQA) Workstream on Ecosystem Restoration. He works to support countries and programmes on lake restoration globally.

Preface

Over many years of consulting about water quality issues in natural and constructed lakes, I have been struck by the close links between internal phosphorus loading and proliferation of cyanobacteria – often called "bluegreen algae". In virtually all projects aimed to combat cyanobacteria, a thorough system analysis (based on the inspection of available limnological information and further monitoring) revealed the presence of occasional or persistent hypoxic bottom waters and phosphorus release from the bottom sediments as internal P loading in stratified and mixed lakes. These applied studies formed the basis of this text. "Ecologists should not be afraid of applied problems, they can tell us much about general principles" (Harris 1994).

My contractual studies have involved more than 35 north-temperate freshwater systems in North America and Europe, including 5 reservoirs and 2 river systems. They included lakes with large ranges of productivity, morphometry, stratification regimes, and other limnological characteristics. Not surprisingly, productive lakes and reservoirs often exhibited cyanobacterial blooms, especially during dry summer seasons. But signs of deteriorated water quality also turned up in "cottage country" lakes that had historically been nutrient-poor, pristine, with clear or tea-coloured water, with only the occasional summer cottage, cabin, or camp along the shores. During cool, wet summers, water quality was generally good, and there were no obvious cyanobacteria. In hot and dry summers, cyanobacteria accumulated into "blooms". All of these events pointed to the possible importance of climate-dependent development of internal P loading.

My PhD work (Nürnberg 1984a) focused on the geochemistry and modelling of internal P loading in deep, stratified lakes and its possible effect on phytoplankton. Shallow lakes seemed more complicated, involving additional processes that are hard to separate. For example, phosphate released from sediments may be taken up by P-limited organisms before it can be measured. Phytoplankton may be light limited. Sediments can be resuspended. The elusiveness of internal P loading, the fact that the classic studies involved mainly stratified lakes with higher productivity, and the general necessity of controlling external P load provided a background of neglect of and even disregard for internal P loading for many years.

Eutrophication has accelerated in recent years, leading to shifts in phytoplankton species including an increased proliferation of cyanobacteria in the wake of climate change (Adrian et al. 2009). A synthesis of the connections between sediment P release, climate change, and cyanobacterial proliferation is overdue. This book aims to provide this synthesis. The work builds on my session at the 33nd NALMS meeting in San Diego, California, USA, in 2013, *Cyanobacteria and internal phosphorus loading: hypotheses and case studies*, and subsequent presentations (e.g., in SIL 2016 in Torino, Italy, and IFHAB 2018 Toronto, Ontario).

With this study, I hope to provide a solid theoretical foundation leading to stronger tools for the management of cyanobacterial proliferation in freshwater. Such a combination of theoretical and applied limnology has been my goal since embarking on graduate studies in Canada. As Cooke et al. (2020) put it: "We define 'success' in applied environmental science as respectfully conducted, partner-relevant research that is accessible, understandable, and shared and that can create opportunities for change (e.g., in policy, behaviour, management)".

I did not use Artificial Intelligence (AI) methods in the preparation of this book.

Acknowledgements

My clients, including lake associations, towns, cities, and municipalities, and other governmental organizations and consulting companies, provided the incentive for looking beyond phosphorus and including phytoplankton in my deliberations and studies. Discussions with fellow attendees at numerous meetings of many organizations, including the North American Lake Management Society (NALMS), the Interdisciplinary Freshwater Harmful Algal Blooms Workshop (IFHAB), and the International Association for Great Lakes Research (IAGLR), provided sounding boards for ideas and studies that led to the synthesis presented in this text. Especially valuable were participants of my nearly annual workshops on internal P loading at NALMS meetings in their quest for understanding the increasing cyanobacteria proliferation in lakes and reservoirs of all types.

I want to acknowledge the Indigenous peoples worldwide who are the original stewards of the freshwater described in this research and many summarized studies. These include First Nations, Métis, and Inuit in Canada; Native Americans and Alaska Natives in the United States; Sámi in Europe; Aboriginals in Australia; Māori in New Zealand, and many other indigenous groups across the world.

I am grateful to the scientific community at large, where the many published studies were crucial for this synthesis to be assembled. But I regret not having been able of citing more and hope for the reader's understanding if I neglected to mention an important publication.

A considerable part of the manuscript has been read by my colleagues and friends, and their comments and criticisms have improved this study in every case. In particular, I would like to thank them for their enthusiasm and encouragement, which means a lot to a lonely writer. Two of my colleagues and friends of NALMS were essential to this undertaking as they read and supported my application to CRC press from the start and reviewed the semi-final product: Jennifer Graham, U.S. Geological Survey, New York Water Science Center, reviewed Chapters 1 and 3 and William (Bill) James, University of Wisconsin-Stout, reviewed Chapters 1, 2, and part of 5. In addition, Olga Tammeorg, University of Helsinki, Finland, looked over the whole text and specifically reviewed Chapters 2, 5, and 6. Francis Pick, University of Ottawa (emeritus), provided comments on Chapters 1 and 3. Andrew Paterson, Ontario Ministry of the Environment, Conservation and Parks, reviewed Chapter 4. Peter Isles, Department of Environmental Conservation, Vermont Agency of Natural Resources, USA, helped streamline early versions of Chapters 1 and 5. Miquel Lürling, Wageningen University & Research, the Netherlands, reviewed Chapters 1 and 6.

The enthusiasm expressed by many colleagues when asked for specific information or photos made writing so much more fun. David A. Matthews, Director of the Upstate Freshwater Institute, Syracuse, New York, USA, kindly provided the raw data for Figure 3.13. Photos were provided helpfully and promptly by Iran E. Lima Neto, Federal University of Ceará, Brazil, and my reviewers Jennifer Graham, Frances Pick, and Olga Tammeorg; additional photos were obtained through previous collaborations.

My long-time friend since student times, David Currie, University of Ottawa (emeritus), graciously provided stylistic and general comments on the Preface and Introduction. I also want to acknowledge the rigorous training by the limnology group of McGill University, Montreal, Quebec, Canada. My late supervisors Robert Peters and Frank Rigler introduced me to empirical thinking and how to look at limnological tasks related to societal problems with the intent of finding solutions.

My family members all provided active support and never-ending encouragement. My son-in-law, Alex Koiter, Brandon University, Manitoba, commented on Chapters 2 and 5; my daughter, Stefanie LaZerte, R programmer and biological consultant, commented on Chapters 1 and 5 and produced most of the beautiful graphics depicting regression plots; and my son, Franz LaZerte, encouraged me all the way. Bruce LaZerte, my husband and partner in life and work (Freshwater Research), provided more technical input, editorial support, and loving encouragement than can be quantified; it will never be forgotten.

Regardless of the assistance, the responsibility for synthesis, interpretations, and conclusions rests with me, and all errors are mine. In this context, I encourage discussion and pointing out errors, which are foreseeable in such an endeavour.

The support and diligence of the editorial stuff of CRC Press, Taylor & Francis Group, from start to finish are gratefully acknowledged.

About the Author

Gertrud K. Nürnberg has been an environmental scientist for more than 40 years, studying and modelling the geochemistry of lakes and reservoirs. She holds a PhD (1984) from McGill University, Montreal, Canada, on "The availability of phosphorus from anoxic hypolimnia to epilimnetic plankton", very much the subject of this book.

As the head of *Freshwater Research*, she has focused on the restoration and modelling of eutrophic lakes and reservoirs. Her main interests include the sediment–water interactions in stratified and polymictic lakes, especially phosphorus release from lake bottom sediments, using several methods to quantify internal phosphorus loading. She has developed theoretical and limnological concepts, most importantly the anoxic factor, which describes the temporal and spatial spread of anoxia in lakes. She is active in science as a previous editor and constant reviewer for numerous journals. As she is particularly interested in the improvement of the status quo, she has contributed to lake restoration by publishing on and investigating several techniques to decrease internal phosphorus loading and hence curtail cyanobacterial blooms. Her efforts have been recognized by several awards from the North American Lake Management Society (NALMS.org).

Symbols, Acronyms, and Abbreviations

A: lake surface area
APA: alkaline phosphatase activity
C: carbon
CO_2: carbon dioxide
DIC: dissolved inorganic carbon
DO: dissolved oxygen
DOC: dissolved organic carbon
Fe: iron
HABs: harmful algal blooms
IPCC: Intergovernmental Panel on Climate Change
LMB: lanthanum-modified bentonite
mcyE: microcystin synthetase gene
MERIS: Medium-Spectral Resolution Imaging Spectrometer
Mn: manganese
N: nitrogen
NH_4: ammonium
NO_3: nitrate
NO_x: the sum of nitrite (NO_2) and nitrate (NO_3)
$[NH_2]_2CO$: urea
P: phosphorus, see total P (TP) and soluble reactive P (SRP) in Glossary
PAC: poly aluminum chloride
PAR: photosynthetically active radiation
q_s (m/yr): *annual areal water load*, annual outflow volume (Q, cubic m) per surface area (A_o, square m), where $q_s = Q/A_o$
τ (yr): *annual water detention time* or *annual water residence time*, lake volume (V) divided by annual outflow volume (Q), where $\tau = V/Q$
V: lake volume
v (m/yr): *settling velocity*, average distance that TP settles downward within one year
$z/A^{0.5}$: morphometric ratio

Chapter 1

Introduction

Freshwater comprises 2.5% of the total water on the planet of which just 1.2% is stored on the surface with less than a quarter unfrozen liquid freshwater of lakes, rivers, and wetlands (www.usgs.gov/media/images/distribution-water-and-above-earth). Water is unevenly distributed, and most of the world's lakes are located above the 40° N northern latitude (Lehner 2024). For example, Canada, the second largest country of the world, has over half of the world's freshwater lakes (https://reviewlution.ca/resources/how-many-lakes-in-canada/, accessed, December 23, 2023). It is estimated to have the most water resources after Brazil and Russia, occupying the third position. There are over 8 500 rivers and more than 2 million lakes (with 1 million lakes having area >10 ha) covering almost 9% of its area (Monk and Baird 2014). Ninety percent of Canada's municipal drinking water supply originates from surface waters (Huot et al. 2019), and much of the rest is also derived from shallow groundwater. Other countries have fewer freshwater resources. Water scarcity is often associated with poor quality, but water quality is often compromised also in regions with sufficient water quantity. It is obvious that an acceptable water quality is an important issue from both ecological and societal perspectives (Vörösmarty et al. 2010). Clean water and sanitation comprise one of the United Nations 17 goals for sustainable development (https://sdgs.un.org/goals, accessed December 12, 2023). Eutrophication has long been recognized to hamper water quality.

The classic definition of eutrophication implies the enrichment of a water body with the nutrients phosphorus and nitrogen, leading to an increased overall productivity. But more recent use of the term implies an increase in the abundance of potentially toxigenic cyanobacteria. "Eutrophication . . . is a syndrome of aquatic ecosystems leading to blooms of nuisance (often toxic) algae, anoxic events, fish mortality and substantial economic losses" (Carpenter and Brock 2006). The accumulation of cyanobacteria on the lake surface as scums or in deeper layers is sometimes called harmful algal blooms (HABs or cHABs). Cyanobacterial blooms are unsightly and potentially toxic (Chorus and Welker 2021). Decaying blooms can increase hypoxia, enrich the sediments with nutrients and organic matter, generally decrease water quality, and impair ecosystem functioning and services (O'Neil et al. 2012; Paerl and Otten 2013).

A striking example of the societal impact of HABs occurred in Toledo, Ohio, where tap water is drawn from western Lake Erie (Bullerjahn et al. 2016).

> In early August 2014, the municipality of Toledo, OH (USA) issued a "do not drink" advisory on their water supply directly affecting over 400,000 residential customers and hundreds of businesses. . . . This order was attributable to levels of microcystin, a potent liver toxin [produced by cyanobacteria], which rose to 2.5 μg/L in finished drinking water.

DOI: 10.1201/9781003301592-1

This level far exceeded the World Health Organization guidelines for long-term exposure to drinking water with microcystin concentration of 1 μg/L (WHO – World Health Organization 2022, 2021, 2020) and closely approached the warning issued for the short-term (<2 weeks) use of water with microcystin concentration of 3 μg/L for bottle-fed infants and small children.

Cyanobacterial blooms have been reported with an increasing frequency in the public news (Van de Waal et al. 2024). Updates can be accessed by a simple Google search (e.g., *cyanobacteria in the news)*. For example, Lake Okeechobee, Florida, USA, has been notorious for its cyanobacterial blooms and has been the subject of investigative reporting: "It's Toxic Slime Time on Florida's Lake Okeechobee" (www.nytimes.com/interactive/2023/07/09/climate/florida-lake-okeechobee-algae.html?smid=nytcore-android-share, accessed July 9, 2023).

Damages caused by cyanobacterial blooms can be costly. They reflect damages to commercial fisheries, public health, provision of potable water, recreation and tourism, coastal management, and much else (Dodds et al. 2009; Smith et al. 2019). For example, the 2011 bloom in Lake Erie was estimated as costing approximately US $71 million of lost economic benefits and another bloom in 2014 as costing $65 million (Bingham et al. 2015).

There is no longer any scientific debate that climate is changing due to human activities (Intergovernmental Panel on Climate Change 2022). Numerous important lake characteristics depend on climate including water temperature, stratification, nutrient regeneration rates, light regime, and water residence time. Cyanobacteria proliferation, in turn, responds to many of these climate-dependent variables. Climate change, the human-caused increase of nutrients, and hydrological alterations have been identified as the main problems leading to eutrophication (Beaulieu et al. 2019; Burford et al. 2020; Paerl and Huisman 2008; Paerl and Paul 2012). Especially worrisome is the likely existence of tipping points with respect to climate change (https://global-tipping-points.org/download/4608/)(Wunderling et al. 2023), which means that problems do not increase gradually but may turn suddenly into extremes (Woolway et al. 2022b):

> We should not expect gradual ecosystem responses to climate change, as is illustrated by global projections . . ., but, rather, tipping point responses that lead to new ecosystem states once ecological thresholds are crossed. . . . The ecological consequences of climate change will be abrupt, and the impacts of extreme climate events are already occurring and will continue to be felt locally, without warning, and without time to adapt. The effects of climate change often occur cumulatively and interact synergistically with multiple environmental stressors. Climate change will exacerbate problems with water quantity and quality, the latter including eutrophication, salinization, contamination, and the spread of invasive species, to name a few.

I suggest that an important tipping point in freshwater systems occurs when a lake's bottom sediments begin to release phosphorus as internal P loading. The development of cyanobacteria and HABs often seems to follow. Such a tipping point, or switch, has been noted previously, even without climate change considerations (Harris 1999) "as the nutrient load is increased it is transferred to the sediments and the deoxygenation slowly builds up. Then, quite suddenly, the internal load increases and the whole system switches over to the new (eutrophic) state."

This book explores the steady development of the occurrences of sediment P release and investigates the effects of climate change. Gradually declining dissolved oxygen concentrations in the bottom water and at the sediment–water interface lead to the release of sediment P. This internal P loading fuels biomass and eventually increases sediment pools of labile organic matter, which results in more oxygen depletion and more P release. Climate change acts as an accelerator of this process in many ways. The most important is probably the enhancing effect of increased air temperature on lake water column temperature and stability, thus increasing release areas and providing longer periods for P release to occur. Other numerous effects include drought with resultant low water levels, wind stilling or enhanced disturbances by wind and rain affecting lake-mixing regimes, increased lake colour by organic acids, and salinization. In addition, and at the same time, many of the recently developing climate conditions favour the proliferation of cyanobacteria, especially by increased temperature and stability, in the context of ongoing cultural eutrophication.

The widespread problem of HABs has prompted a call to the scientific community for the "examination of extremes in water quality" asking for definitions concerning cyanobacterial blooms and pointing to the need of studies that combine climate change with changes in water quality and increased eutrophication (Michalak 2016). This book seeks to answer this call.

1.1 Contribution of internal P loading and climate change to the recent increases in cyanobacteria blooms – a hypothesis

The main hypothesis underlying this book is that internal P loading has increasingly become a major driver of cyanobacterial growth in many lakes and that climate change is intensifying this effect. This would mean that:

A major contributing factor to the recent increase in cyanobacteria bloom expansion and frequency is the initiation and persistence of internal P loading as intensified by climate change.

If this is so, options for dealing and abating the accelerating cyanobacteria proliferation become clearer and can facilitate evidenced-based and pro-active stewardship of freshwater systems in the wake of climate change.

This book examines the potential causes of cyanobacterial proliferations, including human-caused nutrient increases, hydrological alterations, and climate change. A systematic sorting and organizing process involving hundreds of individual publications has helped to clarify the potential influences by internal P loading on the increase of cyanobacteria blooms. In the end, the goal of this text is to provide a conceptual framework for future studies that further explore the relationship between internal P loading, climate change, and cyanobacteria.

This book is addressed to aquatic scientists interested in the management of cyanobacterial proliferation, as well as more generally to biologists who study the consequences of climate change on eutrophication and toxigenic cyanobacteria blooms. By gathering much of the related scientific work, I hope to simplify the task of addressing these substantial water quality problems that the global community is facing currently.

There is a vast number of studies and articles that explore the reasons for cyanobacteria proliferation, and I relied on many excellent reviews and meta-analyses published in the scientific literature. None of these literature compilations are complete. I offer my sincere apologies for not considering relevant work that I may have overlooked. Older studies may only be cited in the reference list of newer, included papers.

I here concentrate on findings that are relevant to the hypothesis that climate change increases internal P loading, resulting in the proliferation of cyanobacteria. To address this subject, I include descriptions of the P cycle, cyanobacteria characteristics, and climate change evidence. Once relationships are established and conclusions drawn, suggestions of risk assessment (estimation of external versus internal P load in P mass balance analysis) and remediation techniques are presented.

The text is organized in six separate chapters that can be read individually. After the introduction in Chapter 1, Chapters 2, 3, and 4 describe the characteristics of phosphorus, cyanobacteria, and climate change and their effects on freshwater systems separately. Chapter 5 presents the conclusion in a summarized assessment of internal P loading effects on cyanobacteria (and vice-versa), showing the synergy between many of the previously explored water quality determinants. I also consider competing interpretations that have been published elsewhere. After all, testing of hypotheses implies an attempt at their falsification. The knowledge accumulated mainly for lakes is then applied to other freshwater systems, including reservoirs, rivers, wetlands, and estuaries.

For a quick summary addressing the presented hypothesis, Chapter 5 can be read separately, or even just Sub-section 5.8, with occasional reference to previous chapters, depending on the reader's interest and previous knowledge. The final Chapter 6 presents an overview of options to reduce cyanobacterial abundance, concentrating on techniques that address internal P load remediation (since reduction of external loading is already well understood) – this section can also be read separately.

1.2 Method of analysis

It is important to test theoretical (mechanistic) thinking with quantitative assessments. To do this, I draw on observations from in-situ monitoring studies and from experimental results. Ideally, such numerical results are used to construct predictive models in an attempt of quantification.

In purely empirical analysis, data can be analyzed without any theoretical underpinning (Peters 1986), so that results document correlations (i.e., the likelihood of two things occurring together). But clearly, correlations are not necessarily causal. Neither can science "produce logically indisputable proofs about the natural world. At best it procures a robust consensus based on a process of inquiry that allows for continued scrutiny, re-examination, and revision" (Oreskes 2004). This is based on the logical fact that causal relationships cannot be proven – only disproven (Popper 1959). Causal hypotheses can describe our perception of the natural world in a wider context as a model. But they need to be tested. As pointed out by Popper (1972, *Objective knowledge: An evolutionary approach. Oxford*, cited by Peters (1986)): "The only objective test for understanding is predictive success and therefore any dispute between . . . approaches must be based on their respective predictive powers." While I have tried to present as many quantitative relationships as I could find to test my hypothesis, I also compare my hypothesis to relevant theoretical deliberations, understanding, and limnological consensus.

The "Principle of Parsimony", or Occam's Razor (William of Ockham, c. 1287–1347), proposes that *among competing hypotheses, the hypothesis with the fewest assumptions should be selected* even though "parsimonious" does not necessarily imply "truth". I suggest that the hypothesis of *internal P loading as one of the main drivers of cyanobacteria in the context of climate change*, as discussed in this book, is a parsimonious consolidation of many relationships (between cyanobacteria and climate and climate-change-related variables) identified in the extensive literature reviewed here.

Chapter 2

Phosphorus transfer and cycle in freshwater systems

Phosphorus (P) is such an important nutrient in fresh waters that it could be called the *freshwater currency*. The water in lakes, reservoirs, and rivers is characterized by its turbidity, colour, and content of plankton and abiotic particles, all of which are related to P. Most importantly, productivity from prokaryotes to eukaryotes, from bacteria, phyto-, and zooplankton to fish, is somehow related to the concentration of P. Quantitative studies have found correlations between P and numerous limnological variables in surface waters. P is typically the nutrient most often limiting productivity in freshwater systems and dictating the trophic status (Section 2.3), besides the "other", potentially critical, macronutrient, nitrogen (N, as discussed in Sections 2.5 and 3.3). P manifests itself in many chemical forms in nature (Section 2.1). Careful differentiation between bioavailable, inert, and other distinct P forms help detect patterns and relationships with water quality variables, including cyanobacteria biomass (Sections 2.2 and 3.3).

The P cycle in freshwater starts with external P input from terrestrial (main contribution) and atmospheric sources as *external P load* (Figure 2.1). While some of the external P directly affects the water quality of the lake, much of it can settle and be retained as it accumulates on the bottom sediments (*accumulation*). Transformation (geochemical and microbial) of accumulated P turns part of the settled material into a potential *internal P source*. *Mobilization* that occurs under certain conditions (especially affected by redox potential, temperature, pH, and geochemical characteristics, i.e., sulfur versus Fe prevalence, and mineralization) then provides the *internal P load* to the freshwater system as determined by release or diffusion rates, bioturbation effects, and resuspension. Thus, this step includes a transport mechanism as realized by Søndergaard and Jeppesen (2020) in their three aspects driving internal P loading (mobile P in the sediment, its mobilization, and its transport). Part of this internal load can settle and is retained again in the sediment after being distributed in the water column. The final *impact* on the system's water quality depends on temporal, physical, chemical, biological, climatic, and other limnological conditions (Figure 2.1). The P cycle as outlined here extends the *P transfer continuum* concept for catchment basins (Haygarth et al. 2005). It is our guideline in the following explorations of internal P load influences.

2.1 Chemistry and bioavailability of P

In the freshwater environment, many chemical P species can be separated by chemical and biological analyses (Table 2.1). Total P (TP) is obtained after chemically digesting a water sample to convert all P into orthophosphate (e.g., Andersen 1976) that then is readily analyzed by a well-established method (usually molybdenum-blue, spectroscopic analysis (Nagul et al. 2015)). The TP analysis of lake water includes all chemically identifiable P in a water sample,

DOI: 10.1201/9781003301592-2

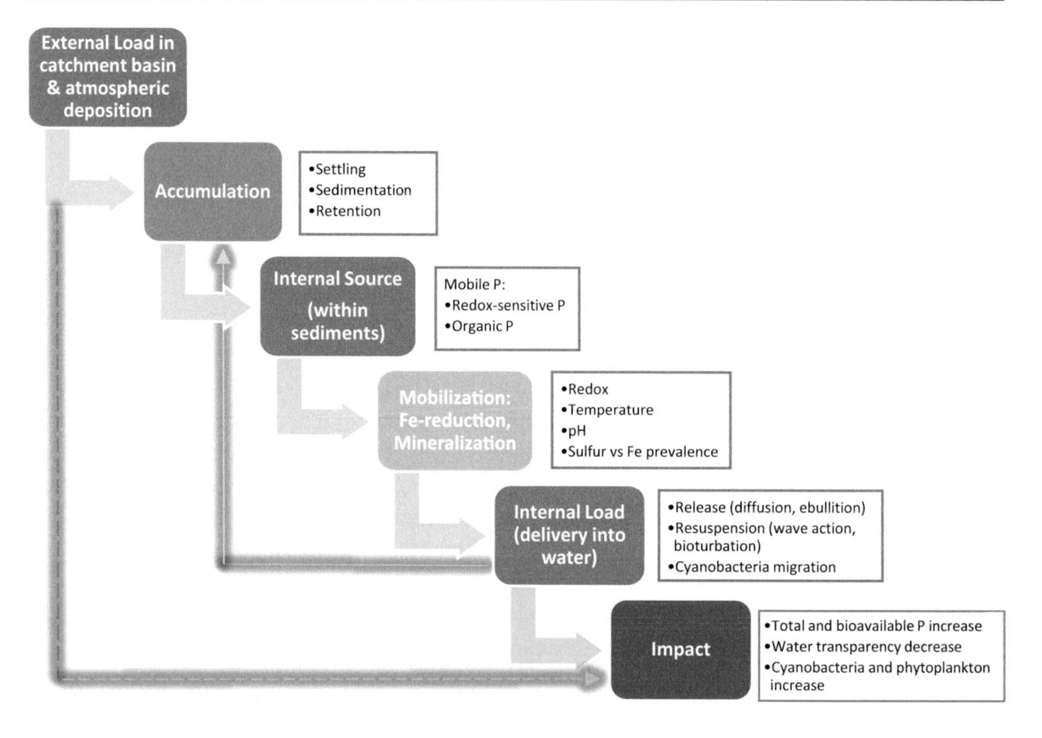

Figure 2.1 Transfer (arrows) of phosphorus (P) in freshwater systems from external load to impact on the water quality via internal P load, including a loop back to accumulation. The direct impact of external P load on the water quality is indicated by broken lines.

hence represents the "total" amount. Phosphorus located in other lake compartments including water plants (macrophytes), fish and other macrofauna, as well as the lake bed sediments and its macrobenthos, is not included.

Some chemical distinctions depend on whether the P compound is dissolved and soluble or not, defined as filtrable by a certain (usually <0.45 μm) filter pore size. Usually, the particulate fraction (not filtrable) is determined indirectly as the difference between TP and dissolved P (Table 2.1). Only a small fraction of TP may be in the form of free soluble orthophosphate (PO_4^{3-}) that is "bioavailable" and can be taken up immediately by the phyto- and bacterioplankton. Because many surface waters are nutrient limited in the summer, bioavailable inorganic concentrations of the main nutrients, P or nitrogen (N), are low. (The question about which nutrient, P or N, is more important and drives eutrophication is discussed in Section 3.3.)

2.2 Spatial and temporal variability

Spatial and temporal variability of the various P compounds is large between and within lakes and is much controlled by a lake's mixing status. A lake can be frequently wind-mixed and unstratified throughout the year from top to bottom (polymictic) as occurs in shallow lakes or

Table 2.1 Chemical phosphorus (P) species in freshwater and their analysis or computation. MB refers to the routine molybdenum-blue analysis for phosphate (Nagul et al. 2015; Nürnberg 1985).

P fraction	*Analysis/computation*	*Meaning*	*Comment*
TP – total P	Digested, then MB (unfiltered).	All P in water sample.	Most reliable.
TDP – total dissolved P	Filtered through 0.45 μm pore size filter, then digested, then MB.	Can be organic in lake water with variable, mostly long-term bioavailability; in rivers, considered instead of TP to avoid sedimentary P.	Often mixed up with SRP but not necessarily bioavailable.
PP – particulate P	PP = TP – TDP	Seston, plankton	Depends on TP and TDP analysis.
SRP (DRP) – soluble reactive P (dissolved reactive P)	Filtered through 0.45 μm pore size, then MB.	Immediately biologically available, mostly orthophosphate.	Difficult handling and analysis; possible contamination or adsorption; can overestimate phosphate especially in presence of organic acid compounds containing iron (Nürnberg 1984b).
TRP – total reactive P	MB on raw water, compared to blank (unfiltered).	Fe–P complexes, include SRP and PRP.	Easy, but colour and transparency blanks are needed for comparison (Nürnberg 1984b).
PRP – particulate reactive P	PRP = TRP – SRP	P adsorbed onto iron oxyhydroxides (Fe-P).	Naturally found at anoxic–oxic boundaries; artificially induced by the aeration of anoxic water samples containing Fe (Nürnberg and Peters 1984a).
BAP – biologically available P	Bioassay	Availability depends on bioassay method, test organisms, time period considered.	Time consuming, difficult (Nürnberg and Peters 1984a).

it can be seasonally thermally stratified (mono- or dimictic) or always stratified (meromictic). Thermally stratified lakes typically stratify during the warm-water-growing period (i.e., May/June to September/November in the northern hemisphere and equivalent months in the southern hemisphere, depending on the location). There are no clear borders between these lake types as the classic concept assumes. In a thermally stratified lake, there is an upper mixed layer that fluctuates in depth depending on morphometric and geographical (e.g., latitude) characteristics as well as wind, weather, and other short-term disturbances (Figure 2.2).

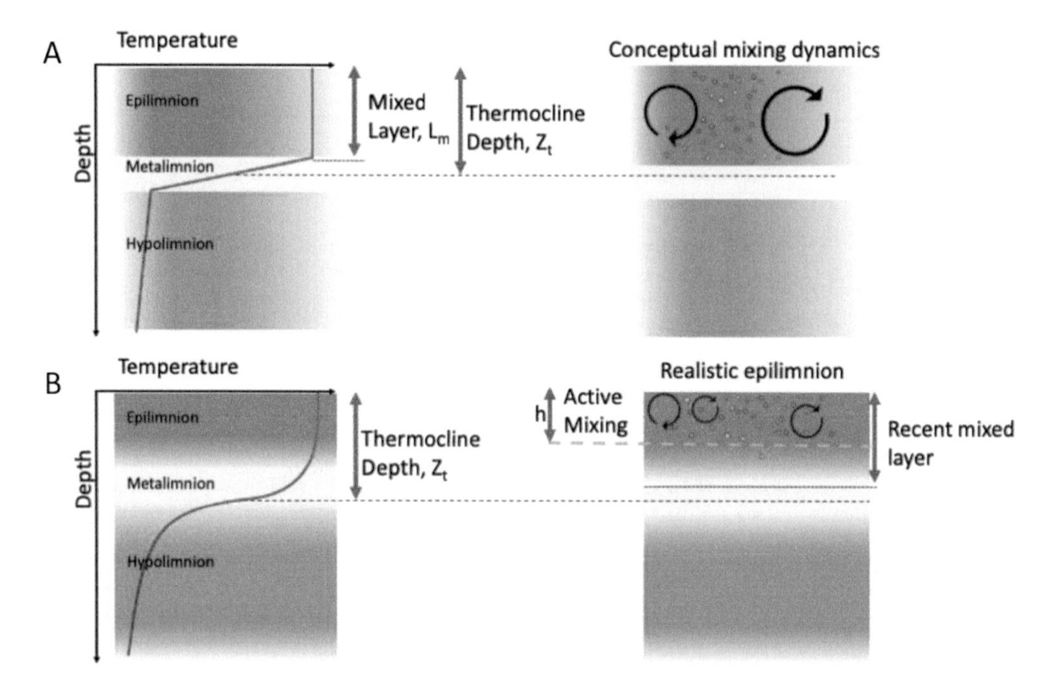

Figure 2.2 Conceptual mixing regime in stratified lakes assuming uniform mixing (A) compared to more realistic observations (B), where the active mixing depth is much shallower than the thermocline or the recent mixed layer. The differences result in a shallower distribution of buoyant cyanobacteria. *Figure 2 from Wells and Troy (2022) with permission.*

Monomictic lakes stratify only in one period annually, usually the summer, dimictic lakes have two periods of stratification, one in the summer and one in the winter under ice cover in lakes at higher latitude or elevation. While stratified lakes mix at least once during the year (holomixis), meromictic lakes do not mix completely each year, so that there is a consistent stagnant layer at the bottom. Meromixis occurs in deep lakes with a relatively small area like in kettle and alpine lakes or in tropical and other lakes with specific temperature and salt conditions.

The extent of mixing (or stratification) dictates the distribution of P compounds spatially and temporally in the water column. In addition, the mixing state controls mechanisms that can increase P, such as oxygen depletion during stagnant periods that trigger phosphate release from enriched, anoxic lake bed sediments (Section 2.4.2). Other effects of stratification include decreased particulate P in the photogenic zone by enhanced settling. Accordingly, a measure of mixing is a useful additional variable in many relationships involving P and other limnological characteristics (Section 2.3).

The extent of mixing or the strength of stratification depends on morphometric characteristics easily expressed as the deviation of lake shape from an ideal cone, called the *morphometric ratio* (m/km, Equation 2.1, where z is mean depth in units of m, and A is lake surface area in units of km^2) (Nürnberg 1995a).

Equation 2.1: Morphometric ratio $= z/A^{0.5}$

This ratio controls the duration of the annual stratification period for lakes at similar latitude and altitude. A larger ratio indicates that fall turnover at the end of summer stratification occurs later (Nürnberg 1988a). The ratio, sometimes called *Osgood Index*, was initially used to determine the influence of hypolimnetic P on the epilimnion, with smaller ratios indicating more influence due to enhanced mixing (Osgood 1988). The inverse of the morphometric ratio is especially useful in shallow polymictic lakes. It is called *dynamic ratio* (Håkanson 1982) and was initially used to illustrate the extent of sediment disturbances by erosion compared to accumulation processes and later of wave action (Bachmann et al. 2000).

For example, the morphometric ratio is an important variable in the relationships of lake TP or external TP load with the hypolimnetic oxygen depletion rate and with annual anoxia (Nürnberg 1995a). The ratio significantly improves the prediction of extent and duration of hypolimnetic anoxia (quantified by the anoxic factor, Box 2.1) by the growing period average TP concentration of the mixed layer in a multiple regression model (Figure 2.3). Such relationships are important in the estimation of internal P load (Section 2.4.2).

Box 2.1 Anoxic factor (*AF*) and its model, active area release factor (*AA*) (Nürnberg 2019, 1995a)

The active area involved in redox-related P release in stratified lakes resembles the sediment area that is overlain by anoxic water (Figure 2.8A). By integrating the time period for different depths of the oxycline (1–2 mg/L dissolved oxygen, DO) and subsequent division by lake area, an anoxic factor can be calculated from oxygen profiles. For applications other than determining the active sediment release area, higher DO thresholds (e.g., 3.5 mg/L or 5.4 mg/L depending on the fish species preference or regulation) can be used to compute hypoxic factors, HF, which are useful in habitat quantifications.

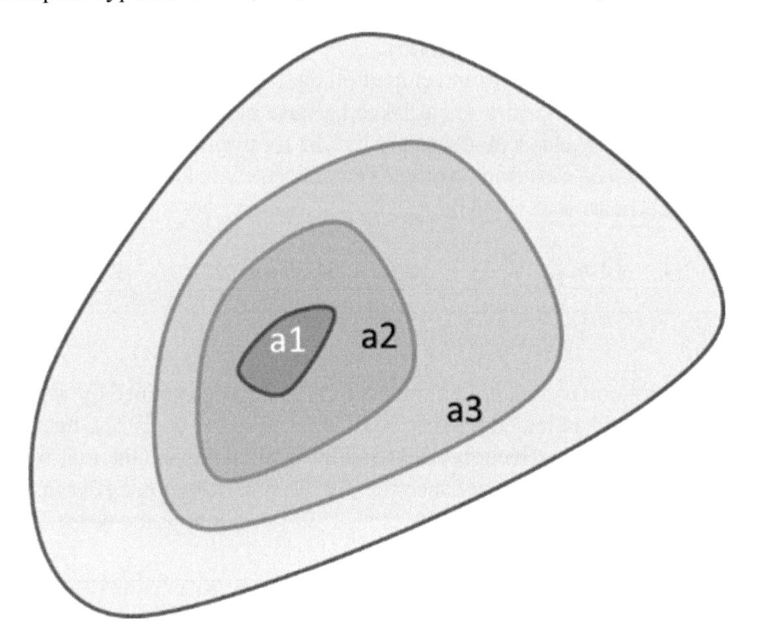

These factors represent the number of days in a season (summer-fall) or year when a sediment area equal to the lake surface area is anoxic or hypoxic.

The anoxic or hypoxic factor is computed from DO profiles according to Equation 2.2 (Nürnberg 1995a):

Equation 2.2: $$AF\ or\ HF = \frac{t_1 \cdot a_1}{A_1} + \frac{t_2 \cdot a_2}{A_2} + \frac{t_3 \cdot a_3}{A_3} + \ldots + \frac{t_n \cdot a_n}{A_n} = \sum_{i-1}^{n} \frac{t_i \cdot a_i}{A_i}$$

where t_i is the period of anoxia or hypoxia (days), a_i is the corresponding area (m^2), A is the lake surface area (m^2), and n is the number of periods with different oxycline depths (for *AF*) or different threshold-DO depths (for *HF*).

Based on *AF* values from DO observations in stratified lakes, models were developed from the average TP concentration of the mixed layer during the growing period and the morphometric ratio (Equation 2.1). The predicted factor is also called the active area release factor (*AA*, Equation 2.3, in units of days per summer) (Nürnberg 2020a, 2019). *AA* describes the active release area and period (when the sediment–water interface is anoxic) also in polymictic lakes (Figure 2.8B) because it adequately predicts internal loading in these lakes when combined with a P release rate (Nürnberg 2005). As *AA* is based on a regression model, it does not necessarily represent only redox-related information but may summarize all conditions that support P release related to trophic state (TP concentration) including the mineralization of organic matter.

Equation 2.3: $$AA = -36.2 + 50.1\ log_{10}\ (TP_{summer}) + 0.762\ z/A^{0.5}$$

Where, TP_{summer} is the growing period ("summer") average composite (epilimnetic or mixed surface layer) TP concentration (µg/L); $z/A^{0.5}$ is the morphometric ratio (m/km, Equation 2.1); and z is the mean depth (m).

The model (Equation 2.3) was developed on data from temperate lakes and underestimates *AF* values in semi-arid tropical lakes because of their higher temperatures. Different parametrization yields a model to predict *AA* for tropical stratified lakes during the wet period (*AA-tropical-wet*) from average TP concentration during the wet period (TP_{wet}, Equation 2.4, (Carneiro et al. 2023)).

Equation 2.4: $$AA\text{-}tropical\text{-}wet = -8.90 + 64.50\ log_{10}\ (TP_{wet}) + 9.87\ z/A^{0.5}$$

Besides lake morphometry, water column stability is also determined by weather influences including storm and wind action (Bocaniov et al. 2014; Reiss et al. 2022), further discussed in the context of climate change (Section 4.2.3) and by internal waves. Internal waves or *seiches* (Nielson and Henderson 2022) can enhance the distribution of hypoxia (Bernhardt et al. 2014; Flood et al. 2021) and nutrients (Ostrovsky et al. 1996) throughout the water column in otherwise stratified lakes.

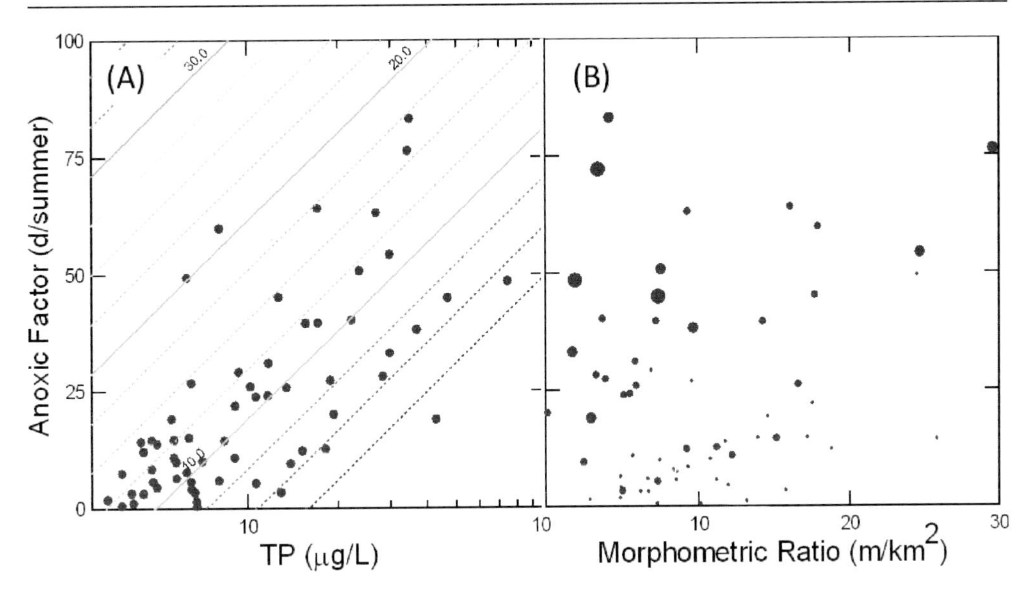

Figure 2.3 Relationships between TP or trophic state with anoxic factor and morphometric ratio. (A) *AF* versus TP, where diagonal stripes indicate the influence of the morphometric ratio (m/km^2, Equation 2.3). (B) *AF* versus morphometric ratio, where symbol size indicates the four trophic states (oligo-, meso-, eu-, and hyper-eutrophic, Table 2.2). *With data and permission from Nürnberg (2019, 1996).*

2.3 Total P influence on trophic state

The water quality of the mixed surface layer during the growing period is used to indicate a lake's trophic state or productivity. Trophic state is a classification approach to quickly sort lakes into several acknowledged states of nutrient enrichment that is associated with water quality (Table 2.2). The trophic state of a particular lake is typically determined by the (growing-period) average concentration of the main nutrients, P and N, the algal pigment chlorophyll *a*, and by the water transparency measured as Secchi disk depth.

When determining a trophic state value, it is important to consider the exact definitions, that is, average or median concentration of a growing period or stratified summer season in the surface mixed layer (Nürnberg 1996). Individual values of unspecified depth and time, water-column-integrated measures or integrated areal expressions do not qualify for trophic state determinations (Nürnberg 1999).

TP estimates are more robust than phytoplankton-based variables. Thus, TP's relationships with other variables have been used to set most trophic state limits, where previously proposed thresholds were generally similar (Nürnberg 1996). Reasons that make TP the ideal candidate as a trophic state identifier are three-fold: (1) TP is probably the most often measured variable in lake studies, and estimates are reliable (Table 2.1), so that quantitative TP values are available in many lakes for long periods of time. (2) TP is less variable in lakes with respect to space and time compared to the patchiness of algal biomass indicators, for example, chlorophyll and Secchi transparency (Haldna et al. 2013). (3) TP is correlated with the other trophic state variables

Table 2.2 Trophic state classification based on the growing season averages of the mixed surface layer. The thresholds between the four states are based on identical intervals of log-transformed variables (except for *AF* that was not transformed) (Nürnberg 1996).

Trophic state variables	*Oligotrophic*	*Mesotrophic*	*Eutrophic*	*Hyper-eutrophic*
Total phosphorus (mg/L)	< 0.010	0.010–0.030	0.030–0.100	> 0.100
Total nitrogen (mg/L)	< 0.350	0.350–0.650	0.650–1.200	>1.200
Secchi disk transparency (m)	> 4.0	2.0–4.0	1.0–1.9	< 1.0
Chlorophyll *a* (µg/L)	< 3.5	3.5–9	9.1–25	>25
Anoxic factor (d/yr)*	0–20	20–40	41–60	>60

*See Box 2.1

including chlorophyll, Secchi transparency, total N (Nürnberg 1996; Wurtsbaugh et al. 2019), colour, and lesser studied productivity measures (Nürnberg and Shaw 1998) under various geochemical conditions (Figure 2.4).

The impact of variables connected with organic acid concentration and colour in tea-stained, brown lakes (Section 2.6) has recently been used to further differentiate the trophic state classification. In this scheme, lakes with TP concentration below 30 µg/L (oligo and mesotrophic lakes, Table 2.2) were divided into clear (blue) lakes and dystrophic (brown) lakes; lakes with higher TP concentration (eutrophic and hyper-eutrophic lakes, Table 2.2) were sorted into eutrophic (green) lakes and mixotrophic (murky) lakes (Meyer et al. 2024).

2.4 P sources, loads, and TP concentration modelling

The importance of P is further reflected by the numerous studies on P sources and P cycling in the freshwater environment. Total P mass balances (Box 2.2) have been assembled for many lakes for a long time. Many decades ago, Vollenweider (1969, 1976) introduced the concept of dependencies between P input, P lake concentration, and phytoplankton biomass by quantifying the impact of P loading on lake TP and chlorophyll concentration, which has been followed by numerous further studies (Bachmann and Jones 1974; Golterman and Oude 1991; Lowe and Steward 2011). In a mass balance, P is not a conservative or an inert chemical (such as the chloride ion, for example), and its loss rates (retention) depend on hydrological and other environmental variables and are highly variable between lakes.

To help quantify various pathways of the lake P cycle without unattainably detailed monitoring efforts, P fluxes have been modelled in numerous ways. In particular, empirical models (similar to Vollenweider's 1969 approach) have been developed to predict (model) P retention separately in lakes (Brett and Benjamin 2008; Nürnberg 1984c) and reservoirs (Hejzlar et al. 2006; Straskraba 1994) for P mass balance models (Box 2.2). Detailed and data-intensive models (sometimes called *mechanistic* or *process-oriented* models) are used to consider the various P paths in a more or less complex fashion (Mooij et al. 2010; Robson 2014). More recently, a combination of machine-learning and process-oriented modelling was conducted on a well-studied lake (Lake Mendota, Wisconsin, USA (Hanson et al. 2020)) and included external and internal P fluxes from the watershed and the lake sediments, respectively.

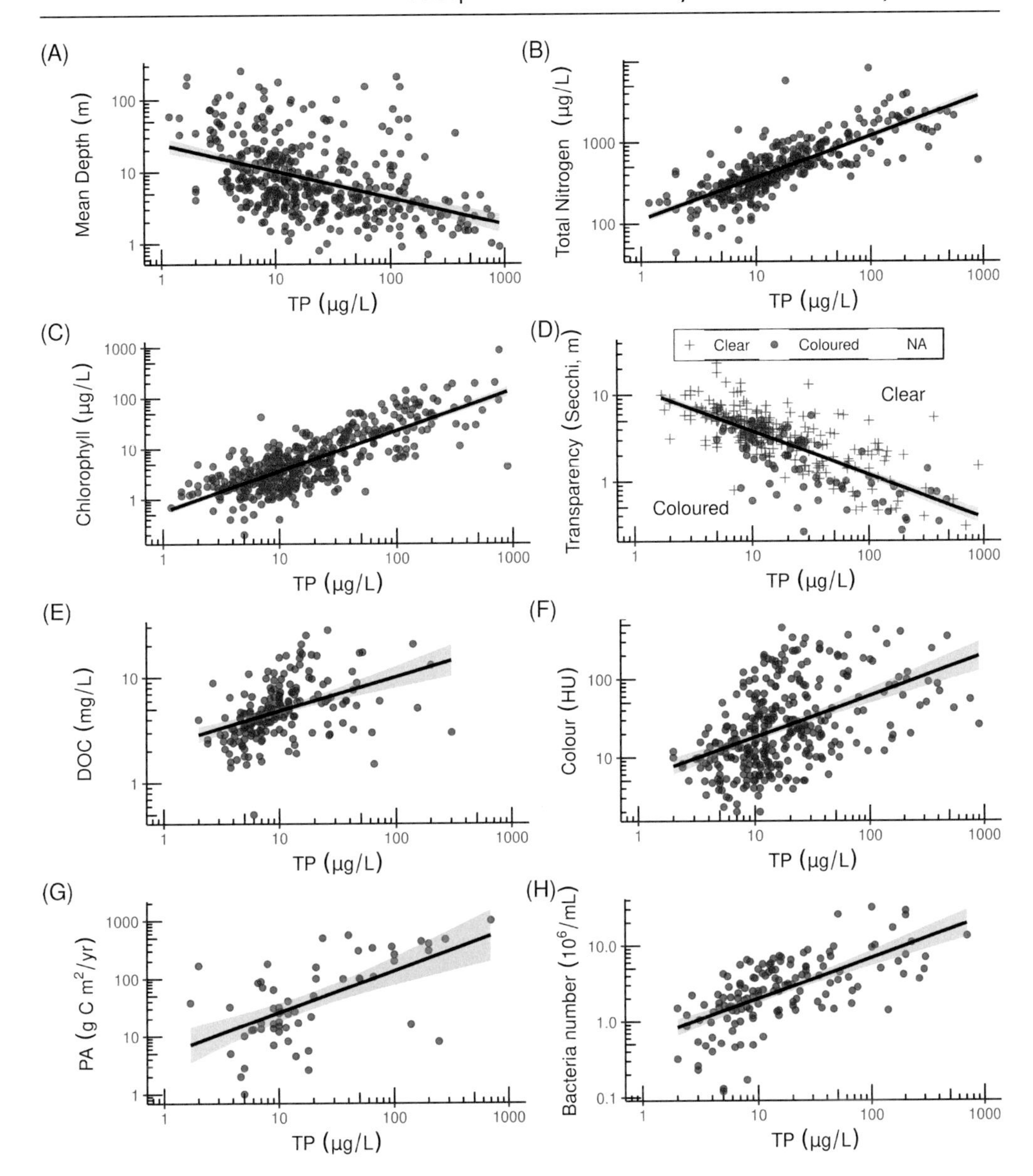

Figure 2.4 Relationships with mixed-layer growing period TP concentration averages from worldwide lakes (A, mean depth; B, total nitrogen; C, chlorophyll; D, Secchi disk transparency, E, dissolved organic carbon (DOC); F, colour; G, annual integral primary productivity (PA); H, bacteria abundance). Significant regression lines with 95% confidence bands (grey shading) are shown. *With data from Nürnberg (1995a, 1996), Nürnberg and Shaw (1999).*

Box 2.2 Phosphorus mass balance in a lake

Within the lake, two fluxes oppose each other: settling or sedimentation that transports P in particulate form (in seston) downwards to the bottom sediments and P release or diffusion from the bottom sediments. This upward flux constitutes the internal P load. Because the upward flux can also settle out (just like the externally derived P), time and location of measurements must be consistent so that reported fluxes are comparable and expressed as a gross mass. Annual average lake TP concentration can change, and that change has to be considered in an annual mass balance (*Change* in equation given in Figure 2.5). This P mass balance approach simplifies P fluxes in, within, and out of a lake so that they can be quantified as is shown in Sections 2.4.1, 2.4.2, and 2.4.3.

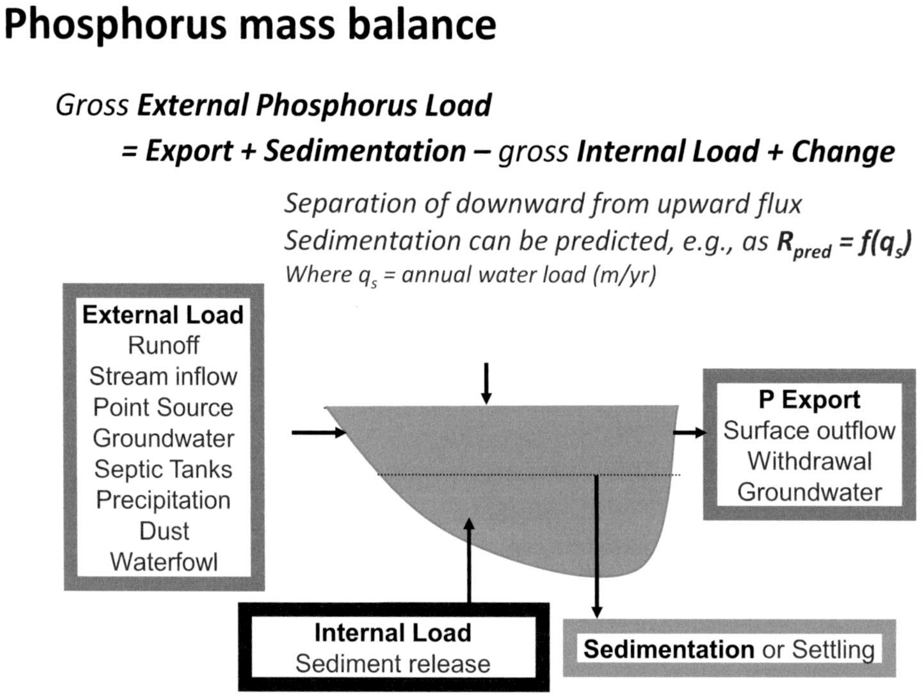

Phosphorus mass balance

Gross ***External Phosphorus Load***

= Export + Sedimentation *– gross* ***Internal Load + Change***

Separation of downward from upward flux

Sedimentation can be predicted, e.g., as $\mathbf{R_{pred}} = f(q_s)$

Where q_s *= annual water load (m/yr)*

External Load
Runoff
Stream inflow
Point Source
Groundwater
Septic Tanks
Precipitation
Dust
Waterfowl

P Export
Surface outflow
Withdrawal
Groundwater

Internal Load
Sediment release

Sedimentation or Settling

Figure 2.5 Schematic of the annual phosphorus mass balance in a lake. External load includes all total P mass that enters the lake via air, surface, and underground in a year (Section 2.4.1). Internal load includes all internally derived P, here shown for the most important source, that is, sediment-released P (Section 2.4.2).

2.4.1 External P sources

External sources of bioavailable P include point sources from waste water treatment, runoff from fertilized fields and urban areas, natural sources like runoff from fertile catchment soils, inflow from productive upstream lakes, beaver ponds, wetlands, and groundwater, atmospheric deposition, and waterfowl droppings (Devito and Dillon 1993; Nürnberg and LaZerte 2016a; Savenko and Savenko 2022; Stackpoole et al. 2019). Upstream water systems that experience sediment P release and have deep-water outlets (e.g., in reservoirs, during hypolimnetic withdrawal treatment) can fertilize downstream waters (Nürnberg 2020b). In many developed regions with advanced waste water treatment and best management practices, external P load is only partially available phosphate, where P is attached to silt and clay particles or associated with Ca, Al, and minerals (James and Barko 2005; James and Larson 2008). The median-dissolved P portion in the global hydrographic network has been estimated as about one-third (Savenko and Savenko 2022). Also, runoff volume and bioavailability are not constant throughout the year but often decrease during the growing period when water and P uptake by terrestrial plants is enhanced.

External P input is determined in various ways depending on the study system and available information. In a fast-flushed reservoir where the main river is regulated and gauged, river inputs and outputs can represent close to 100% of the water budget. Here, most of the P load is easily computed as the product of flow and TP concentration (Nürnberg 2009). In other systems including natural lakes, several contributing inflow loads must be determined separately as they can show high variability among each other and between years (e.g., Lake Winnipeg, Figure 2.6).

When P sources are diverse and difficult to monitor, sub-models are used to determine external P load. Sub-models can be based on previously published TP export coefficients for specific land use in the catchment basin, perhaps considering impervious versus pervious land cover ratios; regional atmospheric P input; septic system additions based on soil characteristics and usage number; and other specific inputs such as contributions from waterfowl, macrophyte senescence, and contamination sites (Nürnberg and LaZerte 2016a, 2004). More detailed watershed models use water flow, land use or soil characteristics, and combinations to estimate external P input with variable accuracy (e.g., Arhonditsis et al. 2019; Huttunen et al. 2016; Mooij et al. 2010; Neumann et al. 2023).

Often, P load estimates are available as an annual average or for certain time periods only, because there are large seasonal variations. Water density differences between the stream and

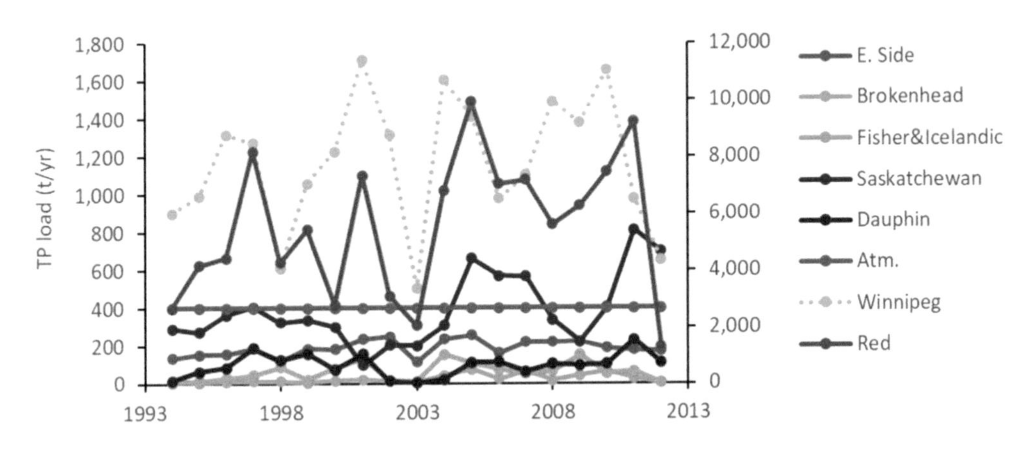

Figure 2.6 Annual TP loads of all monitored rivers and atmospheric deposition (Atm.) in large shallow Lake Winnipeg, Manitoba, Canada. The large load of the Red River is scaled on the right axis. *With data from Nürnberg and LaZerte (2016b).*

lake vary seasonally and can dictate where the P load enters the mixed layer of the lake or reservoir and where it settles (Carmack et al. 1979; James et al. 1987). In temperate climates, the summer growing period usually coincides with low flows so that external inputs are generally lower in the summer period compared to the cooler months (Figure 2.7). This seasonality is important for evaluating the different impacts of external relative to internal sources on cyanobacteria.

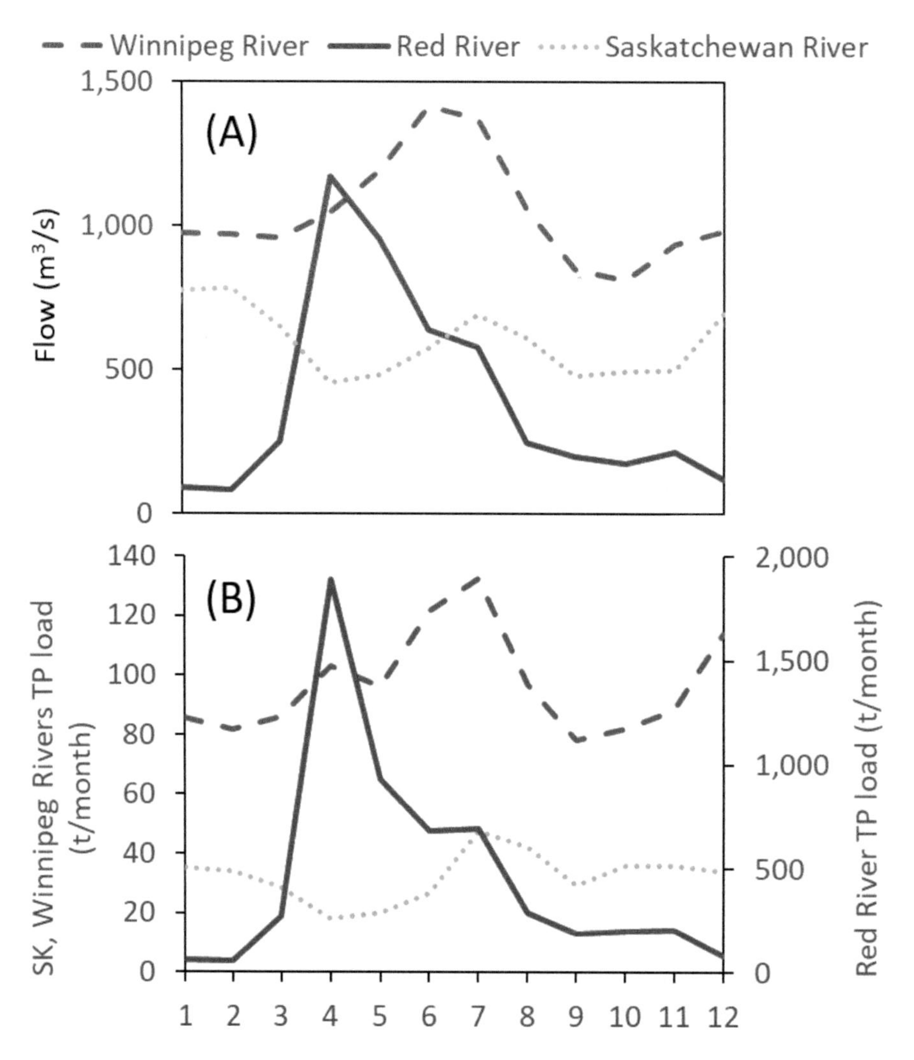

Figure 2.7 Monthly average flow (A, top) and external TP input (B, bottom) of the main inflows to Lake Winnipeg, the Red, Winnipeg, and Saskatchewan (SK) River (1994–2012). *With data from Nürnberg and LaZerte (2016b).*

Note to Figure 2.7: Because of the rivers' large differences in TP concentration, their impact on Lake Winnipeg varies. While Red River's large P input increases lake TP concentration, Winnipeg River and Saskatchewan River actually dilute the lake water. Because external TP inputs from the Red River are smallest in the summer and early fall, increases in TP concentration and deteriorating water quality at that time can be attributed to internal P loading.

TP is usually the measure of P that is considered in mass balances and models that support loading restrictions and reduction strategies for water quality protection. But low bioavailability and lower flows during the important growing period mean that only a portion of the annual external TP input directly affects the productivity of the receiving waters, compared to internal P load that is almost totally bioavailable and occurs during summer and fall (Section 2.4.2).

The settled portion of external load that accumulates in the lake forms a potential internal P source (Figure 2.1). After transformations, the new and former settled P is transformed into mobile P compounds (Rydin 2000) that can be released from the sediments under certain conditions (including low-redox state, high temperature, high tropic state and depending on sediment P and organic matter enrichment). The released part of this mobile P then constitutes the internal P load to a lake's P budget.

2.4.2 *Internal P sources*

When P does not originate externally within the timeframe under consideration, but enters from sources within the lake, it is called internal P load (Albright 2022; Liu et al. 2022; Nürnberg 2009, 1984c; Orihel et al. 2017; Rocha and Lima Neto 2022; Søndergaard et al. 2003; Søndergaard and Jeppesen 2020; Steinman and Spears 2020a). Much of the internal load can be traced back to external sources from previous years and decades, especially in anthropogenically enriched, eutrophic lakes. This is because a large part of the previous P input is retained and accumulates on the lake bed sediments where it presents a *legacy* source of internal load. (The term *legacy phosphorus* was developed for the large portion of P, 70–80%, which has accumulated throughout the watershed, when a large part of previous P input, often by agricultural activities, enters stores in soil, river sediments, groundwater, wetlands, riparian floodplains, and finally in lakes and estuaries (Jarvie et al. 2013; Kleinman et al. 2011; Sharpley et al. 2013).) The legacy load in a lake turns into internal load as it moves from the lake sediments into the lake water for timesteps of individual years or less (Figure 2.1). Using this definition, principles of annual TP mass balance models can be applied that improve the predictions of average seasonal P concentrations (Section 2.4.3) and related water quality variables (Section 2.3).

Different mechanisms for internally derived P are listed in general order of importance, indicating that sediments are usually the primary internal source representing the largest internal P input, plants can be a temporary source, and the water column is negligible in comparison. The actual relative importance is highly lake-specific.

a. *Anoxic or redox-related P release* from the lake bed sediment after dissolution of P adsorbed to iron oxyhydroxides upon decreases in redox potential. This is the classic internal loading mechanism considered ubiquitous. The process involves long-term diagenetic changes in the sediment including upward migration of releasable P. Reductive dissolution releases simultaneously P and Fe into the water, unless Fe forms iron sulfide (Fe_2S) in a sulfur-rich sediment environment (sulfuretum) where sediment P release is uncoupled from Fe.
b. *Microbial-mediated P release* from the sediment of organic P compounds previously settled on the sediment. In addition to mechanism a, the more recently settled material in enriched (eutrophic and hyper-eutrophic) systems and sedimentary bacteria contains organic P that can be mineralized by microbes and released into the overlaying water.
c. *P release from decaying macrophytes.*
d. *P release from particles settling within the water column.*

e. *Sediment resuspension*, due to wind mixing or bioturbation (benthivorous fish, benthic macroinvertebrates), can adsorb (decrease) or contribute to available P fractions in the water. Resuspension of sediments liberates P in the water column, when porewater is high in P but low in material that adsorbs phosphate, such as aluminum and oxidized iron compounds (Fe-oxyhydroxides). If P-adsorbent material is high in the sediment and water P concentration low, it can decrease P in the water.
f. *Bioirrigation* (sometimes included in bioturbation) from burrowing mayflies, chironomid larvae, and oligochaetes.
g. *Translocation* of sediment P by upward migrating cyanobacteria or other motile algae.

Much of the accumulation of the sediment involved in P release is influenced by physical properties including lake depth, wind exposure, and inflow proximity. A process called *sediment focusing* (Blais and Kalff 1995) leads to a higher accumulation of fine-grained sediment particles at deeper depths with associated higher accumulation of releasable P (Tammeorg et al. 2022b). Further, sediments close to P-rich inflows (James et al. 1987) and in macrophyte beds can retain fine particles enriched in releasable P (James and Barko 1990).

The mechanism of sediment P release related to anoxia and low redox potential at the sediment–water interface (mechanism *a*) has been known for almost a century in stratified lakes (Einsele 1936; Mortimer 1941) and has been tested and supported by much subsequent work (Amirbahman et al. 2013; Nürnberg 1988b; Tammeorg et al. 2020a). At anoxic conditions of approximately 200 mV, iron reduction releases any P adsorbed to ferric oxyhydroxides. For example, sediment fractionation (Table 2.3) has been used successfully to determine reductant-soluble P that is released under anoxic conditions from iron oxyhydroxides as the bicarbonate dithionite P fraction (BD-P) (Psenner et al. 1988; Psenner and Pucsko 1988).

Because of the redox-sensitivity of manganese (Mn) that is released at higher oxygen concentration and earlier than iron at a redox potential of approximately 400 mV, Mn is sometimes also thought to be involved in P cycling. This assumption is theoretically unlikely and was not supported by a high-resolution spatio-temporal sampling study on eutrophic sediments of Lake Taihu, where P was controlled by Fe but not Mn compounds (Chen et al. 2019a).

At very low redox potentials (~0 mV) in often eutrophic, hardwater systems, a sulfuretum is established where reduced gases, including hydrogen sulfide and methane, develop. Under these conditions, iron is removed from the P cycle by the formation of iron sulfide (FeS and FeS_2) (Jensen et al. 1995). Therefore, lake sediments can be grouped into iron-releasing versus H_2S-producing sediments during the internal P-loading process (Nürnberg 1994).

When the mineralization of settled organic matter during eutrophication was increasingly investigated (mechanism *b*), studies described an organic P fraction in sediments that can be responsible for some of the sediment P release and is analyzed as the non-reactive P of the NaOH fraction (Hupfer et al. 2009; Reitzel et al. 2007; Rydin 2000) (Table 2.3), by specific enzymes (Tu et al. 2020), or by modified ^{31}P NMR spectroscopy (Ni et al. 2022). Such release is sometimes supported by anoxic conditions (Fleischer 1978; Rydin 2000).

Release from iron oxyhydroxides still occurs in productive systems, even when organic P in the sediment has accumulated and contributes to internal load (Alam et al. 2020; Liu et al. 2022; Søndergaard et al. 2023b). Redox-related Fe-P fractions contributed 70% of TP in the sediment (1959–2017, mesotrophic to hypertrophic status) of stratified, recently eutrophied Lake Lugano, where the surface sediment layer (0–5 cm) showed increased organic P forms. Similarly, the potentially releasable organic P components constituted only a small fraction of TP compared to Fe-P after 11 years of high nutrient loading in an experimental mesocosm study (Saar et al.

Table 2.3 Phosphorus sediment fractions and assumed involvement in lake P cycle ("Meaning") in the modified Psenner extraction scheme (Lukkari et al. 2007; Paludan and Jensen 1995; Rydin 2000).

Name	*P fraction*	*Analysis – extraction*	*Meaning*
"Mobile" fractions contributing to internal P load			
Porewater-P	Loosely bound, porewater P	Water or NH_4Cl	Porewater immediately available during resuspension
BD-P	Redox-sensitive, reducible, reductant-soluble P (Fe-bound)	Bicarbonate-buffered dithionite	P released into the water during anoxia at the sediment–water interface
nrNaOH-P	Organic P (includes short-chain polyphosphates)	Non-reactive P in sodium hydroxide extraction	P released from organic compounds, including settled phytoplankton; especially important in highly eutrophic conditions
Relatively "permanent" fractions, usually not contributing to internal P load			
rNaOH-P	Al-bound, pH-sensitive P (P bound to Al oxides and non-reducible Fe oxides)	Reactive P in sodium hydroxide extraction, w/o the humic-P fraction	pH-sensitive, can be released from sediments under extreme pH conditions
Humic-P	P bound to humic and fulvic acids	Acidification of NaOH-extracted P	Unavailable (refractory) P in the short term
HCl-P	P bound to carbonates and apatite	Hydrochloric acid extraction	Non-releasable P but responds to pH changes
Residual-P	Leftover, not extracted P	Aqua Regia	Non-releasable residual-P

2022). Mineralization contributed only a sub-dominant amount to the porewater SRP in two eutrophic Finnish lakes (Zhao et al. 2024). But total organic P fractions contributed 40–75% to the mobile P fraction in comparison to 10–36% Fe-P and 3–22% loosely bound P in seven hyper-eutrophic shallow lakes in Iowa, USA (Albright et al. 2022).

Redox-related P release also occurs in polymictic lakes and lake regions, which was not considered in the classic internal loading model mentioned above (Einsele 1936; Mortimer 1941). Observations indicate that oxyclines follow the bottom contours so that they are closer to the surface at shallower sites (Molot et al. 2021b, 2014), and there is evidence for widespread hypoxia and sediment oxygen demand at shallower mixed lake sites (Tammeorg et al. 2020b). A study on 436 Danish lakes (area range: 0.01–40 km^2, maximum depth range: 1.3–45 m) observed at least one mixing event during the summer stratified season with oxygen concentrations below 1 mg/L, despite only very small (0.5–1.0 C) differences between the top and bottom temperature in lakes with maximum depths of 4–10 m (Søndergaard et al. 2023a). This means that redox-related release of P and metals is not restricted to deep stratified lakes. It can occur in shallow lakes and in the nearshore productive areas of large lakes in addition to their deeper accumulation areas and provide nutrients to the mixed layer (Figure 2.8).

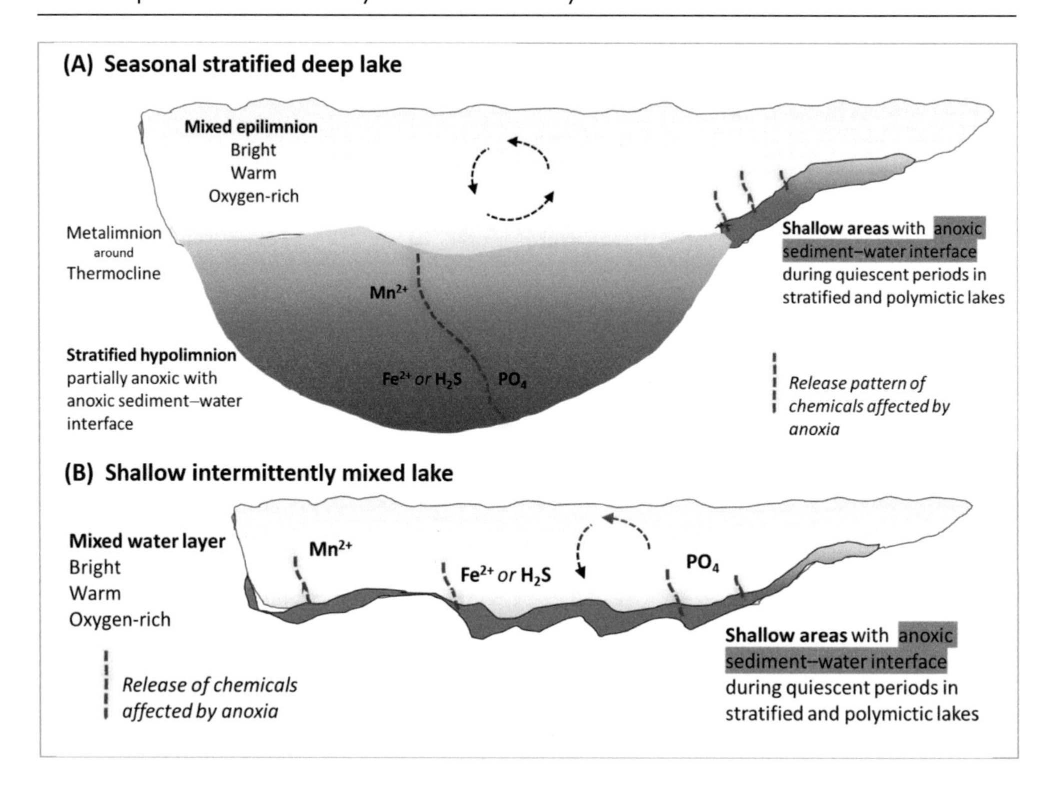

Figure 2.8 Differing mixing regimes leading to different oxygen depletion and redox-related release of phosphorus, metals, and reduced gases in stratified lakes (A) and in polymictic shallow lakes (B). (Stratified lakes have variable shallow areas with anoxic sediment interface.)

In lakes with extensive macrophyte cover, senescing plants can contribute to the internal P pool (mechanism *c*) (Carpenter 1980). But lakes with macrophytes cover are usually not prone to have extended cyanobacteria because they are still relatively clear (van Nes et al. 2007; Scheffer et al. 1993). Chlorophyll was substantially lower in Danish lakes where macrophyte covered at least 20% of the area (Søndergaard et al. 2017). Water hardness and calcium concentration can also affect P release in macrophyte-covered lakes. The dissolution of calcium-bound P contributed to P release in a calcium-rich, macrophyte (*Chara*) covered lake (Kisand and Nõges 2003). The interactions between macrophytes, water, and sediment are complex; they include slowed movement with decreased resuspension, increased sediment accumulation, and pH changes in macrophyte beds (James et al. 1996; James and Barko 1990; Madsen et al. 2001).

Phosphorus in settling particles (seston) in the water column (mechanism *d*) does not represent a separate (net) P source but originates from external or internal P sources. P associated with settling particles may not be easily available to lake plankton. Studies have shown that seston adsorbs rather than releases phosphate, leading to the conclusion that increases in hypolimnetic SRP (a measurement of phosphate, Table 2.1) must have originated from anoxia-related sediment P release in a mesotrophic, stratified lake instead (Gächter and Mares 1985).

In contrast, sediment resuspension (mechanism *e*) can increase TP concentration in the overlying water (Li et al. 2019; Tammeorg et al. 2013). Depending on the nutrient enrichment of the particles, they can adsorb or desorb phosphate, while increasing TP. In shallow oxic parts of oligotrophic lakes, sediment resuspension does not liberate phosphate but rather adsorbs it from the surrounding water because of P-adsorbing components that increase the P adsorption capacity (Cyr et al. 2009; James 2017a). In contrast, in shallow unstratified eutrophic lakes, the mixing of porewater that is enriched with P from redox-related desorption mechanisms leads to P increases in the water column throughout the warm period, hence contributing to redox-related internal load, source *a* (Tammeorg et al. 2020b, 2020a). It is also possible that dissolved P from organic compounds or aluminum compounds is liberated from sediments at certain conditions such as changes in pH. For example, P desorption from suspended particles is related to pH, and studies have shown that P release from sediments increased at pH above 9 (Istvánovics and Petterson 1998; James et al. 1996; Koski-Vähälä and Hartikainen 2001; Penn et al. 2000) especially in hardwater lakes, due to calcium interaction with P.

Bioirrigation (mechanism *f*) by burrowing mayflies, chironomid larvae, and oligochaetes can have varying effects (Hölker et al. 2015; Steinman and Spears 2020a). It can increase sediment P release (Chaffin and Kane 2010) or aerate the sediment, possibly increasing sediment P retention rather than contributing to an internal load (Lewandowski and Hupfer 2005).

Phosphorus translocation from the sediments into the open water by cyanobacteria (mechanism *g*) is possible for specific cyanobacteria taxa (Gervais et al. 2003), but it rarely constitutes a large consistent amount throughout the growing period (Section 5.4.1) (James 2017b; James et al. 1992).

Redox-related P released from bottom sediments (mechanism *a*) is potentially the most important internal source (compared to other sources) because it is released as phosphate, which is an immediate bioavailable form of P (Table 2.1). Studies on SRP and bioassays (15–60 min short-term bioassays with P-limited natural lake water) demonstrated that indeed 80–100% of the TP in anoxic hypolimnia of stratified anoxic lakes is orthophosphate and can be taken up by the natural phyto- and bacterioplankton (Figure 2.9, Nürnberg and Peters 1984a). The SRP proportion increases with concentration so that at higher hypolimnetic TP, almost all is SRP.

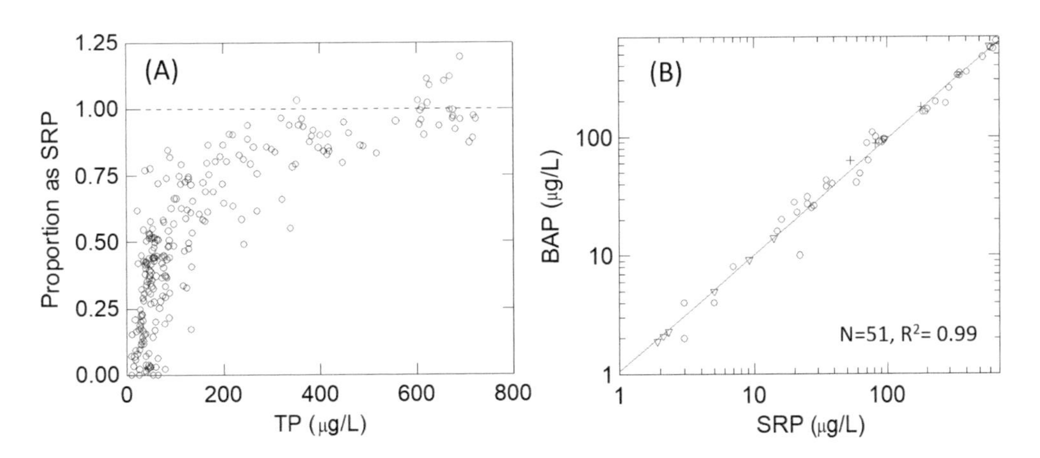

Figure 2.9 Proportion of TP that is SRP in anoxic hypolimnia (A) and the comparison of biologically available P (BAP, by 3 different bioassays) with SRP from anoxic hypolimnia (B). *Based on data from Nürnberg and Peters (1984a).*

Dilution is an important controller of the availability of P released from resuspended or bottom sediments. When a small amount is suspended into a large volume of water, proportionally more P is in a bioavailable form (Cyr et al. 2009; Koski-Vähälä and Hartikainen 2001), similar to the enhanced availability of P in diluted hypolimnetic water (Nürnberg 1985; Nürnberg and Peters 1984a). Only when the dilution of hypolimnetic P-rich water is small, less P is taken up by lake plankton and more is adsorbed by metals and settles out.

Indications of internal loading from the bottom sediments (mechanism *a*) differ between lakes that stratify and those that do not and are polymictic (Table 2.4). In stratified lakes, redox-related P release occurs when the hypolimnion becomes oxygen depleted, leading to large P increases close to the sediment–water interface and into the hypolimnion (Figure 2.10). Hypolimnetic P can diffuse into the epilimnion during wave action from unsettled weather (Niemistö et al. 2012) and seiche activity (Nielson and Henderson 2022) and during late summer thermocline erosion and becomes distributed into the whole lake at turnover that typically occurs in the fall (Nürnberg and Peters 1984b) as controlled by lake location, morphometry, and hypolimnetic temperature (Nürnberg 1988a).

Depending on the degree of enrichment, polymictic lake sediments can occasionally be anoxic at the surface despite overlaying oxygen-rich water, so that the mechanism of redox-related P release occurs here as well (Tammeorg et al. 2020b). Variable redox conditions occur when turbulence creates relatively high DO concentration above the sediment, while sediment oxygen demand rapidly decreases oxygen concentration during quiescent conditions (Bryant et al. 2010), at high temperatures, and elevated trophic state, resulting in internal loading (Søndergaard et al. 2003). Different from stratified lakes, where sediment-released P typically remains in the hypolimnion, released phosphate can

Table 2.4 Indicators of internal load in stratified (A) and polymictic lakes (B) (based on Nürnberg 2009).

A. Stratified, mono-, or dimictic (deep) lakes

- Severe hypolimnetic anoxia, low redox potential (<200 mV)
- Increasing TP and SRP with depth
- Increasing hypolimnetic TP and SRP throughout summer
- Concomitant iron, manganese, or reduced gas development
- Increased turbidity and phytoplankton blooms during fall turnover
- With respect to P mass balance:
  a. More TP leaving the lake than entering (negative retention)
  b. Less TP retained than predicted (from morphometric–hydrological characteristics)
  c. Higher TP concentration than predicted

B. Polymictic (shallow) lakes

- Seasonal: Increasing TP and potentially, SRP, throughout summer, even in upper water layers
- Increasing turbidity and phytoplankton blooms during summer
- Thin oxic sediment layer; occasional anoxia in weed beds and open water during quiescent conditions (early morning)
- Occasional iron, manganese, or reduced gas development during quiescent conditions
- With respect to P mass balance:
  a. More TP leaving the lake than entering (negative retention)
  b. Less TP retained than predicted (from morphometric-hydrological characteristics)
  c. Higher TP concentration than predicted

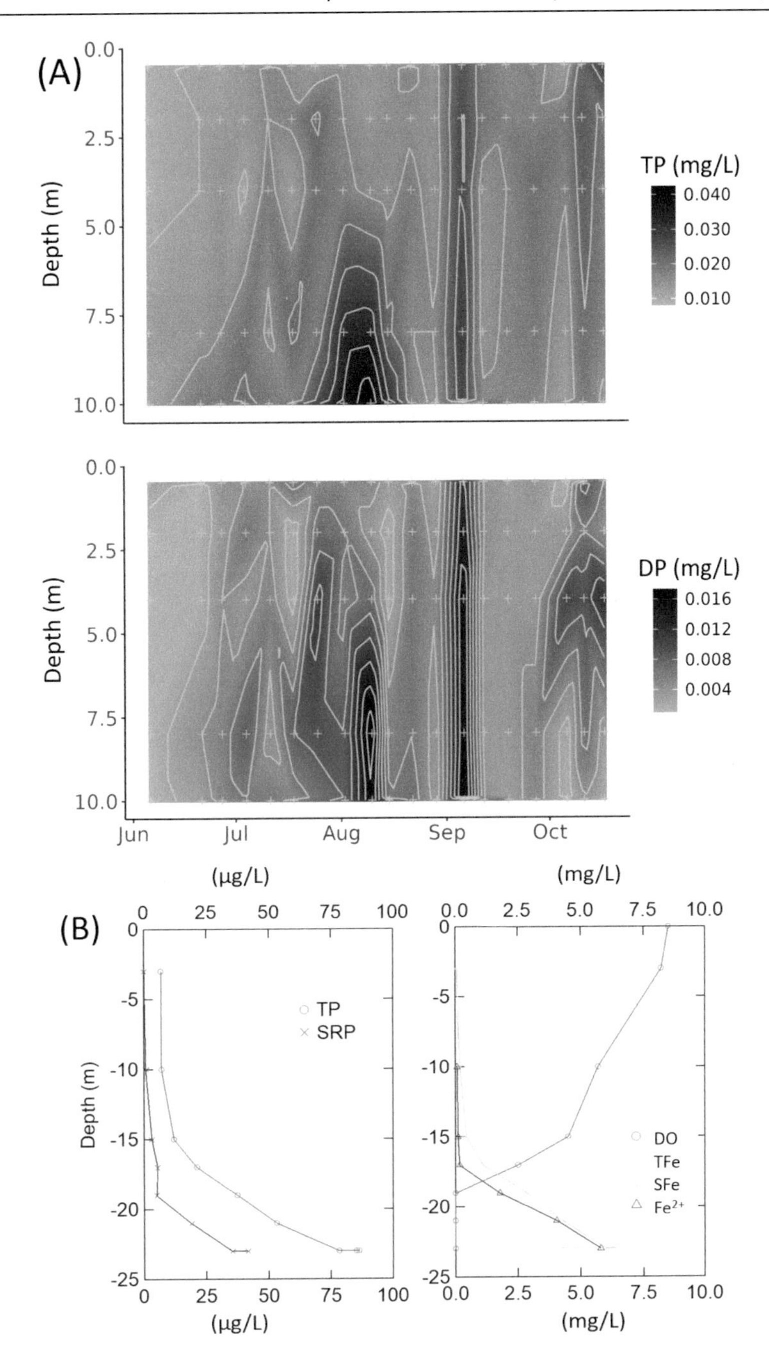

Figure 2.10 Changes of phosphorus and related variables in stratified lakes. (A) Total P (TP) and total dissolved P (DP) concentration contour lines in mesotrophic Tower Road Reservoir, New Brunswick, Canada (deep site B3, 2022, with data from City of Moncton, New Brunswick, Canada) and (B) fall profiles of TP, soluble reactive P (SRP), dissolved oxygen (DO), total iron (TFe), soluble iron (SFe), and ferrous iron (Fe^{2+}) in oligotrophic Chub Lake, Ontario, Canada (September 13, 1982).

be immediately distributed throughout the water column in polymictic lakes so that whole-lake P concentration increases and fertilizes the phytoplankton (Figure 2.11) (James et al. 2015; Nürnberg 2009).

The earliest indications of imminent P release in polymictic lakes may be Mn rather than Fe, because Mn is released at a redox potential higher than Fe (~400 compared to 200 mV) and

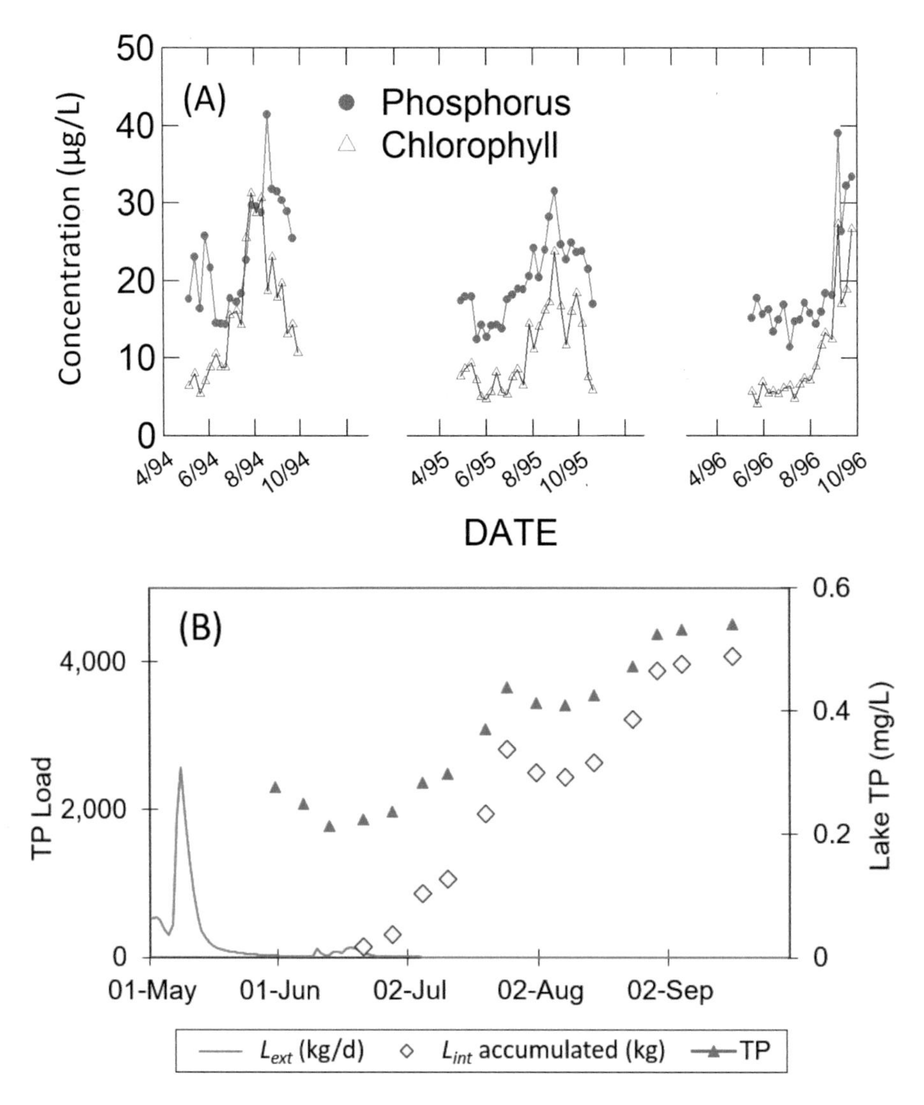

Figure 2.11 Changes of phosphorus and chlorophyll in polymictic lakes. (A) Variable timing of increases of TP and related chlorophyll concentration indicate different stratification regimes in three growing periods of Brome Lake, Quebec, Canada (1994–1996, 1. summer stratified then mixed, 2. summer mixed, 3. summer stratified). (B) Increases in TP concentrations were used to estimate internal P load (L_{int} accumulated) in the polymictic reservoir Lake Mitchell, South Dakota (2001), with L_{ext}, external P load. *Figures 3 and 4 from Nürnberg (2009), open access.*

is released earlier and prevails longer after re-aeration. Although Mn is not directly involved in P release, the Mn concentration increase was suggested as an indicator for the beginning of cyanobacteria blooms in shallow eutrophic Missisquoi Bay of Lake Champlain, Vermont, USA (Giles et al. 2016).

Approaches to quantify internal load depend on data availability, lake characteristics, and the extent of scientific and financial commitment among other aspects (Nürnberg 2020a). Detailed process-based models (also called, mechanistic models) that are data intensive have sub-models to indirectly incorporate internal load in water quality models for specific lakes and reservoirs. In comparison, empirical, cross-system models developed on data from many lakes over a wide range of conditions provide simpler and more testable approaches to predict internal load, as illustrated below.

A direct determination of internal load is relatively simple in stratified lakes and can be based on the TP mass increase in the hypolimnion or the whole water-column during the stratified anoxic period (i.e., summer and fall in the temperature zone) (Approach 1 in Box 2.3). Estimating internal load in shallow, mixed systems is more complicated. Here, TP cannot easily be traced back to the source after it is released from the sediments right into the mixed water column where it can be taken up by phyto- and bacterioplankton. Further, the actively releasing sediment area that is involved in redox-related P release cannot easily be determined because the depth of oxygen penetration that controls P adsorption and release is rarely known. However, Approach 1 and models developed on stratified lakes were tested and, with specific constraints, are also applicable to polymictic lakes.

When a TP mass balance can be established from available TP input and export, internal load can be computed, considering a predicted P retention that accounts for sedimentation (downward flux) (Approach 2 in Box 2.3). The separation of internal load into its components, which are the active release area and the P release rate from such area, provides a third approach to estimate sediment-related internal load (Approach 3 in Box 2.3). The temporal and spatial extent of the area involved in P release can be estimated from dissolved oxygen profiles or predicted (*AF* or *AA*, Box 2.1, Nürnberg 2009, 2020a). Areal P release rates (*RR*, also called diffusive flux, Figure 2.12) are often defined as P release per actively releasing (anoxic) sediment area per day ($mg/m^2/d$). *RR* can be determined in the lake or in the laboratory by sediment–water incubations and depend on temperature and oxygen penetration depth (Figure 2.12A, B) among other variables. Models for the prediction of these anoxic release rates include regression equations based on (a) the concentration of specific sediment P fractions involved in redox-related release or just sediment TP (Figure 2.12C), (b) lake trophic state (Figure 2.12D) or just the TP concentration in the water, and (c) watershed characteristics, including land use and geochemistry, as reviewed in (Nürnberg 2020a).

The sediment P adsorption capacity that regulates long-term P retention and release depends on chemical constituents, especially Al, Fe, and Ca relative to the releasable P fractions. Specific thresholds of ratios have been described to improve the predictions of possible P release under specific conditions, including lake and catchment acidification (Kopáček et al. 2015; Nürnberg et al. 2018; Ostrofsky and Marbach 2019), and aluminum enrichment due to reversed acidification (Section 5.1.8) or restoration treatments (Section 6.4.1). Relationships between such sediment-fraction ratios and lake TP concentration were significant predictors in addition to watershed properties (e.g., percent agricultural area) and dissolved organic carbon (representing colour from organic matter, Section 4.3.2) in 126 lakes in Maine, USA (Amirbahman et al. 2022).

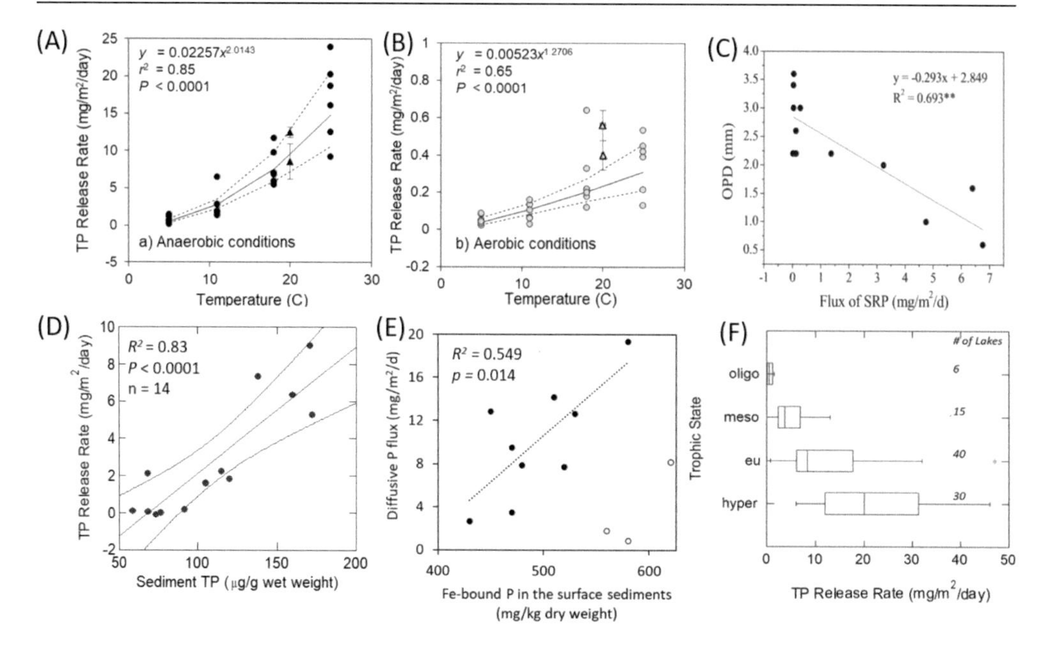

Figure 2.12 Dependency of phosphorus release rates (RR, P release per actively releasing, anoxic sediment area per day, mg/m²/d) on oxygen and temperature in the water overlying the sediment (A, B, *Figure 3 from James 2017a, open access*), on the oxygen penetration depth (OPD in C, *Figure 8 from Ding et al. 2018, with permission*), on sediment TP (D, *redrawn from Nürnberg 1988b*), on redox-sensitive sediment P fraction (E, *redrawn from Tammeorg et al. 2020a, with permission*), and on trophic state (F, *based on Nürnberg (1997) with added data*).

Box 2.3 Approaches to estimate internal P load in stratified and polymictic lakes

Approach 1: A partially net estimate from ***in situ* water-column TP increases**, in stratified and mixed lakes. This approach considers any internal source during the study time period.

In-situ internal load, L_{int_I}, can be determined from the increases of the TP concentration in the whole water column between spring/summer (minimum TP concentration) and fall (maximum TP) (Equation 2.5).

$$\text{Equation 2.5:} \quad L_{int\_I} = (P\_t_2 \times V\_t_2 - P\_t_1 \times V\_t_1)/(A)$$

Where, t_1 is the initial date (early summer) and t_2 date at end of period (late summer and fall), P_t the correspondent whole water-column pro-rated average P concentration of lake representative stations. and V_t the correspondent lake volume. A is lake surface area for each lake in km². In stratified lakes, the increases in the lower layer (hypolimnion) can be used instead of the whole water column average. This approach is expected to yield an underestimate because the calculation does not include internal loading that was redeposited during the considered time period. It therefore supports the other two methods, especially when its (partially net) results are below those of the (gross) values of Approaches 2 and 3.

Approach 2: Net (and gross) estimates from annual P budgets (*mass balance approach*). This approach considers any internal source during the study time period.

The mass balance calculations of approach 2 yield net estimates of internal load, L_{int_2}, which can subsequently be converted to gross estimates to make them comparable to L_{int_3} and external fluxes. L_{int_2} is based on the separation of downward (settling) from upward (sediment release) fluxes of P via the theory of P retention. Measured lake retention (R_{meas}) is the proportion of P that is retained by the whole lake and should include all external input and output fluxes, as *(in – out)/in* or

Equation 2.6: $R_{meas} = (L_{ext} - L_{out})/L_{ext}$

where L_{ext} is the annual areal external P load and L_{out} is the annual export via the outflow, both in units of mg m^{-2} yr^{-1}. According to Equation 2.6, $R_{meas} \times L_{ext}$ equals the net amount of P retained by the lake ($L_{ext} - L_{out}$), or the difference between annual sedimentation to and release from the sediments.

In contrast to measured retention, predicted retention (R_{pred}) represents the downward flux of P due to settling and sedimentation. It is modelled according to Nürnberg (1984c):

Equation 2.7: $R_{pred} = 15/(18 + q_s)$

where q_s is the annual water load in m yr^{-1}, measured as annual average outflow volume over lake area. This model was developed for stratified lakes ($R^2 = 0.63$, $n = 54$) and has been successfully applied to polymictic lakes (Nürnberg 2005), although other R models produce a better fit in some lakes (Nürnberg 2009). Accordingly, the settled portion of external load can be calculated by the term $R_{pred} \times L_{ext}$. In lakes with internal load, predicted retention overestimates measured retention approximately by the net amount of P released from the sediments (net L_{int_2}) (Nürnberg 1984c) so that:

Equation 2.8: $Net\ L_{int\_2} = L_{ext} \times (R_{pred} - R_{meas})$

As it is based on an annual mass balance, L_{int_2} includes both summer and a potential winter internal load. Net and gross estimates of L_{int_2} are related by R_{pred} according to Nürnberg (1998) who showed that internal P released from the sediment settles back down at the same rate as external P on an annual basis. Consequently, gross internal load can be calculated from net internal load:

Equation 2.9: $Gross\ L_{int\_2} = Net\ L_{int\_2}/(1 - R_{pred})$.

The errors associated with L_{int_2} estimates are often large because the estimates are calculated from the differences between observed input and output fluxes and are affected by the errors associated with the determination of these fluxes. Nonetheless, positive multiple year estimates would indicate the occurrence of internal load.

Approach 3: Internal load ($RR \times AA$, $RR \times AF$) determined from measured or predicted P release rates (RR) and anoxia at the sediment–water interface (AA or AF). This approach mainly considers redox-related P release from iron oxyhydroxides and from organic sources as the mechanism and yields a gross estimate of internal load.

Equation 2.10: $L_{int\_3} = RR \times AF$ or $RR \times AA$

Where RR represents the *daily anoxic areal P release rate* (mg/m²/d), and anoxia is measured as AF (d/summer or d/year) or predicted as AA (d/summer or d/year) that represents the *active sediment release area* (Box 2.1). AF only applies to stratified lakes, while AA applies to all, polymictic and stratified, lakes. RR can be measured experimentally or modelled in different ways, e.g., from sediment P fractions, trophic state, lake water TP concentration, and catchment basin characteristics.

Sediment temperature influences RR and AA drastically and contributes to the variability of internal load due to varying air temperature. Sediment temperature fluctuations are especially pronounced in polymictic lakes, and changes due to climate change have to be considered in predictions (Nürnberg et al. 2012; Nürnberg and LaZerte 2016b). To include this variability, both AA and RR can be adjusted with respect to the summer mixed layer or bottom water temperature to follow the Q10 rule of Van Holst (Equation 2.11).

Equation 2.11: $RR_i = RR \times c^{(t\_i - t\_avg)/10}$

RR_i is the average daily release rate for year$_i$ based on a given summer average lake temperature t_i; t_{avg} is the average summer temperature of all study years, and c is the Van Holst constant or Q10 value, usually close to $c = 3$, which can be calibrated.

Approach 3 is based on sub-models that were originally developed for P release from anoxic sediments. Other potential internal sources, for example, those of senescing macrophytes (Graneli and Solander 1988), littoral P release (Cyr et al. 2009; James and Barko 1991), benthivorous fish (Sereda et al. 2008), and groundwater, may have to be quantified independently. However, they are usually small or negligible compared to the portion derived from lake bottom sediments (Sereda et al. 2008).

The different internal load estimation approaches (Box 2.3) deliver slightly different values for several reasons. (1) The approaches do not always yield a gross estimate; e.g., Approach 1, in situ increases, includes sedimentation throughout the period under study, and therefore yields slightly smaller, partially net, estimates than Approach 3. The mass balance, Approach 2, first yields a net estimate that must be converted to gross before it can be compared to external load, a gross estimate. (2) There are errors to be considered, based on sample size, conditions, sub-model applicability, and climate variability. The natural variability of P cycling in lakes can be large (Figure 2.13), which means that the number of years for which loads are calculated needs to be high to obtain a representative estimate.

The size of the observed internal load quantities compared to external P load is variable in the presented examples but is usually substantially higher than other potential internal fluxes, like P input from resuspension (Tammeorg et al. 2016) (unless the porewater is enriched from redox-related P mobilization), decaying macrophytes, or migrating cyanobacteria (Section 5.4.1).

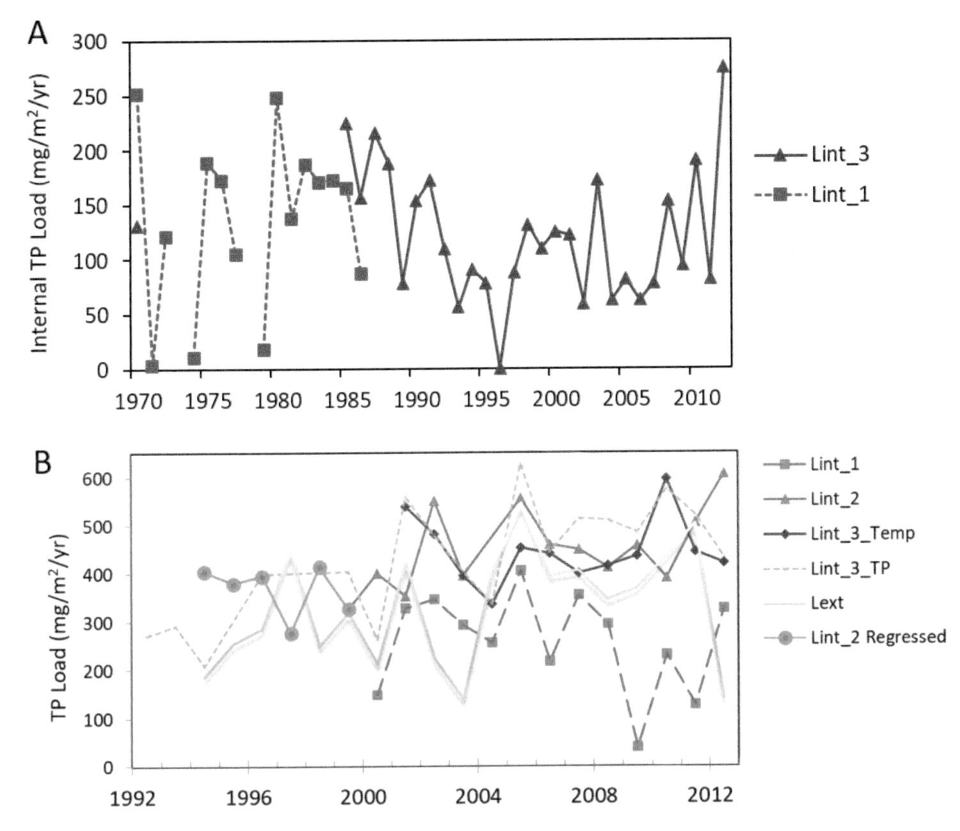

Figure 2.13 Comparison of different internal P load calculations (Box 2.3, with L_{int_1}, in situ internal load, L_{int_2}, mass balance-derived load, and L_{int_3}, internal load predicted as $RR \times AF$ (Erie) and $RR \times AA$ (Winnipeg)). (A) Lake Erie and (B) Lake Winnipeg, Manitoba ($L_{int\_3\_Temp}$, release rates based on temperature, $L_{int\_3\_TP}$ release rates based on lake TP, L_{int_2} Regressed based on a regression to extend values of L_{int_2} backwards in time) also showing the external load for comparison. *Redrawn with permissions from Nürnberg et al. (2019) and Nürnberg and LaZerte (2016b).*

2.4.3 *Dependencies (TP concentration modelling)*

Because P affects lake trophic state and water quality so importantly (Sections 2.2 and 2.3), many data-intensive as well as simpler models have been developed to predict various TP concentration variables (e.g., spatial or temporal averages or medians). In empirical mass balance modelling, the TP concentration is determined by the external and internal P loads and their retention (Box 2.2). Seasonally different TP averages are predicted by either including or not including the internal load term with various amounts of settling (retention, *R*) (Table 2.5).

2.5 Comparison of phosphorus with nitrogen

Total phosphorus in lakes is closely correlated with total nitrogen concentration averages (Figure 2.4). Nitrogen is also considered to be an important macronutrient for phytoplankton and drives eutrophication especially at higher P concentrations (Section 3.3).

Table 2.5 Predictive equations for TP averages for different periods in stratified and mixed lakes (Nürnberg 2020a, 2009, 1998). Examples of empirical (simple) models based on the mass balance concept.

	Formula	*Predicted TP averages for different periods and water layers*	
		Stratified lake	*Polymictic (mixed) lake*
Internal load not explicitly incorporated:			
A	$TP = \frac{L_{ext}}{q_s} \times (1 - R_{meas})$	Annual water column	Annual water column, often close to annual mixed surface layer
Internal load as gross estimate:			
B	$TP = \frac{L_{ext} + L_{int}}{q_s} \times (1 - R_{pred})$	Annual water column (close to stratified period epilimnetic)	Annual water column; minimum of growing period
C	$TP = \frac{L_{ext}}{q_s} \times (1 - R_{pred}) + \frac{L_{int}}{q_s}$	Maximum fall	Maximum of growing period, close to fall average.
Internal load as partially net estimate (in situ increases):			
D	$TP = \frac{L_{ext}}{q_s} \times (1 - R_{pred}) + \frac{L_{int\_1}}{q_s}$	Stratified period epilimnetic	Growing period
Internal load not considered:			
E	$TP = \frac{L_{ext}}{q_s} \times (1 - R_{pred})$	Minimum summer epilimnetic (often early summer)	Not applicable, underestimate unless $L_{int} = 0$

Note: External (L_{ext}) *and internal* (L_{int}) *load values are gross estimates (mg/m²/yr).* L_{int_1} *is determined from in-situ TP increases throughout the growing season and is a partially net estimate (Approach 1 in* Box 2.3*).* $R_{meas} = (L_{ext} - L_{out})/L_{ext}$ (Equation 2.6), $R_{pred} = 15/(18+q_s)$ with q_s being the annual areal water load (m/yr) (Equation 2.7). Formula A is not a model but just reflects the mass balance constraints.

Within a lake, the two main nutrients, P and N, occur in various chemical forms. In a P-limited stratified lake, there is little orthophosphate in the epilimnion, while the dissolution of P from iron oxyhydroxides leads to an accumulation of orthophosphate and ferrous iron in the hypolimnion when it is oxygen depleted, and the sediment surfaces are anoxic (Figure 2.14A).

Nitrogen forms are also affected by oxygen depletion. The exchanges between various N-compounds in lake water and sediment are complex and include differing redox states between oxidized (NO_x, i.e., NO_2 and NO_3) and reduced forms (NH_4 and urea), and other organic forms (Figure 2.14B). Reduced N compounds enhance anoxic sediment P release and are often correlated with P concentration changes during the period of internal P loading. N paths include N-fixation (biotic N uptake from N_2 gas), denitrification (conversion of NO_x into N_2), and the Anammox process (anaerobic ammonium oxidation, conversion of NH_4 into N_2) (Lee et al. 2009). Specific N-forms affect cyanobacterial growth in addition to specific P-compounds and regulate varying nutrient limitation throughout the year (Section 3.3).

The P mass balance is considered closed because there is no gaseous state of the element under natural conditions (Figure 2.14A). In contrast, the element N has several gaseous phases (Figure 2.14B), with 78.1% N_2 in the atmosphere from where it can be incorporated into the water and living organisms, often in form of the reduced gas ammonium (NH_3 or NH_4). NH_4

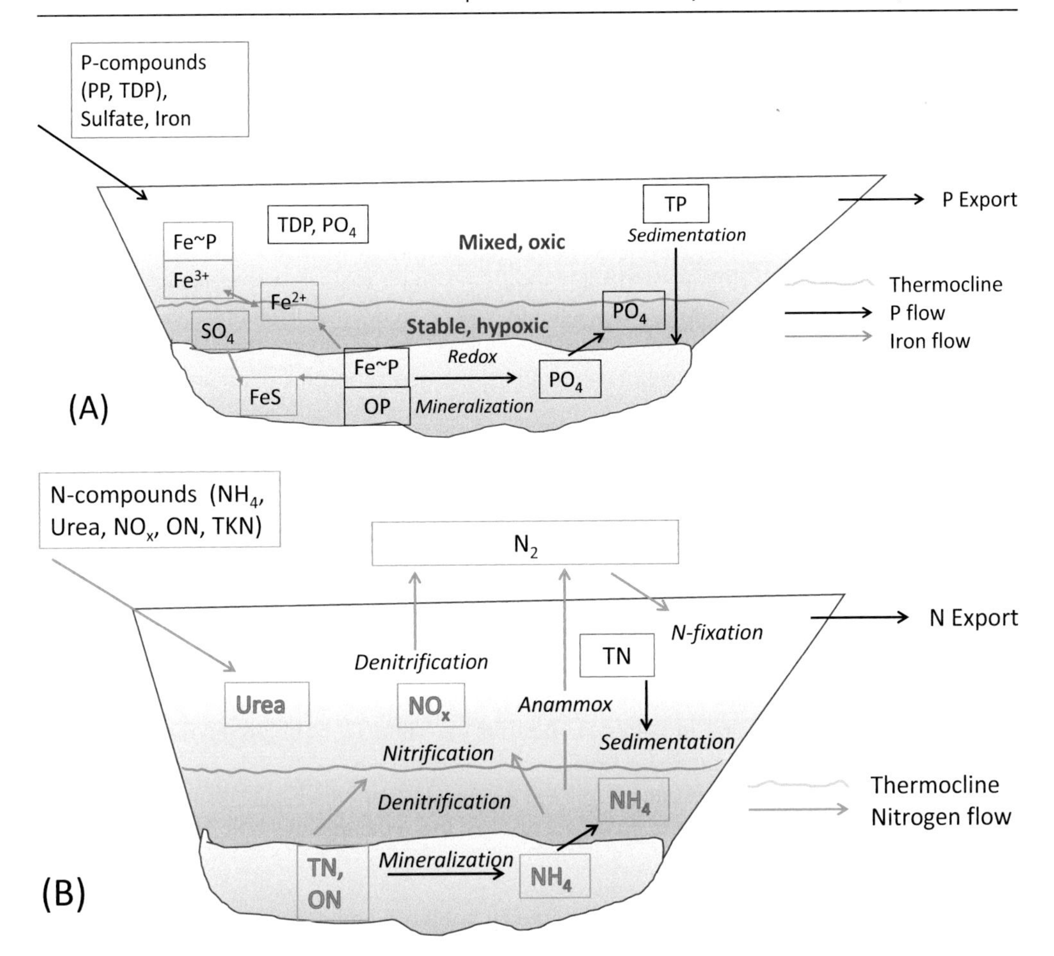

Figure 2.14 Potential fluxes of phosphorus (A) and nitrogen compounds (B) in stratified lakes (TDP, including organic P in the water; OP, sediment organic P; Fe~P, reductant-soluble P from Fe-oxyhydroxides; TP, total P including particulate P, Table 2.1; N_2, nitrogen gas; NOx, nitrate and nitrite; ON, organic N; TKN, total Kjeldahl N, consisting of NH_4 and ON). Sulfur can interact with iron by forming FeS. Most mechanisms occur in the summer in shallow (with micro-stratification close to the bottom) and deep productive lakes.

can form from the bacterial mineralization of organic matter, abiotically under reduced conditions or biotically in living cells (Wetzel 2001). Its accumulation in the hypolimnion or at the sediment–water interface is sometimes called internal N load.

The freshwater N mass balance is further complicated by industrial fixation of N_2 to NH_3 in the Haber Bosch process ($N_2 + 3\ H_2 \rightarrow 2\ NH_3$). This reaction is responsible for the vast increase in recent N fertilizers that provided the food production to support the human population increase from 2 billion people in the early century to more than 7 billion (Glibert 2020). In contrast, the amount of P fertilizer is constrained, because phosphate has to be mined from the deposits of phosphate minerals (Cordell et al. 2009; Gilbert 2009).

2.6 Chemical interactions of P with iron, dissolved organic matter (DOM), and browning

When iron originates from sources in the catchment (including wetlands), its transport from soils to waters is typically associated with dissolved organic matter (DOM, including DOC and humic and fulvic acids) that increases iron solubility in oxic circum-neutral waters. This process increases the colour in the receiving water and can lead to lake *browning* or *brownification*. Phosphorus and iron are often correlated in the anoxic bottom water because of P desorption from iron oxyhydroxides (Figure 2.10B and Figure 2.14A). Consequently, iron and associated processes including brownification are potentially related to internal P loading.

Coloured lakes have specific characteristics compared to clear lakes (Table 2.6) (Ohle 1935; Wetzel 2001). These special characteristics affect the many general relationships involving nutrients, productivity, phytoplankton biomass, transparency, bacteria biomass, as well as the abundance and bloom potential of cyanobacteria (Havens and Nürnberg 2004; Meyer et al. 2024; Nürnberg 1996; Nürnberg and Shaw 1998).

A generally higher trophic state compared to clear lakes was determined in an analysis of 600 lakes spanning a latitude of 39°S to 82°N (Nürnberg and Shaw 1998). Higher trophic state was manifest as higher volumetric primary and secondary productivity, where lakes with "browner" colour and higher DOC had higher TP and chlorophyll concentration and shallower Secchi disk depths. Because coloured lakes have smaller euphotic depths, the browning effect on productivity (PP) differs between areal versus volumetric measures. PP expressed on an areal basis (e.g., $C/m^2/yr$) was lower in humic lakes, but, when expressed volumetrically (e.g., mg $C/m^3/d$), PP was similar or slightly higher in coloured lakes for the same TP or chlorophyll concentration (Nürnberg 1999; Nürnberg and Shaw 1998).

In contrast to an increased productivity (and a constant ratio of TP to DOC concentrations), an increase in light limitation due to browning resulted in a reduction in whole-lake productivity,

Table 2.6 Lake characteristics that differ between coloured and clear lakes, based on a 600-lake study (Nürnberg and Shaw 1998).

Lake characteristics	*Coloured compared to clear lakes*	*Variables used by Nürnberg and Shaw (1999)*
Organic acids, organic matter, DOC	+	DOC
Colour	+	Colour
Transparency	-	Secchi disk depth
Light	-	*n.d.*
Euphotic depth	-	*n.d.*
Total P	+	TP
Total N	+	TN
Fe	+	*n.d.*
pH	-	pH
Conductivity	-	Conductivity
Hypolimnetic anoxia	+	Anoxic Factor
Phytoplankton	+	Chlorophyll
Primary productivity (PP)	+	Volumetric PP
	-	Areal PP
Bacterial productivity	+	Volumetric bacterial productivity
Stratification, stability	+	*n.d.*

Note: + (−) means significant higher (lower) in coloured lakes compared to clear lakes; *n.d.*, not determined.

especially in oligotrophic lakes (Stetler et al. 2021). Here, the ratio of TP to DOC decreased during browning and was explained by catchment recuperation from acid precipitation and the increased growing period of terrestrial vegetation, which would increase DOM export but decrease TP export.

The differing effects on productivity can perhaps be explained by colour intensity. In 102 Finnish lakes with a colour gradient, phytoplankton biomass increased up to a specific threshold of DOC and water colour after which it decreased (Horppila et al. 2023). At high DOC (>14.4 mg/DOC in this study), intense colour did not only reduce access to photosynthetic radiation but also led to a decrease in the illuminated part of the mixing zone, all of which caused light limitation.

Because of the complex chemical and biological interactions, the effect of browning on water quality is not always consistent. Dissolved organic matter (mostly DOC) cycling includes biotic reactions such as microbial transformation of particulate to dissolved organic matter and abiotic reactions including redox cycling among others as reviewed in (Lau and del Giorgio 2020). Humic–metal–P complexes can form and bind part of P, making P unavailable for bioproduction under nutrient-poor conditions, as observed in a microcosm experiment (Saar et al. 2022). But the complex formation was hindered by high pH due to enhanced photosynthesis in the P-enriched treatment.

Studies demonstrating an increase in browning and lake water iron concentrations also lend support to increasing internal P loading because of the geochemical connection of iron oxyhydroxide–phosphate complexes (Section 2.4.2., Figure 2.14). Many of the lakes experiencing Fe enrichment are nutrient-poor, and it is not always obvious (because of the low P concentration and the immediate uptake by P-starved phytoplankton and bacteria) that along with redox-related Fe mobilization, also phosphate becomes available (Nürnberg 1985).

Sedimentary organic content can have opposing effects on internal P load. Organic matter has been found to be negatively correlated with anoxic P release rates because it can bind P (Chen et al. 2023; Nürnberg 1988b; Tammeorg et al. 2022b), but organic matter positively affects oxygen depletion (Schwefel et al. 2023). That means that internal load, which is controlled by P release rates and the spread and duration of anoxia (Section 2.4.2), may be affected in opposite directions during the process of brownification (i.e., decreased P release rates but increased duration and spread of anoxia).

2.7 Conclusion concerning phosphorus

It is widely accepted that phosphorus is the most important nutrient in freshwater systems (besides nitrogen that becomes important under P-rich conditions, Section 3.3, Box 3.2) and controls lake water quality including the trophic state. There are two basically different P loads that contribute to the P concentration in lakes, external and internal, which have conspicuously different impacts on the lakes' P concentration and related water quality. These differences support later arguments on internal load effects on cyanobacteria. A simple approximation of the relative importance of external versus internal inputs is available as the watershed to lake area ratio. The pollution history further dictates the amount of accumulated P as internal P load (Figure 2.1).

The most important differences between external and internal load are:

External load

- P input originates in the catchment basin (above- and underground) and the atmosphere.
- Typically, only a small portion is in a biologically available chemical form (~30% on average).

- Often scarce during the growing period (e.g., dry summer period).
- Creates legacy sediment P that drives internal P loading.

Internal load

- Typically in a highly biologically available chemical form (80–100% orthophosphate)
- Mostly provided during the growing period and distributed directly into the mixed layer (polymictic lakes) or into the epilimnion as entrainment during episodes of internal wave action and late summer thermocline erosion (stratified lakes).
- Responds to external load after a lag time on the order of years to decades (faster in highly productive systems).
- Can also create legacy sediment P after being recycled into the lake food chain (from settled phytoplankton and detritus).

Because of the generally high quantity of P derived from anoxic sediments compared to that from other internal sources, sediment-derived redox-related P can be considered as the prevalent source of internal load. In addition, mineralization and release from organic settled matter (including cyanobacteria) contribute to internal load in eutrophic and hyper-eutrophic lakes.

A collection of detailed discussions of various aspects of internal P loading is presented in *Causes, Case Studies, and Management* (Steinman and Spears 2020b).

Chapter 3

Cyanobacteria in freshwater

Cyanobacteria are primarily autotrophic prokaryotes (except for a few lesser-known genetically determined groups of non-photosynthetic bacteria (Soo et al., 2017)). Different from other prokaryotes (bacteria and archaea), they perform oxygenic photosynthesis with the green pigment of chlorophyll *a*. Cyanobacteria have been on the earth since the Archean and Proterozoic time periods, are ubiquitous, and inhabit a wide variety of environments with extremes of temperature, salinity, mineral and nutrient availability, and redox conditions in terrestrial and watery habitats.

This chapter concentrates on the characteristics of cyanobacteria inhabiting mainly lentic systems (lakes, reservoirs, and slow-moving rivers). Many freshwater cyanobacteria taxa have evolved to gather on a lake surface in the summer in large floating masses ("scum") to form (potentially toxic, Section 3.2) *harmful algal blooms* (HABs) (Figure 3.1). Several genera can control their buoyancy by the formation of gas vesicles, have a low sensitivity to UV-light, are adapted to elevated temperature (high maximum growth temperatures), and are less edible compared to eukaryotic phytoplankton (Sections 3.5 and 3.6). Other cyanobacteria form layers at specific depths in the water column or are distributed throughout the photogenic zone. Several taxa migrate between the open water and the bottom sediments daily to reach nutrients or light, and seasonally, as vegetative or specially formed resting cells (akinetes). Cyanobacteria are also adept at assimilating macronutrients (P and N, Section 3.3) in low supply and micronutrients (Section 3.4), which provides an advantage over competing phytoplankton.

Cyanobacteria appear to be increasing in recent years (Ho et al. 2019; Huisman et al. 2018; Paerl and Huisman 2009; Pick 2016; Scholz et al. 2017; Taranu et al. 2015) even in nutrient-poor lakes (Favot et al. 2023; Reinl et al. 2021; Winter et al. 2011). Global mapping (30-m-resolution Landsat satellite images between 1982 and 2019) revealed expansion of areas and frequency of HABs over the past decade (Hou et al. 2022) (Figure 3.2). Paleolimnological sediment analysis of 108 time series and 18 decadal-scale water monitoring records demonstrated that "(1) cyanobacteria have increased significantly since 1800, (2) they have increased disproportionately relative to other phytoplankton, and (3) cyanobacteria increased more rapidly post 1945" (Taranu et al. 2015). This ubiquitous increase in cyanobacteria is especially worrisome, because many bloom-forming cyanobacteria species have toxic strains that may produce cyanotoxins (Section 3.2).

In most, especially eutrophic lakes, the summer and fall chlorophyll *a* pigment concentration is mostly contributed by cyanobacteria. For example, growing period chlorophyll *a* was one of the three strongest metrics to assess the variability of the ecological status of European lakes (besides cyanobacterial biovolume and Phytoplankton Trophic Index, PTI), (Carvalho et al. 2013). Because of relative ease of determination, often only the chlorophyll *a* concentration was

DOI: 10.1201/9781003301592-3

Figure 3.1 Cyanobacteria blooms in worldwide lakes and reservoirs.

Note to Figure 3.1:

1. Warning sign on mesotrophic Bright Lake, Algoma, Ontario, Canada, September 12, 2009. Photo by Gertrud Nürnberg.
2. Cyanobacteria bloom in eutrophic Fanshawe Lake, reservoir of the Thames River. Ontario, Canada, August 26, 2005. Photo by Karla Young.
3. *Aphanizomenon* dispersed throughout the water column in hyper-eutrophic Mitchell Lake reservoir, South Dakota, June 30, 2001. Photo by Gertrud Nürnberg.
4. *Microcystis aeruginosa* and *Raphidiopsis raciborskii* in Maranguapinho Reservoir, in the state of Ceará, Brazil, February 19, 2021. Photo by Mario Barros.
5. Shoreline accumulation of *Aphanizomenon* on the east end of eutrophic Desbarats Lake, Algoma, Ontario, Canada, August 23, 2007. Photo by Peter Pollard.
6. *Microcystis aeruginosa* bloom in eutrophic Milford Lake, Kansas, July 14, 2016. Photo by Jennifer Graham.
7. *Microcystis aeruginosa* bloom in eutrophic Missisquoi Bay in Lake Champlain, Quebec, August 2023. Photo by Frances Pick.
8. *Microcytsis aeruginosa* and *M. viridis* accumulation in northwest Lake Peipsi, a transboundary lake in Estonia and Russia, June 26, 2022. Photo by Olga Tammeorg.
9. *Planktothrix rubescens* blooms under ice in mesotrophic Lake Wilcox, Ontario, Canada, in winter 1999. Photo by Dan Olding.
10. *Planktothrix rubescens* blooms after ice-out in Lake Wilcox, Ontario, Canada, March 1999. Photo by Dan Olding.

monitored historically to determine general phytoplankton biomass, while other, more specific chlorophylls (Chl *b*, *d*, and *f*) have not often been analyzed. (Planktonic chlorophyll *a* is meant whenever referring to chlorophyll in this text.) Recent technological advances, including in-situ sensor deployment and the determination of cyanobacteria-specific pigments (e.g., phycocyanin and phycoerythrin), have permitted more direct routine cyanobacteria monitoring.

There are no generally accepted quantitative definitions to describe cyanobacteria accumulations large enough to constitute a bloom. Such HABs are sometimes defined for a specific study area but often just address the accumulation of cyanobacteria on the surface or

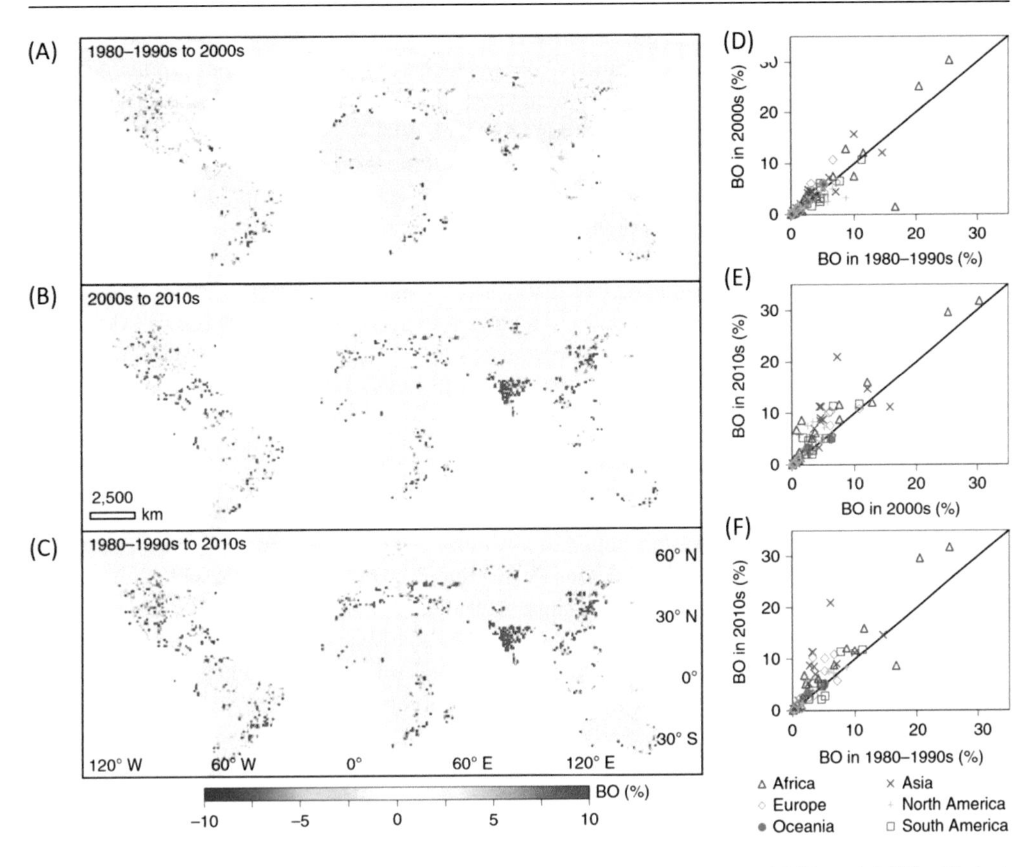

Figure 3.2 Decadal changes of lacustrine algal blooms (BO) between 1980 and 2010. Global, A to C, and country-level (n = 65), D to F. *Figure 4 from Hou et al. (2022) with permission.*

close to shore that is unsightly, brightly coloured, and odorous. The criterion of harmfulness that was originally expressly included (Smayda 1997) has now often been assumed. Most of these measures reflect cyanobacterial abundance or dominance and are not a direct indicator of harm, although increased cyanobacterial abundance implies an increased potential for cyanobacteria-related problems.

For example, cyanobacteria blooms in the western arm of Lake Superior were defined, "as an event when dense phytoplankton populations were observed either suspended in the water or floating on the surface" (Sterner et al. 2020). Others calculated an index, the summer bloom frequency index, as the percentage of summer chlorophyll values above 20 μg/L for six New Zealand lakes, where chlorophyll was mainly contributed from cyanobacteria (Waters et al. 2021). Medium Resolution Imaging Spectrometer (MERIS) imagery was used to quantify the intensity of cyanobacterial blooms in Lake Erie as cyanobacteria index (CI) (Stumpf et al. 2012).

A recent detailed framework of bloom categories that connects temporal patterns in bloom time series with underlying theories of process was developed from the review of the scientific literature (Isles and Pomati 2021): "A period of net phytoplankton biomass accumulation within a defined area or volume, resulting from growth rates that exceed loss rates, followed by eventual decline to near baseline concentrations". When such blooms occur in the summer and fall, they usually consist of cyanobacteria, as compared to spring blooms that typically comprise eukaryotic phytoplankton, especially diatoms (Xie et al. 2020).

The second edition of *Toxic Cyanobacteria in Water* of the World Health Organization (Chorus and Welker 2021) provides the following (qualitative) definitions for accumulated cyanobacteria: A *bloom* consists of a high average phytoplankton (i.e., algae and/or cyanobacteria) cell density in a waterbody, and a *surface bloom* consists of cyanobacteria accumulation near or at the water surface, forming visible streaks that may be discernible in remote sensing images. Surface blooms can be caused by concentrating mechanisms despite low average cyanobacterial cell density in the water column. Surface blooms can lead to surface *scum*, a massive accumulation of buoyant cyanobacteria, sometimes leading to cell lysis staining the water blue by phycocyanin.

3.1 Classification of cyanobacteria

Cyanobacteria (cyanoprokaryota) belong to the phylum of photosynthetic bacteria (cyanophytae) even though they are often put together with algae and are sometimes called "bluegreen algae" or just "bluegreens" (Wehr and Janse van Vuuren 2024). Prior to the advent of modern microscopy, all algae were considered plants. But algae are a diverse group of microorganisms that span four kingdoms: the prokaryotic Eubacteria (cyanobacteria) and the eukaryotic Chromista (diatoms, dinoflagellates, cryptomonads, and golden algae), Plantae (green algae), and Protozoa (euglenoids). The unifying traits of this diverse group of organisms traditionally called algae are an aquatic habitat, a relatively simple structure without a vascular system, and the ability to conduct oxygenic photosynthesis using the pigment chlorophyll *a* (Reynolds 1984; Wehr and Janse van Vuuren 2024).

Cyanobacteria contribute to the phytoplankton in fresh and marine waters (Bonilla and Pick 2017; Chorus and Welker 2021). Their simple, prokaryotic cells consist of pigments also present in "real" (eukaryotic) algae, such as chlorophyll *a*, but have different pigments in addition, including phycocyanin and phycoerythrin (the latter is also found in the eukaryotic cryptophytes). These pigments increase photosynthetic efficiency. Another compound specific to cyanobacteria, cyanophycin, is used for N storage. These pigments can be used to distinguish cyanobacteria from eukaryotic algae, in addition to more elaborate methods (genetics and optical and electron microscopy).

Besides the taxonomic classification related to genetic and evolutionary principles, a classification based on functional characteristics in the phytosociological context of freshwater phytoplankton was developed (Reynolds et al. 2014, 2002; Rigosi et al. 2014). The still-evolving system of *Reynolds Functional Groups* (also called *Codons* or *Traits*, Table 3.1) was more effective and potentially more permanent in describing cyanobacteria response to environmental and other influences than the classic biodiversity index based on taxonomic classification (Kruk et al. 2020). A review of known cyanobacteria characteristics suggested that "a trait-based perspective should improve our understanding of the mechanisms that drive HABs" (Litchman 2023).

Using functional groups enables the grouping of individual cyanobacteria with respect to responses to environmental challenges and effects on ecosystem processes, instead of considering genetic and evolutionary constraints. In this way, closely related species with different demands of favouring different conditions can be differentiated, which improves the evaluation of a freshwater environment's suitability for specific phytoplankton groups. In the context of internal P load influences on cyanobacteria, such grouping by traits can facilitate examination

Table 3.1 Selected Reynolds Functional Groups (*Codon* or *Trait*) of cyanobacteria (based on the review and update by Padisák et al. (2009)).

Codon or Trait	*Lake characteristics*				*Cyanobacteria**	
	Trophic state	*Depth/Size*	*Mixing state*	*Turbidity*	*Characteristics*	*Species*
H1	Eu	Deep, shallow	Strat or mixed		N-fixer, gas vesicles, akinetes	*Aphanizomenon flos-aquae, A. gracile, A. klebahnii, A. issatschenkoi, A. ovalisporum* *Dolichospermum affinis, D. circinalis, D. flos-aquae, D. planktonica, D. perturbata, D. solitaria, D. spaerica, D. spiroides, Anabaenopsis arnoldii, A. cunningtonii, A. elenkinii, A. tanganykae, Nodularia*
H2	Oligo, meso	Deep, shallow	Strat, mixed	Clear	N-fixer, akinetes	*Dolichospermum lemmermannii Gloeotrichia echinulate*
K					Small-celled. colonial, w/o gas vesicles	*Aphanocapsa, Aphanothece and Cyanodictyon*
L	Oligo	Deep			Late summer	*Aphanocapsa and Aphanothece colonies*
L_M	Eu-hyper	Sm-med			Obligatory co-occurrence	*Microcystis* co-occurring with *Ceratium hirundinella (Dinoflagellate)* and/or *C. furcoides*
M	Eu-hyper	Sm-med			Vegetative benthic forms	All *Microcystis* species
MP	Any	Shallow	Mixed, stirred up	Turbid by inorganic particles	Includes periphytic, epilithic	*Pseudanabaena catenata, P. galeata, Oscillatoria sancta, Cylindrospermum cf. muscicola*
R	Oligo-meso	Deep	Strat		Meta, upper hypolimnion	*Planktothrix rubescens* any filamentous cyanobacteria species in a deep layer maximum
S1			Mixed	Turbid	Shade-adapted	*Planktothrix agardhii, Pseudanabaena limnetica* *Planktolyngbya limnetica, Isocystis pallida, Leptolyngbya tenue, L. antarctica*
SN			Mixed warm		Shade-tolerant, N-fixer, akinetes	*Cylindrospermopsis raciborskii, Anabaena minutissima*
T_C	Eu	Standing water, slow flowing river			Epiphytic on emergent macrophytes; benthic mats	*Oscillatoria spp., Phormidium spp., Microseira (formerly Lyngbya) spp., Microcoleus spp. (formerly Phormidium spp.), Rivularia spp., Leptolyngbya cf. notata, Gloeocapsa punctata*

Note: *Dolichospermum*, formerly called *Anabaena*
Trophic state: oligo-, meso, eutrophic, and hyper-eutrophic and combinations (oligo-meso; eu-hyper)
Strat: thermally stratified
N-fixers have specialized cells called heterocytes where the N fixation occurs, a process that transforms gaseous N_2 by reduction to NH_4 and incorporation.
Akinetes: thick-walled resting cells triggered by unfavourable growth conditions

*Taxonomic groupings:
Chroococcales: *Aphanocapsa, Aphanothece, Microcystis, Snowella,* and *Woronichinia*;
Nostocales mainly belong to Groups H1 and H2: for example, *Aphanizomenon* and *Dolichospermum*;
Oscillatoriales: *Limnothrix, Oscillatoria, Planktolyngbya, Planktothrix,* and *Pseudanabaena*;

and analysis of dependencies. Classifying phytoplankton by functional traits instead of taxonomic units significantly improved the predictability of a plankton metric by 50% in shallow eutrophic Müggelsee, Berlin, Germany (Kakouei et al. 2022).

For example, heterocystous N_2-fixing (diazotrophic) cyanobacteria (*Aphanizomenon, Dolichospermum*) thrive up to a threshold determined by P-availability during N-limitation (Section 3.3.2), so long as other conditions allow (lake-mixing state, light availability, temperature). This association is called trait H1 (Table 3.1) and contributes to cyanobacteria blooms in many lakes. The ability of these cyanobacteria (N-fixers) to incorporate N compounds from the N_2 gas in the air has been used in theoretical speculations that they are independent of N, which has led to an oversimplification in determining their nutrient dependency (see Section 3.3). *Microcystis*, another widespread surface-blooming cyanobacterium, is classified as a separate group, trait M (Table 3.1). While there is affinity to nitrogen, *Microcystis* does not "fix" N. It has a distinguishing life cycle that includes vegetative cells overwintering on the sediment with subsequent migration to the surface water.

Another distinct association forms trait R, represented by filamentous species including *Planktothrix rubescens* which favours the cooler metalimnion, is low-light adapted, and is non-diazotrophic. Thus, surface blooms often consist of species belonging to trait H1 and M, while metalimnetic phytoplankton layers involve trait R. Other functional groups include species of trait SN in warmer mixed and alkaline waters compared to those in S1 or H2 in mesotrophic lakes with high light conditions and many more (Mantzouki et al. 2016) (Table 3.1).

In addition, there are cyanobacteria of the non-freshwater, temperate, and subtropical environments including saline lakes and brackish waters. These systems were not included in Reynolds' original concept, and the taxa resemble closely trait H1 (Table 3.1). These cyanobacteria taxa include *Nodularia* species (planktonic *N. spumigena*), notorious in large blooms of the Baltic and around Australia, in the Dead Sea, in marine aquaculture ponds in Brazil, and in many lakes in semi-desert environments including the Great Salt Lake, Utah, USA (Chorus and Welker 2021).

As the functional group approach is more broadly adapted, it will become more refined and continue to be useful in the study of cyanobacteria.

3.2 Toxicity of cyanobacteria

Cyanobacteria can reproduce explosively with low loss rates and create "blooms" that are not only unsightly (Figure 3.1) but can also be toxic (HABs) to humans, pets, and wild animals upon contact or ingestion (Rastogi et al. 2015, 2014; Trevino-Garrison et al. 2015) and inhalation (Sun et al. 2023b). A review of studies of 2000–2018 found 1 118 recorded identifications of cyanotoxins in 869 freshwater ecosystems from 66 countries throughout the world (Svirčev et al. 2019).

The different types of cyanotoxins include microcystins, cylindrospermopsins (Scarlett et al. 2020), nodularins, anatoxins (Christensen and Khan 2020), saxitoxins, and further unidentified toxins, still being discovered. Various toxins affect the liver (hepatotoxins), the nervous system (neurotoxins), the skin (dermatotoxins), and may be tumour promotors (Chorus and Welker 2021; Zurawell et al. 2005). Relationships with certain diseases are less established in other cyanobacterial metabolites including nonproteinogenic amino acid, β-methylamino-L-alanine (BMAA), and lipopolysaccharides (LPS, also called "endotoxin") (Chorus and Welker 2021). There are also hundreds of other cyanopeptides with unknown effects and toxicity, including cyanopeptolins, anabaenopeptins, aeruginosins, aerucyclamide, and microginins (Janssen 2019; Zastepa et al. 2023b).

Lethal effects on birds, fish, macroinvertebrates, and zooplankton besides mammals and humans were discovered in freshwater, estuarine, and marine systems. Mass mortalities, especially of fish, were deemed to be often the result of multiple stress factors that co-occur during cyanobacterial blooms (including hypoxia and elevated temperature). Bioconcentration of cyanotoxins (direct biotic uptake and accumulation of toxins from the water) was found to be more important than biomagnification (uptake and accumulation via feeding) in some studies (Ibelings and Havens 2008), but biomagnification can also occur (Chorus and Welker 2021). Biotransport to non-aquatic food webs also exists (e.g., via emerging aquatic insects eaten by birds) and was reported in riparian habitats (Moy et al. 2016).

The most studied of the microcystin variants is MC-LR (Rastogi et al. 2014; Spoof and Catherine 2016), a hepatotoxin that often occurs in *Microcystis* assemblages and for which recreational and drinking water guidelines have been available since at least 1998 (WHO – World Health Organization 1998), followed by more recently established guidelines for additional cyanotoxins (Table 3.2).

Table 3.2 Common toxins of freshwater cyanobacteria, distribution, and toxicity guidelines *(not a complete listing).*

Toxin (global proportion)[a]	*WHO guideline (µg/L)*[b]	*Function*	*Species*
Microcystin (63%)	1/24	Hepatotoxin	*Microcystis, Dolichospermum (Anabaena), Planktothrix, Nostoc, Oscillatoria, Anabaenopsis, Pseudoanabaena*
Anatoxin (9%)	30/60	Neurotoxin	*Dolichospermum, Aphanizomenon, Planktothrix, Cylindrospermum, Microcystis*, benthic *Oscillatoria, Tychonema* (tychoplanktic, i.e., planktonic or attached to seston and macrophytes), *Phormidium*
Saxitoxins, including neosaxitoxin Gonyautoxins C-toxins (8%)	3/30	Neurotoxin Paralytic shellfish poisoning (PSP) toxins	*Dolichospermum, Lyngbya, Aphanizomenon, Raphidiopsis, Microseira (Lyngbya)*
Nodularin (2%)	(na)	Hepatotoxin	*Nodularia, Nostoc* (symbiotic in lichen)
Cylindrospermopsin (10%)	3/6	Hepatotoxin Cytotoxin	*Cylindrospermopsis raciborskii (C. raciborskii), Aphanizomenon flos-aquae, A. gracile, A. ovalisporum, Umezakia natans, Anabaena bergii, Anabaena lapponica, Anabaena planctonica, Microseira (Lyngbya) wollei, Rhaphidiopsis curvata, R. mediterranea, R.raciborskii*
Lyngbyatoxin	(na)	Dermatotoxin Cytotoxin	*Microseira (Lyngbya), Oscillatoria*

Note: Dolichospermum formerly called *Anabaena*
Raphidiopsis is similar to *Cylindrospermopsis* but does not form heterocysts.
na, not available.

Source: (Chorus and Welker 2021; Zurawell et al. 2005) and other references cited throughout this text, including (Poirier-Larabie et al. 2020) for *Microseira (Lyngbya)*

a Global proportion (%) is based on 1 118 global records in 869 freshwater ecosystems from 66 countries mainly for 2000–2018 (Svirčev et al. 2019).
b WHO guidelines are indicated for long-term exposure to drinking water (first value) and recreational water (second value) (WHO – World Health Organization 2021, 2020).

There are more than 19 different genera of cyanobacteria that have been shown to produce toxins, often with several different toxins per species, and Bernard et al. (2016) lists more than 11 pages of different species and taxa. The main toxin-producing genera include *Dolichospermum* (formerly *Anabaena), Aphanizomenon, Cylindrospermopsis, Gloeotrichia, Hapalosiphon, Lyngbya, Microcystis, Nodularia, Nostoc, Oscillatoria, Schizothrix, Spirulina*, and *Synechocystis* (Table 3.2). Of these, *Aphanizomenon, Dolichospermum, Microcystis*, and *Planktothrix* are listed as the four taxa most commonly associated with toxic incidents in the United States (Patiño et al. 2023).

Toxicity is not omnipresent, even in the same species, and varies with environmental factors, nutrient conditions, and genetic strains. Therefore, many biochemical and chemical methods have been used to determine the varying toxicity of specific strains as reviewed in (Picardo et al. 2019). In addition, the increasingly used method of gene analysis (Fortin et al. 2010; Willis and Woodhouse 2020; Zastepa et al. 2023b) is especially effective for identifying cyanotoxins, thus determining potential toxicity over a large range of conditions and locations.

An early study determined the microcystin synthetase gene (mcyE) in 70 Finnish lakes with various amount of potential MC-producing *Microcystis, Planktothrix*, and *Dolichospermum (Anabaena)* species (Rantala et al. 2006). This study also detected a pattern with trophic state, as the occurrence of all three genera increased from oligotrophy (low proportion, one genus) to mesotrophy (*Microcystis and Planktothrix)* to hyper-eutrophy (75%, all three genera). Microcystin gene analysis can detect the location of hotspots in large lakes (e.g., Peipsi, Estonia/Russia (Panksep et al. 2020)), help determine the toxin-producing potential of *Microcystis* blooms in the Laurentian Great Lakes (Dyble et al. 2008), and distinguished between metalimnetic red strains of *Planktothrix rubescens* and *P. agardhii* in decade-old samples from Lake Zürich, Switzerland (Kurmayer et al. 2016).

Microcystis contributed up to 42% to cyanobacteria in Lake Erie as a mixed population of potentially toxic and non-toxic genotypes in 2003–2005 (Rinta-Kanto et al. 2009). Further, microcystin concentrations, total *Microcystis* abundance, and the abundance of microcystin-producing *Microcystis* were significantly positively correlated with TP concentration, while another study found correlations with N (Gobler et al. 2016). Other genetic studies determined the spatial differences for three isolates of *Microcystis* in Lake Erie and upstream Lake St. Clair, also finding a significant presence and expression of genes from worldwide bacterial cultures (Meyer et al. 2017).

Cyanotoxins are also distributed in lotic systems. Synthetase genes were detected at all 11 large river sites but one, located throughout the United States, and the various proportions indicated the prevalence of mcyE and saxitoxin (sxtA) (Graham et al. 2020). In an expanded study, these genes originated mainly in *Microcystis* and in *Planktothrix*, when abundant, were diverse, and positively correlated with nutrients (Linz et al. 2023).

3.3 Macronutrient (N and P) effects on cyanobacteria

A fitting statement summarized the general understanding of previous work (e.g., Smith 1985; Watson et al. 1997) a decade ago: "As bodies of freshwater become enriched in nutrients, especially phosphorus (P), there is often a shift in the phytoplankton community towards dominance by cyanobacteria" (O'Neil et al. 2012).

There is no doubt that cyanobacteria proliferation is related to increased nutrient availability (Sections 2.3 and 2.5) emphasizing the importance of "bottom-up" influences in the food web as compared to "top-down" controls by biological conditions. The major macronutrients, P and

N, are needed by cyanobacteria in different ways compared to eukaryotic algae and affect their abundance differently. Process-based arguments and studies, as well as empirical observations, highlight the predominant effect of the two macronutrients on cyanobacteria biomass in general and their toxicity in particular. For example, the overall importance of controlling nutrients for managing blooms, no matter the functional group, was established in a review that considered specific environmental sensitivities of individual groups (Mantzouki et al. 2016).

Nutrient requirements for specific cyanobacteria and other phytoplankton are well-known from laboratory incubation studies, and growth media have been formulated to provide all needed ingredients in abundance. Growth of cyanobacteria requires a certain amount and composition of specific chemical nutrient forms and micronutrients (Section 3.4.) in addition to specific conditions of temperature (Section 3.5.1, e.g., ~20°C), light (Section 3.5.2, e.g., 60 μmol quanta/m^2/s continuous or intermittent), and water column stability (Section 3.5.3) (Lürling and Beekman 2006).

The question about which of the two macronutrients, N and P, is most important in driving eutrophication and hence should be controlled has been debated in the limnological literature for at least half a century. The theory of the *limiting nutrient* acknowledges that either or both, N and P, are needed by the phytoplankton and that available forms are sometimes in short supply. Related to the limitation theory is the concept of N:P ratios (Box 3.1). The theory of nutrient limitation and the N:P ratio concept are paradigms that inspired many studies, but some critiques have been questioning the general validity, pointing out oversimplifications (Glibert 2020).

Box 3.1 Concept of N:P ratios

The ratio of the total compounds (TN:TP) is decreased when the lake trophic state is high and cyanobacteria are abundant. When and where the bioavailable form of P (SRP, bioassay-P, Table 2.1) is low or below the analytical detection limit, indicating P limitation, bioavailable N-compounds (i.e., inorganic N including, nitrate, nitrite, or ammonium) are typically elevated and vice versa (indicating N limitation at low inorganic N compounds). This teeter-totter between N and P compounds is so common throughout freshwater systems that the concept of N:P ratios has been used to determine nutrient limitation (either P or N) and to explain many water quality conditions, including the prevalence of cyanobacteria (Smith 1983).

Various thresholds of N:P ratios were proposed that coincided with specific nutrient levels in specific systems. These thresholds generally indicate that at lower N:P ratios, signifying relative P abundance, cyanobacteria tend to prevail (Smith 1983). Meanwhile, the exact ratios that were originally based on the marine phytoplankton stoichiometry (cell quota or structural ratio) and specified as the *Redfield ratio* (C106:N16:P1 in molar units (Redfield 1958)) were replaced with newer ratios (C166:N20:P1) based on the analysis of more than 2 000 observations from lakes and marine environments (Sterner et al. 2008).

However, several studies determined that the N:P ratio concept does not always apply, and it is generally acknowledged to be an oversimplification (Glibert 2017). Statistical examination revealed that the nutrient concentrations themselves (TP and TN) were better predictors of average cyanobacteria dominance than the TN:TP ratios (Downing et al. 2001). It is also more illuminating to know whether P has increased or whether N has decreased when the N:P ratio decreases, which is often the case in late summer and fall due to increased sediment-released P (Section 3.3.3 and (Isles et al. 2017b) in Section 4.3.2). In addition, nitrogen can form a gaseous state that is easily lost to the air (e.g., denitrification during anoxia, Section 3.3.3) rendering quantification unreliable, and data analysis involving "raw" N:P ratios is misleading because of statistical problems with respect to the distribution of natural data; instead, the logarithm of molar ratios was suggested as more appropriate (Isles 2020).

Further, total chemicals (TP, TN, Section 3.3.2) do not adequately represent the available fraction of the nutrients so that nutrient limitation cannot always be specified by simple comparison of total nutrient concentrations. To avoid the errors inherent to total chemicals, sometimes biologically available forms have been used in ratios.

As studies showed that the structural N:P ratio at optimal growth is species-specific and quite variable (from 8.2 to 45.0) dependent on ecological conditions, the classic N:P ratio of 16 by Redfield (or the newer ratio of 20 by Sterner) can be deemed an average of phytoplankton species-specific ratios that may have little relevance for phytoplankton in general (Klausmeier et al. 2004) and cyanobacteria in particular (Pick and Lean 1987).

While the nutrient ratio concept is still used to quickly determine the importance of the nutrients N versus P in a lentic system, it is not useful for process-based understanding. The idiosyncrasies with this concept were most clearly expressed by Reynolds (1992):

> The explanation becomes even more implausible if the critical condition "that the cells are nutrient-limited" is ignored. For the growing cell, the only factor of relevance is whether there is enough of each of its various nutrient requirements available to sustain its next cell division. It is immaterial whether the N:P ratio is 50:1 or 3:1 if both nitrogen and phosphorus supersaturate the growth requirements for the current generation.

The controversy about P versus N limitation is of major socioeconomic importance because limitation determines which and how many nutrients must be managed (decreased) to achieve the control of freshwater eutrophication in general and cyanobacteria in particular. The debate is apparent from a large amount of scientific papers since Vollenweider's suggested connections between P loading, P concentration, and phytoplankton standing stock (Vollenweider 1968). An article by Conley et al. (2009a) advocating for a dual nutrient reduction strategy to address eutrophication in lakes, estuaries, and coastal areas provoked an extensive discussion (Bryhn and Håkanson 2009; Conley et al. 2009c; Dodds et al. 2009; Jacoby and Frazer 2009; Schelske 2009; Schindler and Hecky 2009). A review (Sterner 2008) determined that nutrient limitation changes from P to N through the summer, contrary to the concept of persistent P limitation. Since then, the debate has continued (Paerl et al. 2020, 2018; Smith and Schindler 2009), as summarized in Chorus and Welker (2021). Much of the following discussion on macronutrients provides further details and includes the consideration of internal P loading.

Most temperate freshwater lakes have been deemed P-limited (Schindler et al. 2008), meaning that little or no phosphate is available for an easy incorporation into phytoplankton most of the growing period. The P limitation concept also explains the long-term success of external P input control on lake TP and chlorophyll concentration (Vollenweider 1976) and was experimentally verified by whole-lake fertilization experiments in north-central Canada (Schindler 1974), which continue to this day. In a more recent experimental P fertilization in the same lake region, large nitrogen-fixing cyanobacteria blooms were initiated in two oligotrophic lakes without any N addition (Molot et al. 2021a). N content of the mixed water layers doubled and P content temporarily increased during the bloom, probably as a consequence of increased phytoplankton biomass. Similarly, P, not N, was found to control cyanobacteria biomass in Mid-US reservoirs (Petty et al. 2020).

The importance of nitrogen is evident from physiological constraints. Many characteristics and components of cyanobacteria involve nitrogen and are affected during N limitation (Duan et al. 2021). The photosynthetic pigments phycocyanin, cyanophycin, and chlorophyll are N-rich, and buoyancy-controlling gas vesicles contain N_2-gas. Cyanotoxins, for example, microcystin, are N-rich compounds and were found to generally decrease under N limitation and increase under P limitation when N is replete in a meta-analysis of 67 data sets (Brandenburg et al. 2020). However, relationships are complex, and more toxic microcystin congeners (Chaffin et al. 2023) and other cyanotoxins, e.g., anatoxins, increased under N limitation in western Lake Erie (Barnard et al. 2021). This means that N starvation often creates pigment loss, hinders buoyancy control, and affects toxicity (Evans and Saleh 2015).

Cyanobacteria can produce higher biomass per unit P than other phytoplankton (Hosper 1997; Dobberfuhl 2003) and are strong competitors for N (Chorus and Spijkerman 2020). Some cyanobacteria groups can fix N (diazotrophic genera including *Dolichospermum* and *Aphanizomenon*, Table 3.2) but many cannot (non-diazotrophic genera including *Microcystis* and *Planktothrix*). Consequently, differences in species composition between lakes are sometimes explained by N concentration (Dolman et al. 2012) in addition to P, noting that the specific chemical forms affect bioavailability. However, the effects of either macronutrient on the phytoplankton in lake water bioassays also depend on the nutrient's supply rate, indicating the complexity of the involved nutrient drivers (Frost et al. 2023).

Tentative nutrient concentration thresholds for limitation are difficult to infer. In Danish lakes, phytoplankton biomass (chlorophyll) was N-limited only at TP concentrations above 50 μg/L (mainly from internal load) and remained P-limited at TN concentrations above 500 μg/L (Søndergaard et al. 2017). Specific thresholds for the inorganic (bioavailable) nutrient forms were proposed for P limitation, SRP less than 3–10 μg/L and for N limitation, and DIN (NO_x and NH_4) less than 100 μg/L (Camarero and Catalan 2012). Besides the effects of environmental and physiological conditions, these thresholds depend on the analytical methods (that can overestimate low nutrient concentrations) and may be lower (Nürnberg and Peters 1984a).

The principle of mass balance dictates that internal P sources increase lake TP (Section 2.4.3). In lakes where internal P load prevails, external load alone underestimates the predictions of ambient lake TP concentration (Nürnberg 2009, 1984c). Under these conditions, lake restoration by external P input abatement is delayed (Larsen et al. 1981), and nitrogen can become the growth-limiting factor for the phytoplankton, especially the cyanobacteria (Conley et al. 2009a). Acknowledging that internal P load causes a seasonal N limitation, P management (instead of the management of N or both nutrients) has still been proposed as the most promising approach, because external load abatement eventually decreases internal legacy load (Schindler and Hecky 2009). But the importance of N control because of increasing N limitation has also been promoted in shallow eutrophic lakes (Gardner et al. 2017; Scott et al. 2019).

In summary, it is not always clear whether N or P is the nutrient that controls cyanobacteria abundance and toxicity the most. An in-depth discussion aimed at explaining the various observations is presented in the following sections and includes the importance of seasonality in cyanobacteria proliferation related to growth-limiting factors (Section 3.3.3).

3.3.1 Relationships of total nutrient compounds (TP and TN) with cyanobacteria

The total of the nutrient compounds (TP and TN) is the analytically simplest way for comparing the macronutrients P and N. In general, growing period average concentrations of the two main nutrients, TP and TN, are tightly correlated with each other in the illuminated, mixed layer of freshwater systems (Figure 2.4). This co-linearity creates statistical hurdles in the application of simple empirical analysis, such as comparisons (e.g., regression) of one of these nutrients with cyanobacteria characteristics. Therefore, results of such analyses can be hard to interpret, and conclusions about causality can be misleading. In other words, the often-found correlations between summer average phytoplankton biomass (chlorophyll *a* concentration, Secchi disk transparency, or phytoplankton biomass measures) and TP concentration may also hold for TN and visa versa, because growing period average concentrations of TP and TN are often correlated with each other. Further, as biomass accumulates, these nutrients are bound up in cyanobacteria and other phytoplankton, confounding mechanistic understanding of total nutrient effects on any one specific group or taxa. Subsequent conclusions, as to which nutrient controls the phytoplankton and specifically the cyanobacteria, can be misleading unless relationships for both nutrients with cyanobacteria can be compared.

Empirical relationships (positive significant regressions and non-linear relationships) between cyanobacteria abundance measures and TP or TN have been observed in numerous studies. But the underlying conditions that lead to these relationships are complex and can include trophic state, nutrient form, and timing. Cyanobacteria biomass was positively correlated with lake TP concentration growing period averages (Watson et al. 1997) or individual samples (MacKeigan et al. 2023; Vuorio et al. 2020) in studies on north-temperate lakes (Figure 3.3A), as well as with cyanobacteria-specific measures, like the Bluegreen Index (Downing et al. 2001) and individual taxa (Dolman et al. 2012).

In many studies where both TP and TN relationships with cyanobacteria were evaluated, TP had a stronger positive correlation with cyanobacteria biomass than TN, except at extremely high nutrient concentrations as in hyper-eutrophic lakes, where TN became the better correlated nutrient. For example, in a survey of multiple year, late summer data of 102 German lakes, the "positive relationship between total cyanobacteria biovolume and TP concentration disappeared at high P concentrations, cyanobacteria biovolume increased continually with N concentration, indicating potential N limitation in highly P enriched lakes" (Dolman et al. 2012).

Similar relationships were determined between cyanotoxins (usually microcystin) and the nutrients, TP and TN (Graham et al. 2004; Patiño et al. 2023; Yuan et al. 2014) and SRP (Davis et al. 2010), while toxin concentration is often independent of cyanobacteria biomass.

The most numerous studies compare chlorophyll concentration with TP and sometimes TN growing period averages and are available since the seventies (Dillon and Rigler 1974; Jones and Bachmann 1976; Sakamoto 1971). As explained in the introduction to this Chapter 3, chlorophyll is not specific to cyanobacteria but present in all phytoplankton. However, chlorophyll correlates well with cyanobacteria in the summer and fall in eutrophic and hyper-eutrophic lakes. Positive correlations between chlorophyll concentration and TP and/or TN have been described in many studies (Davidson et al. 2023; Jargal et al. 2023; Nürnberg 1996; Quinlan et al. 2021; Watson et al. 1992; Wurtsbaugh et al. 2019; Zhao et al. 2023), also supporting the dependence of phytoplankton and cyanobacteria abundance on N and P (Figure 3.3B). Taken together, the studies on cyanobacteria biomass measures and macronutrients indicate the importance of N and P for the abundance and proliferation of freshwater cyanobacteria.

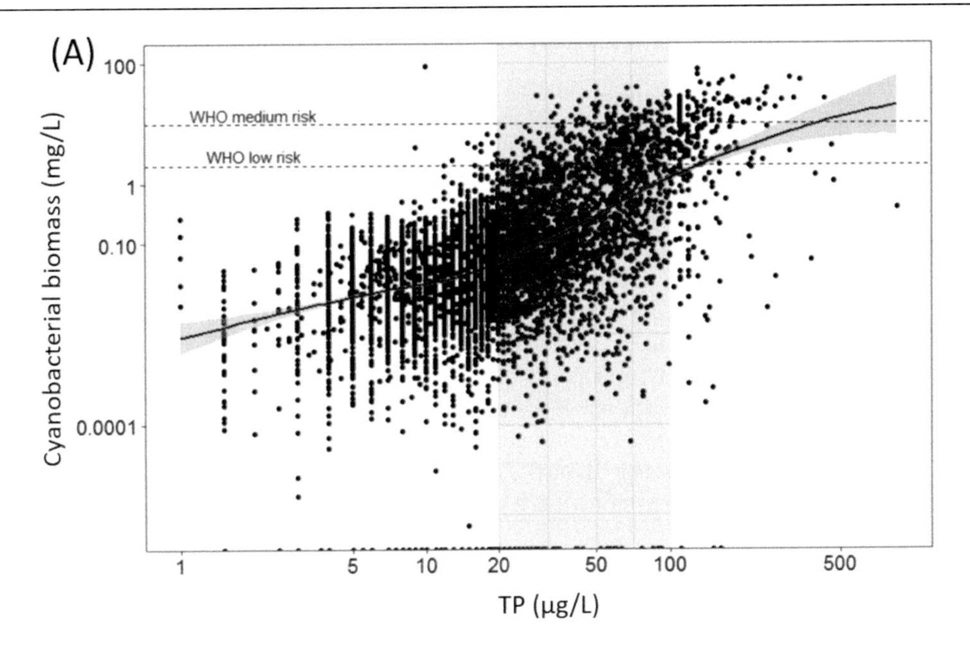

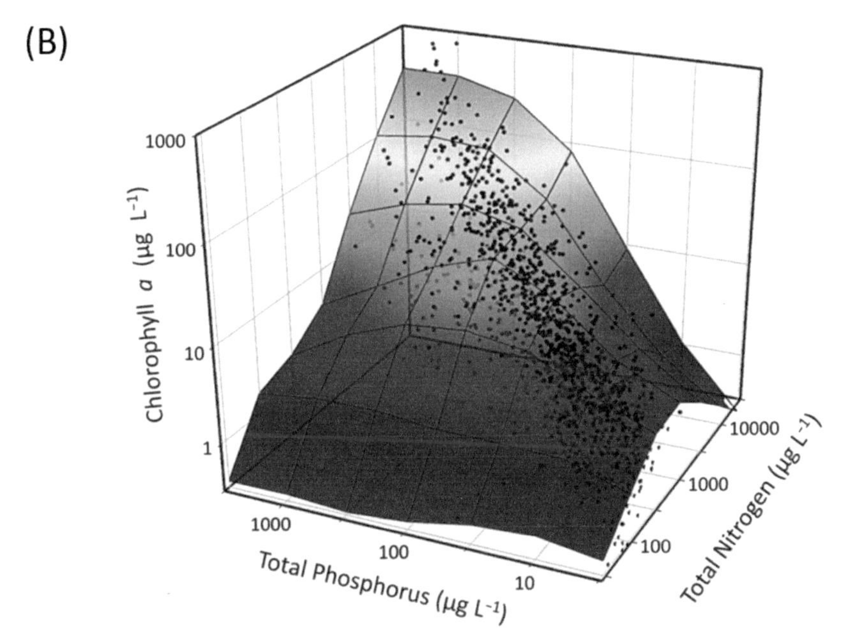

Figure 3.3 Cyanobacteria biomass or chlorophyll versus mixed surface layer nutrient concentration. (A) cyanobacteria biomass versus TP concentration in 2 029 boreal Finnish lakes, showing a linear increase in cyanobacteria biomass between 20 and 100 µg/L TP (shaded) and two WHO thresholds of health risk (low at 2 mg/L and medium at 10 mg/L); *Figure 1 from Vuorio et al. (2020), open access.* (B) Chlorophyll versus TP and TN concentration in 1 264 North American lakes studied in the United States Environmental Protection Agency's National Lakes Survey, fit with local regression smoothing (LOESS); *Figure 10 from Wurtsbaugh et al. (2019) with permission.*

3.3.2 Importance of the chemical nutrient form

"Total" nutrient compounds do not adequately describe the biological availability of the element (see also Box 3.1). Many experimental studies, such as nutrient additions, have strived to determine which of the nutrients and their specific chemical forms are most important for cyanobacteria proliferation and toxicity. Much evidence has been compiled to suggest temporal variation of nutrient limitation of various phytoplankton groups based on specific chemical forms.

Certain chemical P forms, foremost inorganic P (phosphate, measured as soluble reactive P or SRP), but also some short-chained organic P molecules are biologically available with differing long- and short-term availability (Section 2.1). Cyanobacteria prefer phosphate, which they can quickly take up and incorporate as they are highly competitive for inorganic P compared to many other phytoplankton taxa. Some cyanobacteria taxa can take up more P than required for their immediate needs from the P-rich sediment. Such *luxury* P uptake from the bottom sediments and enhanced storage in the form of polyphosphate granules enable *Gloeotrichia echinulata* (trait H2, Table 3.1), for example, to outcompete many other phytoplankton in the P-limited epilimnion after it ascends into the water column (Carey et al. 2008; Pettersson et al. 1993).

In addition, cyanobacteria species exhibit various abilities to access organic P with specific enzymes like phosphatases, for example, alkaline phosphatase, during P limitation (Figure 3.4A) and store luxury *P* as reviewed by Carey et al. (2012). For example, the *Rivulariaceae* (includes *Gloeotrichia*) seems to produce significantly higher P yields from organic molecules than filamentous non-*Rivulariaceae* (*Dolichospermum, Fischerella, Lyngbya*, and *Tolypothrix*); *Anacystis* may be less able to accumulate luxury P than *Dolichospermum*, *Plectonema*, or *Synechococcus*. Differing abilities to process phosphate likely accounted for species variation in western Lake Erie, where *Dolichospermum* and *Planktothrix* were the dominant species at high P concentrations close to P input streams, but *Microcystis* flourished in the offshore, phosphate-scarce regions (Harke et al. 2016).

With respect to nitrogen, the exchanges between various N-compounds in lake water and sediment are complex and include differing redox states between oxidized (NO_x i.e., NO_2 and NO_3) and reduced forms (NH_4 and urea), and organic forms (ON) (Figure 3.4B). Further, N paths include N-fixation (N uptake from N_2 gas), denitrification (conversion of NO_x into N_2), and the Anammox process (anaerobic ammonium oxidation, conversion of NH_4 into N_2) (Lee et al. 2009).

The reduced N-forms, ammonium (NH_4) and urea ($CO[NH_2]_2$), require less energy for incorporation and are preferred by cyanobacteria, especially the non-diazotrophic species (trait M, *Microcystis* and trait R, *Planktothrix*, Table 3.1) (Andersen et al. 2020; Gardner et al. 2017; Wurtsbaugh et al. 2019). Gene analysis found that many cyanobacteria have high-affinity NH_4 transporters (Glibert 2017). In contrast, other phytoplankton groups, including diatoms and chlorophytes, prefer and thrive on NO_x (Andersen et al. 2020; Wurtsbaugh et al. 2019).

For example, the proportion of total reduced N (TKN) to NO_x in the Maumee River, the main contributor of nutrients to western Lake Erie, had a strong positive correlation with chlorophyll and phycocyanin from *Microcystis* in 2009–2015 (Newell et al. 2019). Similarly, in Sandusky Bay of Lake Erie, cyanobacteria relied on regenerated NH_4 for growth and toxin production in late summer (Hampel et al. 2019). Kinetic experiments suggested a competitive advantage for *Planktothrix* over *Microcystis* in Sandusky Bay due to the high affinity of *Planktothrix* for NH_4 at low concentrations, indicating differences among the cyanobacteria taxa.

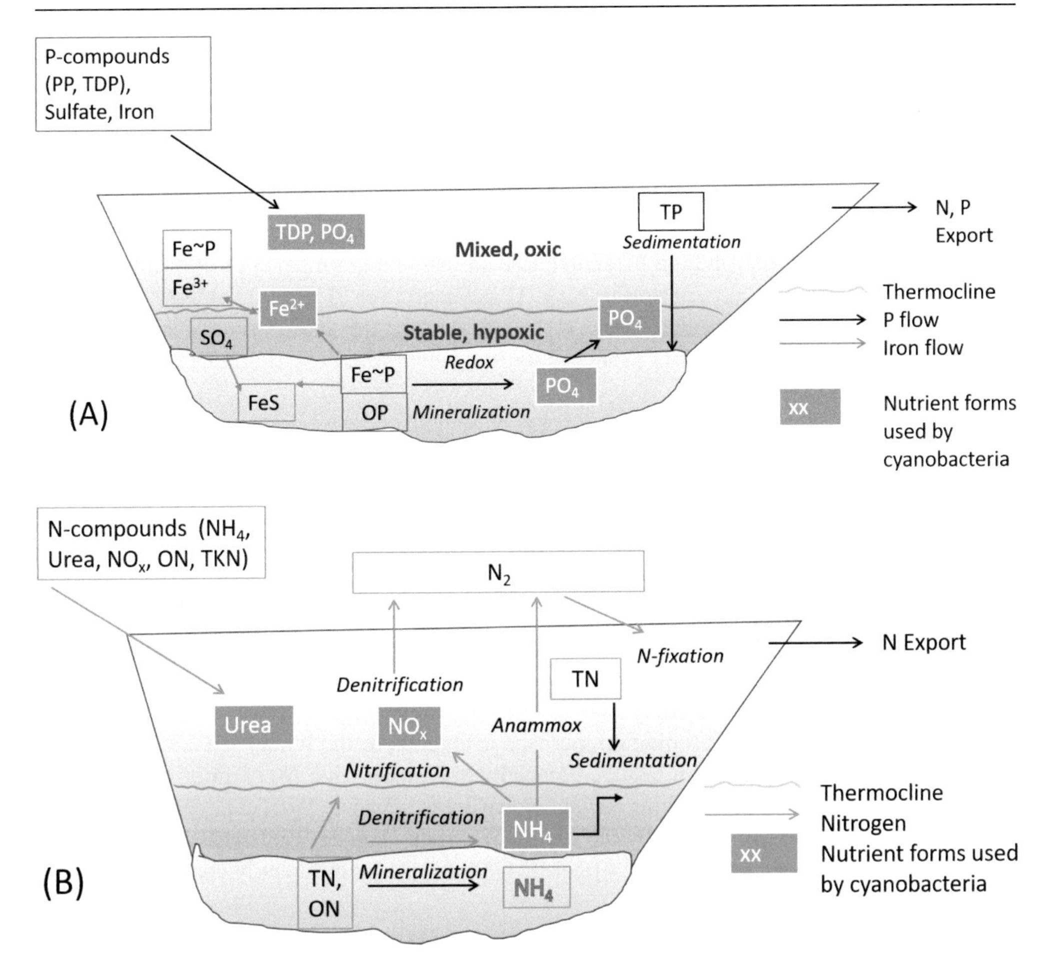

Figure 3.4 Nutrient forms used by cyanobacteria (green) and fluxes of phosphorus (A) and nitrogen compounds (B with affected P fluxes) in lakes (TDP, includes organic P in the water; OP, organic sediment P; Fe~P, reductant-soluble P from Fe-oxyhydroxides; TP, total P including particulate P, Table 2.1; ON, organic N; TKN, total Kjeldahl N) *(see also Figure 2.14)*.

The more recently studied reduced organic N-compound, urea (Bogard et al. 2020; Swarbrick et al. 2020), has been increasing in freshwaters following its growing use as fertilizer throughout the world (Glibert et al. 2014b). Non-N-fixers, such as *Microcystis* and *Planktothrix*, are highly competitive for chemically reduced N-forms, including urea (Hampel et al. 2019). While urea did not enhance growth (compared to NO_x and NH_4) of *Microcystis*, *Dolichospermum*, and *Synechococcus* in experimental studies, it enhanced pigment concentration and therefore photosynthetic capability (Erratt et al. 2018).

The dependence of *cyanotoxins* on nutrient forms, while being different from cyanobacteria biomass, has been described repeatedly. Whether and how much specific cyanobacteria populations are toxic is influenced by environmental factors, including biogeographical (ecoregion and land use), morphological and water quality characteristics (Beaver et al. 2018), and previous nutrient exposure (Hillebrand et al. 2013). Further, specific features of cyanobacteria influence cyanotoxicity, as expressed in different taxa, species, and genotypes (Chaffin et al. 2023; Gobler et al. 2016).

The potential impact of increasing eutrophication, especially of N levels, on toxicity in lakes is evident from 143 New Zealand lakes. Here, the regression of cyanobacteria on trophic state was influenced by location and altitude, so that cyanotoxins were only detected in eutrophic lowland lakes (Wood et al. 2017). Similarly, in a survey of 102 north German lakes, the most abundant potentially toxin-producing Nostocales (*Aphanizomenon* and *Dolichospermum*, Table 3.1) were found in lakes with high N relative to P enrichment (Dolman et al. 2012). In 22 Quebec lakes with various degrees of trophic state, TP concentration (5–130 μg/L) was strongly correlated with phytoplankton biomass, but TN (302–950 μg/L) was more strongly correlated with microcystin concentration (mainly produced by the genera *Microcystis* and *Dolichospermum*) (Giani et al. 2005).

These field observations are supported by experimental studies. Toxic strains (e.g., of *Microcystis*, (Davis et al. 2010)) were preferentially enhanced by inorganic N and P additions in laboratory studies. Elevated NH_4 concentration has been associated with higher cell quotas of microcystin in non-diazotrophic *Microcystis* and *Planktothrix*, particularly under P replete conditions (reviewed by Kelly et al. 2019). Different responses by different genera were reported in nutrient addition studies on Lake Erie sites, where N significantly increased microcystin concentrations (Jankowiak et al. 2019).

In a meta-analysis, N-rich cyanotoxins (including microcystin, cylindrospermopsin, and paralytic shellfish poisoning toxin) increased under P limitation and decreased upon N limitation (Brandenburg et al. 2020). N-rich cyanotoxin production has been proposed as a mechanism to store N compounds during P limitation, perhaps analogous to the luxury uptake of P (Gobler et al. 2016). Toxin concentration may be especially high under high N conditions when cyanobacteria convert N into N-rich toxins during luxury uptake (Van de Waal et al. 2014). However, these dependencies are influenced by the supply of carbon (Van de Waal et al. 2009) as well as specific congeners (e.g., MC-LR, MC-RR, MC-LA) that prefer different N conditions (Chaffin et al. 2023).

3.3.3 Seasonal nutrient limitation changes and the concept of carrying capacity

Much of the seasonal change in temperature, mixing state, light, and nutrient availability helps explain the variability of cyanobacteria in lakes and forms the basis of the concept of carrying capacity (Box 3.2). Typically, there is light limitation in the winter and spring, and then there is P limitation in early summer and N limitation in the fall (Figure 3.5A). There is often an abundance of N in the spring runoff of agricultural catchments compared to P, leading to a P limitation. Over the growing period with optimal temperatures and light, available N increases phytoplankton biomass until there is P and N co-limitation and ultimately N limitation. This change in nutrient limitation is often caused by increased sediment P release that occurs in the summer and fall in meso- and eutrophic lakes (Section 2.4.2) and is supported by denitrification (Figure 3.5). In other words, spring P limitation is overcome by sediment phosphate release later in the summer and fall (provided there is sufficient releasable P in the sediment, including organic P in eutrophic lakes, and favourable physical conditions, low redox potential at the sediment/water interface, and thermocline erosion to facilitate upward diffusion in stratified lakes, Section 2.4.2), leading to simultaneous increases in P and cyanobacteria biomass with potential toxicity (Section 5.1.1).

Box 3.2 Potential carrying capacity for phytoplankton biomass (after *Reynolds (1992)*)

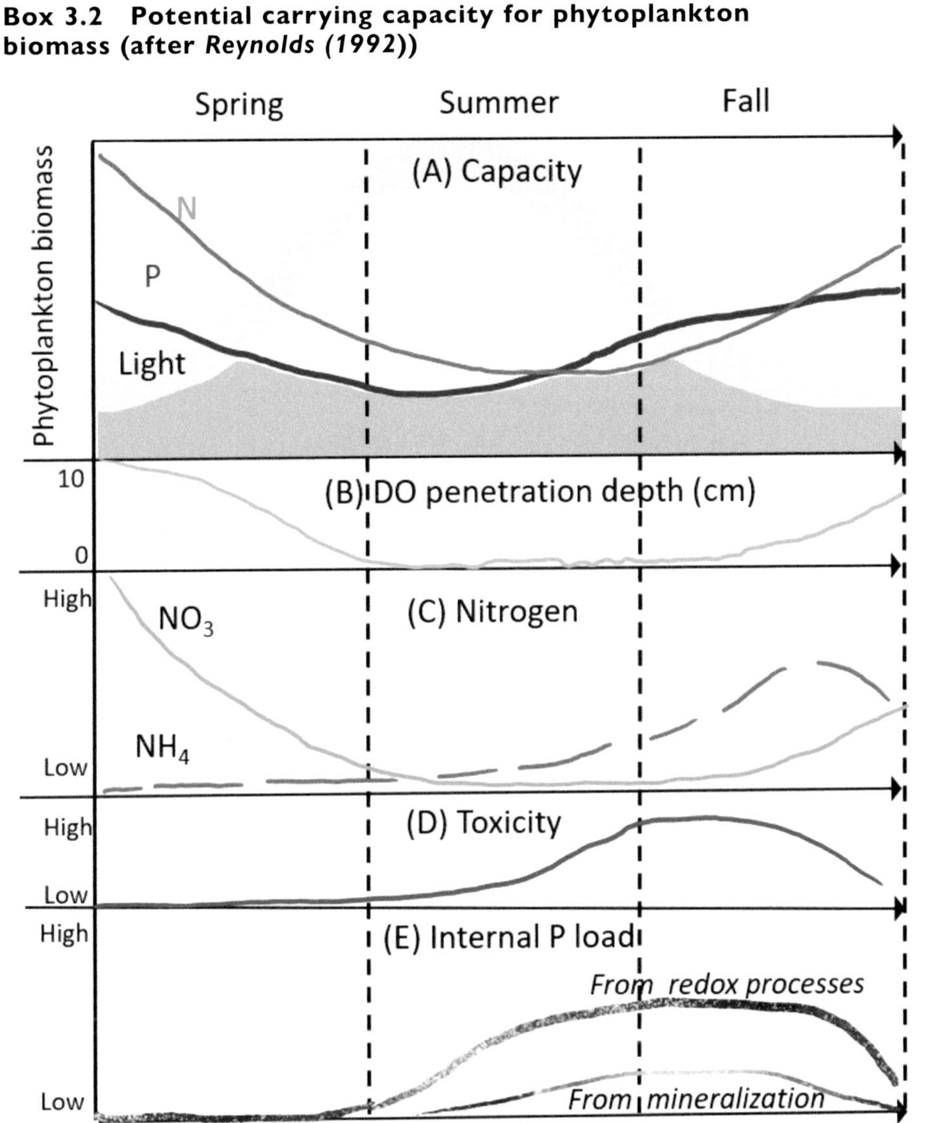

Figure 3.5 In a hypothetical eutrophic lake from spring to fall, seasonal changes in capacity (A) coincide or are related to (B) the depth of dissolved oxygen penetration into the sediment, (C) oxic and reduced nitrogen compounds, (D) potential cyanotoxicity, and (E) internal P loading from redox-related processes and mineralization of organic compounds. Shaded green areas in A indicate conditions with possible growth, when there is capacity due to sufficient P (blue), nitrogen (green), and light (yellow).

The concept of carrying capacity was originally developed by Reynolds (1992) and has been explored and applied to individual lakes since then (e.g., Chorus and Spijkerman 2020). Reynolds expected that the actual chlorophyll concentration (and phytoplankton biomass) may be less than predicted by the capacity from the three variables because of other constraints. Those constraints include temperature, mixing status, and pre-conditions of the phytoplankton.

For example, increased temperature benefits cyanobacteria, oxygen depletion, and sediment P release, and mixing status determines when the maximum biomass of cyanobacteria is expected (throughout the summer in mixed lakes or in late summer and fall in stratified lakes if internal P load is driving the cyanobacteria). In addition, the nutrient status of the phytoplankton (e.g., after luxury P uptake or the incorporation of N into toxins) may dictate the affinity to the nutrient and speed of uptake (Gobler et al. 2016). The exact timing of these processes depends on location and the stratification regime and trophic state as discussed in the text.

Hypoxia in the bottom water and the sediment, manifested as decreased sediment oxygen penetration depth (Figure 3.5B), leads not only to increased internal P loading (Figure 3.5E) but also to a shortage of NO_x (Figure 3.5C). When phytoplankton uptake rates and denitrification (at elevated temperature) are high, the duration of seasonally low NO_x concentration can serve to quantify periods of cyanobacterial proliferation as the measure of Low-Nitrate-Days (LND, Box 3.3, Nürnberg 2007a).

Box 3.3 The concept of Low-Nitrate-Days (LND) as a quantification of cyanobacteria

High NO_x forms in the spring from external inputs during runoff in agricultural catchments are taken up by spring blooms of non-cyanobacteria phytoplankton and decline to low concentrations in the summer (e.g., from 10 mg/L in spring to 1.0 mg/L in summers with cyanobacteria in the Upper Thames River, Nürnberg 2007a). The substantial decrease in NO_x-compounds when cyanobacteria proliferate is consistent in lakes and riverine systems in agricultural catchment basins and can be considered a surrogate for cyanobacteria proliferation, indicating dominance by cyanobacteria (Nürnberg 2007a). A "bloom" index, the period of Low-Nitrate-Days (LND), was developed and defined as the period of time during summer and early fall when nitrate concentration decreased to below a lake-specific threshold (Figure 3.6). The NO_3 threshold was 0.1 mg/L in Lake Okeechobee which was considered a threshold of N limitation (Camarero and Catalan 2012). The threshold was higher, 1 mg/L, in the riverine system with reservoirs of the Thames River in the catchment basin of western Lake Erie. Cyanobacteria blooms also proliferated when NO_x concentrations became low in western Lake Erie (Gobler et al. 2016) and in a eutrophic bay in lake Ontario (Molot et al. 2022). Similarly, NH_4 limitation can coincide with cyanobacteria proliferation and occurred during harmful algal blooms in 10 lakes worldwide (Gardner et al. 2017), when P levels were substantially elevated due to internal P load (Section 5.1).

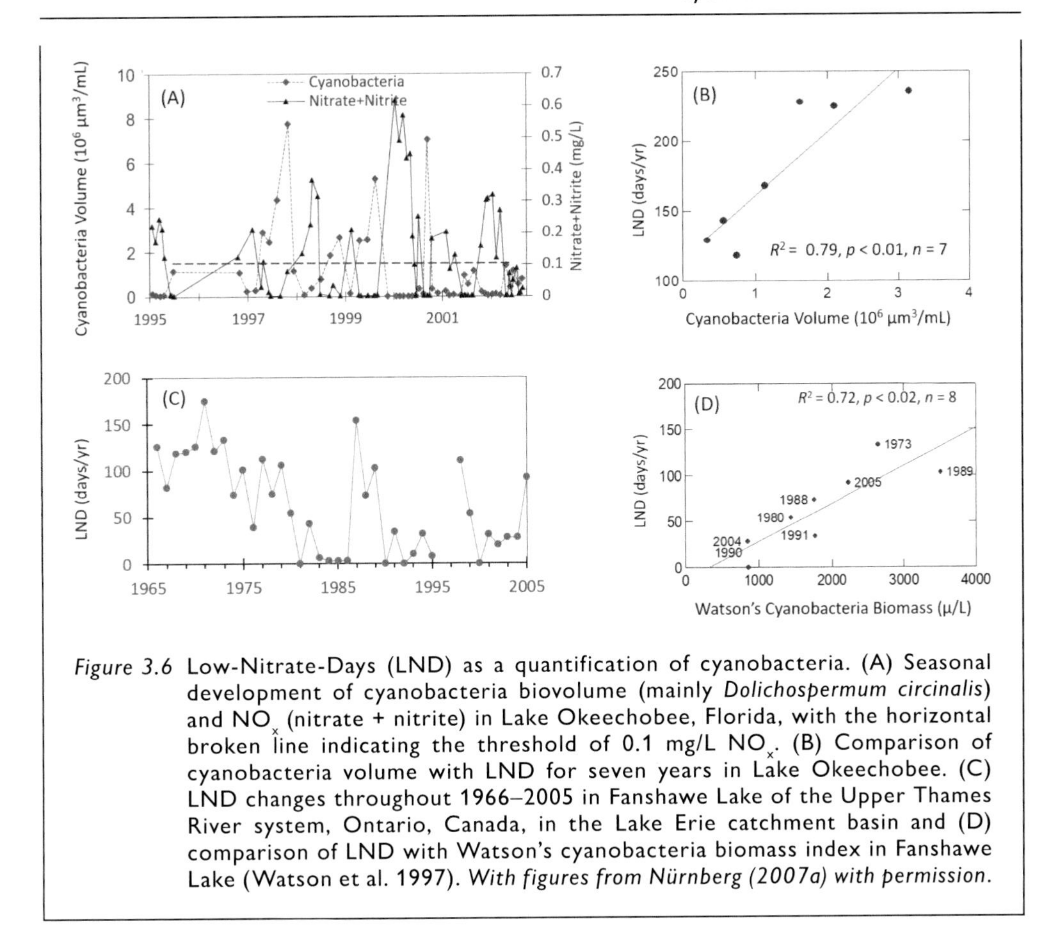

Figure 3.6 Low-Nitrate-Days (LND) as a quantification of cyanobacteria. (A) Seasonal development of cyanobacteria biovolume (mainly *Dolichospermum circinalis*) and NO_x (nitrate + nitrite) in Lake Okeechobee, Florida, with the horizontal broken line indicating the threshold of 0.1 mg/L NO_x. (B) Comparison of cyanobacteria volume with LND for seven years in Lake Okeechobee. (C) LND changes throughout 1966–2005 in Fanshawe Lake of the Upper Thames River system, Ontario, Canada, in the Lake Erie catchment basin and (D) comparison of LND with Watson's cyanobacteria biomass index in Fanshawe Lake (Watson et al. 1997). *With figures from Nürnberg (2007a) with permission.*

Seasonal changes from P limitation to N and P co-limitation and N limitation have been reported throughout the limnological literature (Sterner 2008). Nutrient limitation changes are influenced by mixing state (discussed below) and occur in shallow, nutrient-rich natural lakes (e.g., in Denmark and Florida (Jeppesen et al. 2020; Søndergaard et al. 2017) Figure 3.7) and reservoirs (Lake Austin, Bellinger et al., 2018) and can be an indicator of internal P loading (Isles et al. 2017b).

These field observations are supported by laboratory (microcosm) studies. Nutrient addition bioassays on water samples from two sites of hyper-eutrophic Lake Acton, Ohio, USA, monitoring chlorophyll concentration and cyanobacteria, determined P or N and P co-limitation in early summer, and N limitation in summer and early fall that induced the formation of N-fixing heterocytes after a lag of one to two weeks (Andersen et al. 2020) (Figure 3.8). P limitation ceased in July, coinciding with sediment P release. Similarly, bioassays determined N limitation in late summer in the Maumee Basin of western Lake Erie, causing the reduction of cyanobacteria prevalence and microcystin concentration (Baer et al. 2023; Chaffin et al. 2013). N limitation was attributed to internal P loading in a eutrophic bay of Lake Taihu (Ding et al. 2018), where high-resolution dialysis and diffusive gradients in thin films (DGT) analysis showed seasonal increases in P flux from iron-related P in the sediments below the depth of oxygen penetration (Figure 3.5B).

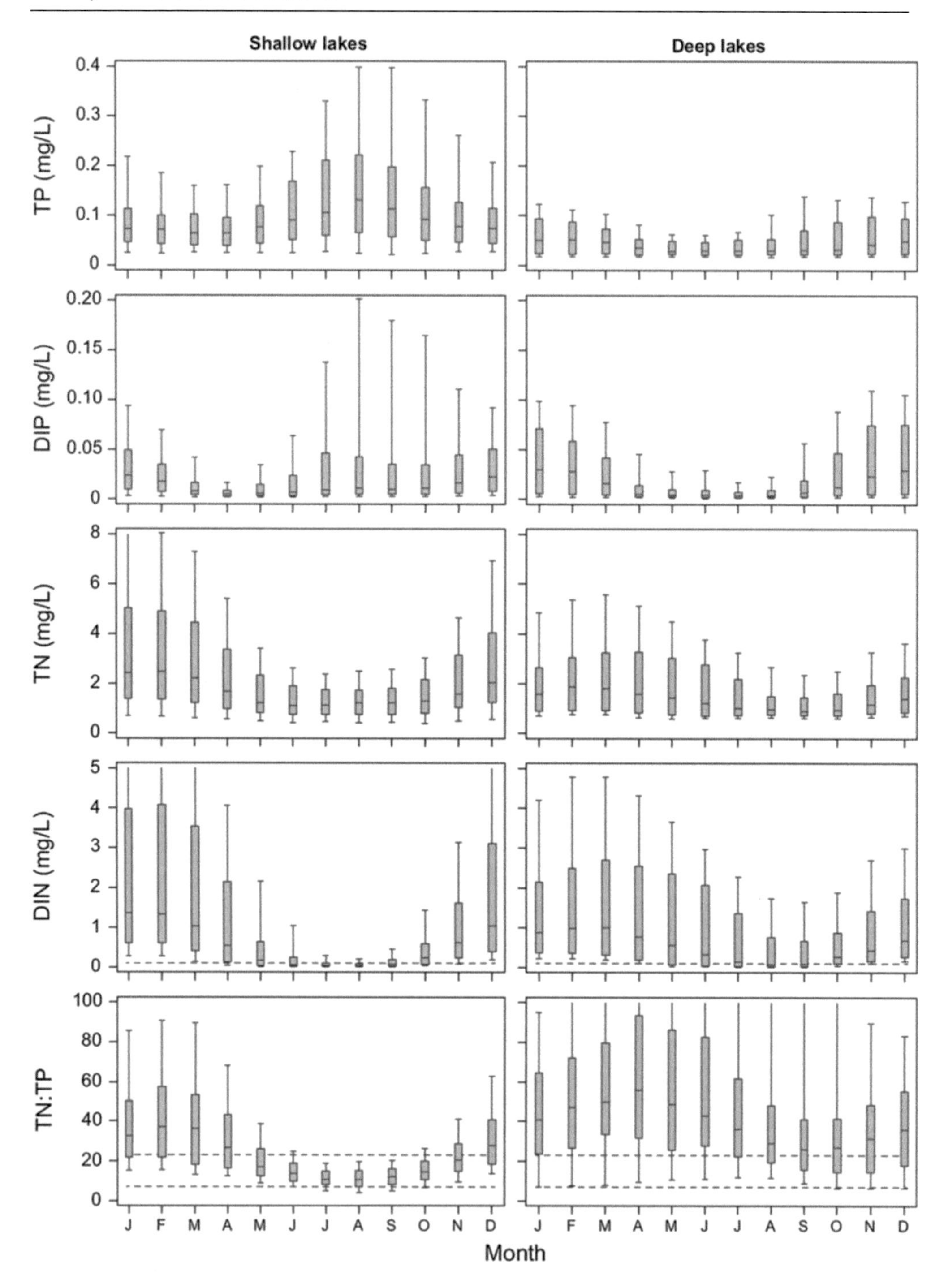

Figure 3.7 Seasonal nutrient variation in 12 shallow and 6 stratified (deep) Danish lakes (DIN = sum of NO_x and NH_4). The boxplots show 10, 25, 75 and 90% fractiles. The line at DIN = 0.1 mg/L indicates the potential threshold for N-limitation. The lines in the TN:TP plots (by mass) indicate potential P (above 22.6) and N limitation (below 7). *Figure 2 from Søndergaard et al. (2017) with permission.*

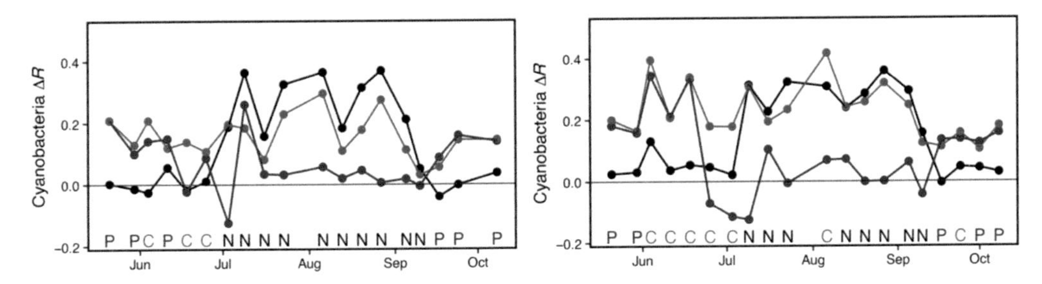

Figure 3.8 Succession of nutrient limitation (P-, N-, and co-limitation) of cyanobacteria, determined by nutrient addition (NH_4 and phosphate) experiments in hyper-eutrophic Lake Acton, Ohio, USA. Left, the shallow mixed inflow site; right, the variably stratified, 8 m deep outflow site. *Figures 3b and 3f from Andersen et al. (2020), open access.*

Note: Limitation is indicated by letters at the bottom of each graph and expressed quantitatively as ΔR, that is, the $\log_{10}$ ratio of the treatment growth relative to the control per day. Values at or below zero (solid horizontal line) indicate no or negative growth in response to nutrient additions.

In highly enriched systems, NH_4 can sometimes increase from near-undetectable levels, supporting cyanobacteria in combination with increased P from internal sources. Such high late summer NH_4 may partially be regenerated from organic settled matter after mineralization (Smolders et al. 2006) contributing to an *internal N load* (McCarthy et al. 2016). Potential NH_4 regeneration was estimated as 60–200% of annual external N load to the western basin of Lake Erie (Hoffman et al. 2022) and contributed 38–58% of potential NH_4 demand for summer–fall *Microcystis*-dominated blooms in hyper-eutrophic Lake Taihu (Xu et al. 2021). Further, the reduced N-compounds, NH_4 and urea, increased the proportion of non-N-fixing cyanobacteria in summer phytoplankton of a hyper-eutrophic polymictic lake at high temperatures and high SRP (Swarbrick et al. 2020, 2018).

The importance of seasonal variation extends to the expression of cyanotoxicity because the prevalence of specific nutrient forms that affect toxicity (Section 3.3.2) changes over the year (Figure 3.5D). In particular, elevated NH_4 concentration has been associated with higher cell quotas of microcystin when P was not limiting (reviewed by Kelly et al. 2019), and luxury N uptake has been proposed to induce cyanotoxin production (Gobler et al. 2016).

Despite seasonal variations in N concentration, there is no evidence that internal N load is as universal as internal P load in eutrophic lakes. None or only a minor TN legacy effect was determined in 23 shallow Danish lakes (Jeppesen et al. 2024a). (1) A TN mass balance model did not require a compartment of upwards flow, as is necessary for TP (Box 2.2). Instead, annual mean lake TN was adequately predicted by external N load, mean depth, and water retention time that is (only) possible for TP in lakes with little or no internal load influence (Table 2.5E). (2) TN retention (in percent) did not change with large N reductions over 30 years of mass balances on ten Danish lakes. Different from lagging TP responses, lake TN concentrations responded immediately to load reductions without considering an internal load.

The prevalent nutrient form (Section 3.3.2) and thus the changes in nutrient limitation of the mixed layer depend on the seasonal stratification regime. In strongly stratified lakes, internal P load remains in phosphate form (SRP) while it is in the hypolimnion, until thermocline erosion and fall turnover (Figure 2.9), so that epilimnetic P does not increase much in the summer. In contrast, in mixed, shallow lakes, sediment-derived phosphate can be immediately taken up by the phyto- and bacterioplankton including cyanobacteria because of the sediment's close proximity to the photogenic zone, removing P limitation. At the same time, oxic N-forms (NO_x)

that are elevated in the spring decline in the summer and fall (Figure 3.5C, Figure 3.6A), while reduced N-compounds form during anoxic stratification but can be quickly incorporated into cyanobacteria biomass during mixed conditions.

Because the stratification regime partially determines the nutrient forms (Figure 3.7), it also influences the seasonal timing of cyanobacterial proliferation. In stably stratified lakes experiencing internal P loading, chlorophyll increases after fall turnover (Figure 3.9), while in weakly stratified and shallow polymictic lakes, frequent thermocline interruptions and internal waves

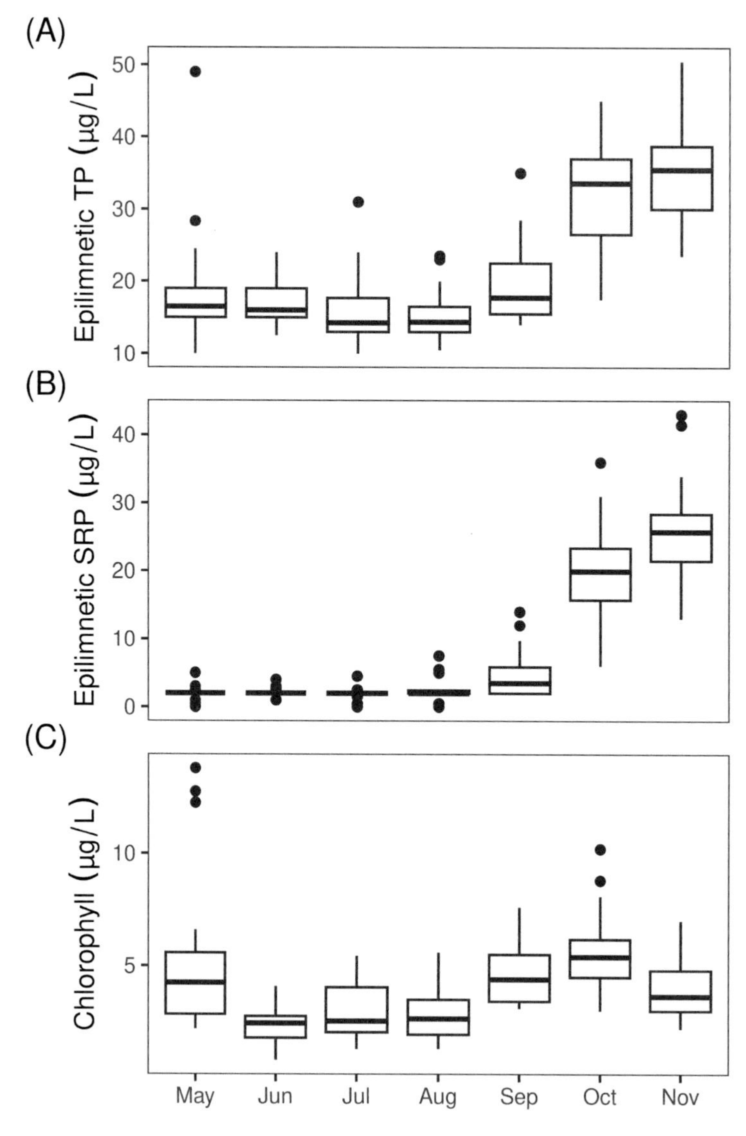

Figure 3.9 Simultaneous increase in TP, SRP, and chlorophyll in the fall in stratified Bornsjön, Stockholm, Sweden (18 m maximum depth, *Nürnberg unpublished study with data from Stockholm Vatten och Avfall AB*) (box-and-whisker plot). Compare to polymictic Brome Lake, Quebec in Figure 2.11.

help entrain sediment-derived nutrients to nurture cyanobacteria throughout the summer. The difference between stratified and mixed lakes depends on their morphometry, especially the morphometric ratio (Equation 2.1). In borderline cases, thermal mixing status can vary between years (Figure 2.11), potentially influenced by lake level differences dependent on precipitation.

Fall blooms were also reported in ten small Canadian lakes located on the Precambrian Shield, adjacent to pockets of lime stone (LeBlanc et al. 2008). Despite mostly low epilimnetic TP concentration ($<<20$ μg/L), cyanobacteria proliferated after the dispersion of hypolimnetic phosphate, accumulated during anoxic hypolimnetic periods, into the photogenic layer in the fall. At the same time, inorganic N compounds (NH_4 and NO_x) were undetectable.

The seasonal changes in carrying capacity (Box 3.2) can also explain observations regarding long-term increases of cyanobacteria from long-term changes in P and N input. The late summer N limitation in many eutrophic lakes, as triggered by an overabundance of sediment-derived phosphate, increases the susceptibility of cyanobacteria to external N input.

In hyper-eutrophic Canadian Prairie lakes (Qu'Apelle lakes, SK), the contribution of cyanobacteria to phytoplankton has been increasing in connection with the long-term rise in reduced N compounds, such as urea and NH_4, from fertilizers (Bogard et al. 2020). The cyanobacteria proliferation was explained by relieving N limitation in the summer, which typically occurred due to high internal P loading rates.

In Lough Neagh, only water temperature was significantly correlated with chlorophyll in 1974–2012, but increases of external N load or lake nitrate concentration were correlated with chlorophyll in addition to temperature in more recent years of 1994–2012 (McElarney et al. 2021). The cyanobacteria (based on chlorophyll concentration) generally increased throughout the period of increasing summer internal P load, but annual averages declined when NO_x (nitrite and nitrate) and external N input declined.

Similarly, decreased inputs of TN and TP from 1980 to 2016 in hyper-eutrophic Müggelsee, Germany, decreased seasonal inorganic N concentration (NO_x and NH_4), but not SRP from internal load, and correlated with decreased summer phytoplankton including N_2-fixing cyanobacteria, also indicating that increased N limitation does not necessarily promote N_2-fixing taxa (Shatwell and Köhler 2019) (Figure 3.10). The authors conclude that "a P-only control would not have been as successful", because of the sustained internal P load.

Conversely, increased N-inputs (as urea from fertilizer) were positively correlated with HABs in Lake Taihu. However, the mere overlap of these increasing variables is no proof for the causal dependencies. In Lake Taihu, as well as in several other Chinese lakes with increased HABs, TN lake concentration significantly decreased and TP concentration tended to increase (Qin et al. 2020a). Typically, phytoplankton and cyanobacteria biomass respond to lake nutrient concentrations rather than external nutrient input, and these variables are not necessarily correlated because inputs undergo change within the lake as they are affected by hydrology, lake morphometry, and sedimentation (retention) (Section 2.4.3).

To summarize Section 3.3, the concept of phytoplankton-carrying capacity depicted in Box 3.2 helps to clarify the nutrient effect on cyanobacteria throughout the year, for different nutrient and stratification regimes, under the consideration of specific phytoplankton traits. Nutrient limitation in freshwater that supports cyanobacteria typically change from P limitation in the spring to N limitation caused by internal load as sediment P release in the summer (in shallow lakes) and fall (in deep lakes) and by denitrification from elevated temperature. That explains why phytoplankton in general and cyanobacteria in particular are increasingly N-limited through summer and fall.

This means that in addition to P control from external load abatement, N load decrease would help reduce the late summer cyanobacteria proliferation as documented in a few lakes (Lough

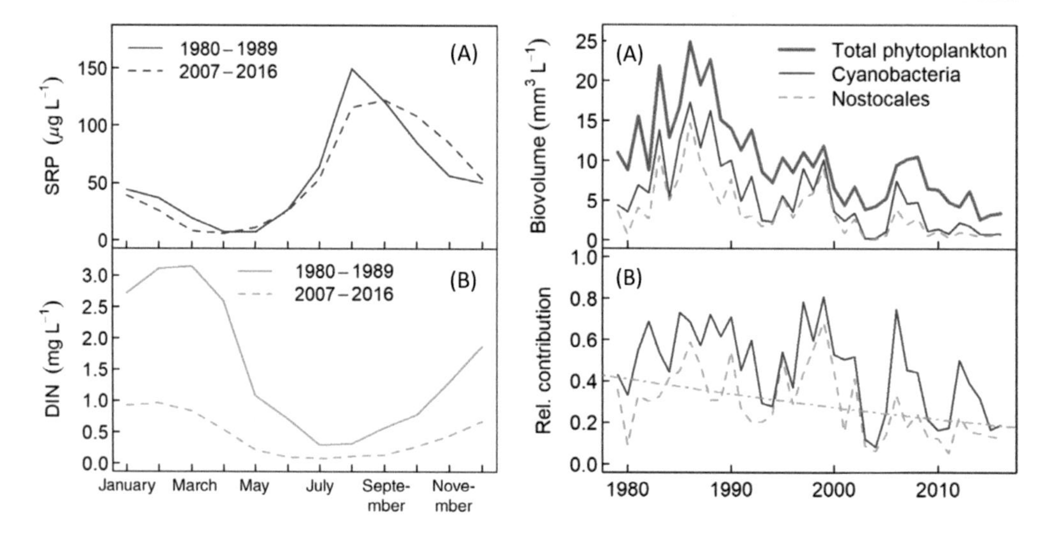

Figure 3.10 Nutrient and phytoplankton changes in hyper-eutrophic Müggelsee, Germany. *Figures 3 and 7 from Shatwell and Köhler (2019), open access.*

Left panel: monthly means of soluble reactive phosphorus (SRP; A) and dissolved inorganic nitrogen (DIN; B) for the decades 1980–1989 (solid lines) and 2007–2016 (dashed lines). Right panel: annual means of the biovolume of total phytoplankton, cyanobacteria, and N_2-fixing taxa (Nostocales) (A) and the relative contributions of cyanobacteria and Nostocales within the total phytoplankton (B). The dot-dashed line in (B) is the regression line for Nostocales.

Neagh and Müggelsee). More studies are needed to further specify the importance of controlling both nutrients P and N. It is conceivable that in situations of extremely large internal P loads that would be difficult to abate (Section 6.2), a strong reduction of N loads may be economically preferable to reductions in P external and internal loads.

3.4 Carbon and micronutrient effects on cyanobacteria

Carbon in carbon dioxide (CO_2) is needed by autotrophs for photosynthesis. While carbon has been historically suggested to be limiting, more recently, one or both macronutrients (P and N) are considered the primary growth-constraining substance. In addition, secondary nutrients (micronutrients) have been suggested to become constraining when the limiting primary nutrient is increased (Sterner 2008). For example, an experimental Fe enrichment of Lake Erie offshore water samples led to increased chlorophyll concentration, mostly from cyanobacteria, only when P and/or N were added, while P enrichment alone resulted in a decrease of Fe (caused by biological uptake of this secondary nutrient) to a level that the N metabolism became co-limited by N (North et al. 2007).

The availability of specific micronutrients can affect cyanobacteria growth, abundance, and species selection. Micronutrients also appear to influence the production and expression of cyanotoxins (Section 3.2), although available studies are not conclusive. Some of the micronutrients have different redox states, so that at conditions of P release due to low-redox conditions, these specific nutrients may be enhanced.

3.4.1 *Carbon*

Physiological characteristics provide cyanobacteria an advantage in comparison to other phytoplankton with respect to carbon fixation needed for photosynthesis. By using a carbon dioxide (CO_2)-concentrating mechanism, cyanobacteria can outcompete other algae at low CO_2 levels

(Badger and Price 2003). In addition, some groups can take up bicarbonate (HCO_3^-) that forms under high pH (Ranjbar et al. 2021). While the high accumulation of cells in blooms increases CO_2 consumption during photosynthetic activity and causes extreme CO_2 limitation, surface-dwelling blooms can further avoid this limitation by taking up CO_2 at the air–water interface (Section 3.5.2).

Carbon as dissolved organic carbon (DOC) contributes to the brown tea-stain in coloured lakes. Cyanobacteria are expected to be positively correlated with DOC because of the positive correlation between average concentrations of DOC and TP in regional and worldwide surveys (Nürnberg and Shaw 1998; Stetler et al. 2021) (Figure 2.4). Indeed, cyanobacteria biomass and DOC were significantly and positively correlated (R^2 = 0.45) in the survey of 640 Canadian lakes (MacKeigan et al. 2023), potentially reflecting enhancing effects of organic matter and trophic state on cyanobacteria proliferation. The influence of DOC on cyanobacteria is further explored in the context of climate-related brownification in Section 4.4.4.

3.4.2 *Iron*

Iron (Fe) is an essential micronutrient and important for N assimilation and photosynthesis by cyanobacteria as it is involved in the reduction of NO_x to NH_4 (reviewed by Wilhelm 1995). In aqueous solutions, Fe exists in two oxidation states: reduced ferrous iron (+II, Fe^{2+}) and oxidized ferric iron (+III, Fe^{3+}). Fe^{2+} is relatively soluble at the physiological pH range and considered to be readily bioavailable, while Fe^{3+} is poorly soluble in oxic freshwater, forming insoluble Fe oxyhydroxides, unless attached to humic and fulvic acids in brown-water lakes. Therefore, iron naturally occurs at low concentration in most oxic waters, although it can be elevated in anoxic hypolimnetic waters.

In softwater lakes, the redox-dependent iron cycle is related to the P cycle, because phosphate is desorbed from iron hydroxides when the redox potential decreases below a certain threshold (Section 2.4.2). Under these conditions (~ 200 mV and below), phosphate and (dissolved) Fe^{2+} are released simultaneously from the sediments and can be detected in the hypolimnetic water overlaying the sediments (Figure 2.10). In hardwater lakes with higher sulfate concentrations, sulfate becomes reduced to sulfide at a low redox potential (~0 mv) and retains ferrous iron in the sediments as the insoluble compound of ferrous sulfide. Consequently, less dissolved ferrous iron is released simultaneously with phosphate in hardwater lakes than softwater lakes. This occurs especially in highly eutrophic lakes and in sedimentary catchment basins (hardwater with high sulfate concentration) and during prolonged hypoxia (Nürnberg 1994; Smolders et al. 2006). When P and Fe are abundant and above saturation, the ferrous iron phosphate mineral of vivianite $[(Fe^{2+})^3 (PO_4)_2 \cdot 8 H_2O)]$ forms a sink for sedimentary P in anoxic sediments (Rothe et al. 2014).

Just as for sediment-released P, the detection of sediment-released Fe is straightforward in softwater lakes under thermally stratified conditions (using anoxic sampling methods) (Nürnberg and Dillon 1993). In mixed conditions, the chemicals are being distributed directly into the aerated water above the sediments, where Fe oxidization leads to Fe oxyhydroxides that can adsorb phosphate and then precipitate it, if that is not prevented by dilution or phytoplankton uptake (Nürnberg 1985).

Cyanobacteria have elevated Fe requirements compared to other phytoplankton, especially under N limitation. They are well adapted to prevent iron starvation because of specific chemicals including siderophores, that is, strong ferric iron chelators (hydroxamates and catecholates), and other iron-regulating proteins (Cheng and He 2020). A study of 30 hyper-eutrophic (N- and P-replete) lakes in Alberta, Canada, observed correlations of Fe-binding ligands (hydroxamates and catecholates) in cyanobacteria with modelled ferric Fe concentration, indicating the potential use of ferric Fe by cyanobacteria (Du et al. 2019).

In theory, Fe limitation should occur especially in hardwater, eutrophic lakes (e.g., the Great Laurentian Lakes, Lake Erken) because of Fe precipitation by sulfide in anoxic sediments. Nonetheless, Fe limitation is not consistent in hardwater lakes, as harmful algal blooms have occurred in many eutrophic and hyper-eutrophic hardwater lakes. For example, in two connected eutrophic prairie lakes (Idaho, USA), cyanobacteria thrived only in the lake that experienced elevated TP concentration from internal loading, despite similar Fe concentrations (Leung et al. 2021). Similarly, addition experiments determined the general independence on micronutrients in Central Basin water of Lake Erie. There were no secondary effects (after nutrients) of Fe, Mo (molybdenum), and B (boron, Section 3.4.4) on phytoplankton biovolume and NO_3 assimilation, and only small effects on chlorophyll (<26% of experiments) (Chaffin et al. 2020), except that combined additions of P and Fe increased chlorophyll during *Dolichospermum* blooms.

Iron additions as treatment of internal P load invoked variable responses in a small hyper-eutrophic hardwater lake in British Columbia, Canada (Section 6.4.3), where the treatment stimulated *Microcystis* blooms when only small quantities were applied (Hall et al. 1994). Similarly, iron additions stimulated the growth of the cyanobacteria *Gloeotrichia echinulata* in hardwater Lake Erken, Sweden. In an in-situ experiment, phosphate and NO_3 were added to all enclosures, but pelagic colonies of *G. echinulata* only increased in abundance in enclosures to which Fe had been added too (Hyenstrand et al. 2001). Similar results were obtained for a *Microcystis aeruginosa* lab culture from a eutrophic lake, where maximal biomass of *M. aeruginosa* was higher with increased Fe availability while it also depended on its initial biomass (Zhang et al. 2017). It appears that the initial cyanobacteria biomass affects the response to Fe addition in nutrient-replete systems.

Vrede and Tranvik (2006) predicted a large potential effect of Fe on lake productivity in 659 oligotrophic un-stained Swedish lakes, based on the experimental enrichment of water from 9 lakes. While P increased chlorophyll in eight lakes, another addition of Fe further increased chlorophyll in seven lakes.

The release of Fe in softwater lakes under anoxic conditions can coincide with increased cyanobacteria biomass (Dengg et al. 2023). These simultaneous events can lead to the hypothesis that iron release from anoxic sediments (iron internal loading) is the primary cause of cyanobacteria blooms (Molot et al. 2014), even though P is released simultaneously. The separation of potential cyanobacteria-biomass-enhancing effects by P internal loading from those by Fe internal loading is discussed in Section 5.6.2. In conclusion, there is no clear evidence about the importance of Fe to pelagic cyanobacteria.

In contrast, bottom dwelling organisms have direct access to the sediment and solutes in its porewater and may directly benefit from sediment Fe and P. The cyanobacterium *Lyngbya wollei* forms mats at the bottom of rivers and lakes. There, steep gradients of redox-dependent metals (Fe, Mn) and nutrients (phosphate, NH_4) occur, caused by sediment anoxia, in addition to pH gradients and light gradients (in shallow lake regions) (Hudon et al. 2014). *L. wollei* requires Fe for synthesizing the enzyme nitrogenase during atmospheric N_2 fixation. It did not grow well without additions of Fe to the growth medium where it thrived at up to 600 μg/L Fe. It is possible that access to Fe supports these taxa also in lakes.

3.4.3 Manganese

Manganese is an essential micronutrient for cyanobacteria and is required for photosynthesis at high light conditions. However, Mn may become toxic at higher concentrations according to earlier studies (Patrick et al. 1969). Since then, a transport protein that regulates and maintains intracellular Mn concentration has been determined that avoids toxicity at least for one species, the genetic model organism *Synechocystis* (Brandenburg et al. 2017). The limitation and toxicity at higher Mn concentrations were reviewed on this species, although potential Mn limitation is rarely studied (Eisenhut 2020). Because Mn is released from sediments at a redox potential (~400 mV) that is higher than for Fe and phosphate (~200 mV), Mn increases are often a precursor of internal P release. Mn can be correlated with P and potentially, Fe, especially in softwater. In hardwater lakes with hydrogen sulfide development, Mn release is the better indicator of potential redox-related P release than Fe, because iron is retained in the sediment by ferrous sulfide (Section 2.4.2).

3.4.4 *Molybdenum, boron, salinity, and other trace elements and vitamins*

Molybdenum (Mo) is an essential micronutrient and especially important for the N assimilation by cyanobacteria due to its presence in nitrogenase, the enzyme that performs N_2 fixation, and in nitrate reductase, the enzyme that reduces NO_3 to NO_2 in the process of NO_3 assimilation (Glass et al. 2012). Mo limited the primary productivity of natural microbial communities and pure cultures in lakes of New Zealand, Alaska, and the western United States (Glass et al. 2012). In their review, the authors concluded that N-fixing cyanobacteria in freshwater periphyton communities have higher Mo requirements than microbial communities in the open water. The simultaneous addition of Mo and nitrate caused an increased activity of proteins involved in N assimilation when NH_4 was scarce, indicating specific responses to the redox state of the environment. Since Mo is less needed when NH_4 is present, it is unlikely to affect cyanobacteria proliferation under conditions that support sediment P release (i.e., during low-redox conditions).

Other addition experiments show only small effects. For example, in Central Basin water of Lake Erie, no or only small secondary effects of Mo on chlorophyll were observed (Chaffin et al. 2020).

Mo, usually known in its oxidized form (+VI), also occurs in the reduced form (+V) in sulfidic mats and other reduced environments in freshwater and oceans as reviewed by Wang (2012). Because of Mo's ability to switch its redox state from +II to +VI, it can be involved in electron and oxygen transfers in reactions of C, N, and S. Reduced forms were deemed to be potentially responsible for cyanobacterial blooms in oceans, but less is known about its involvement in freshwater blooms.

Boron (B) is a micronutrient needed for the synthesis of the cell wall in cyanobacteria heterocytes (specialized cells where N fixation occurs) (Bonilla et al. 1990). B occurs as the highly soluble $B(OH)_3$ at pH less than 9.24 in freshwaters and is not redox-dependent. Under low NO_3 conditions when *Dolichospermum* produces heterocytes, B increased growth by 15.8% when P and B were added (Chaffin et al. 2020). These addition experiments found occasional increases (both as chlorophyll and biovolume) of mostly *Dolichospermum* by the addition of the three secondary nutrients, Mo, B, and Fe, to p-limited water of Central Basin, Lake Erie, throughout the growing seasons of 2014 to 2017 (Chaffin et al. 2020). P and Fe, P and Mo, and P and B enrichments resulted in significantly higher chlorophyll concentration than P enrichment alone in 26.3, 7.6, and 15.8% of the experiments. Similarly, heterocystous *Gloeotrichia* in water from Lake Erken, Sweden, was stimulated by the addition of a mixture of P, N, Fe, and B, compared to one without B (Hyenstrand et al. 2001).

Salinity due to chloride (Cl) ions and resulting salinization (Section 5.6.4) affect species selection and the production of cyanotoxins. In treatment scenarios with N:P ratios indicating

P limitation (N:P = 50 by atom) or N limitation (N:P = 4), growth and toxin production were investigated for salinity increases up to 10.5 g/L (Osburn et al. 2023). *Microcystis* (non-N-fixer) produced more microcystin (MC-LR) toxins at intermediate salinity (treatments 0.35–6.5 g/L salinity) in the N-enriched treatment, while the response by *Aphanizomenon* (N-fixer) to salinity was variable. Based on their experiments, the authors concluded that freshwater cyanobacteria biomass and toxicity may increase when mixed with brackish coastal water downstream (Osburn et al. 2023). Similarly, a field survey determined that cyanobacterial blooms' and cyanotoxins' occurrence (*Dolichospermum, Aphanizomenon, Microcystis*, and *Planktothrix*) was disproportionately higher in brackish than in freshwater lakes in the United States (Patiño et al. 2023).

Vitamins, for example, cobalamin (B_{12}), are organic water-soluble molecules that act as coenzymes impacting cyanobacteria growth and distribution but are rarely studied in freshwater. Recent work in marine systems involves promising genetic techniques that may determine the role of vitamins in freshwater in the future (Helliwell 2017).

3.5 Favoured physical environmental conditions

Environmental effects on cyanobacteria proliferation can be caused by non-nutrient variables (e.g., light, temperature) solely or modified by nutrients. Therefore, most studies either use nutrient-saturated conditions to study the effects of other variables or determine combined effects including specific nutrient levels. The functional groups introduced in Section 3.1 (Table 3.1) indicate the dependence of various cyanobacteria taxa on specific physical environmental characteristics.

3.5.1 Preferred temperature

Cyanobacteria favour elevated water temperatures especially if they are nutrient replete, as in eutrophic and hyper-eutrophic systems. Physiological adaptation to higher temperatures is often presented as the direct reason for the prevalence of cyanobacteria at elevated temperature (Paerl and Otten 2013). Surface or water column average temperature was the second best predictor of cyanobacteria biomass and functional subgroups (bloom-forming, potentially toxic, heterocyst-producing) after nutrient concentration (TN or TP) in a data set of about 1 000 lakes of the continental United States (Beaulieu et al. 2013). Similarly, temperature significantly increased the prediction of cyanobacteria biovolume when added to nutrient concentration (either TP or TN and TP combined) for a data set of 143 shallow (average of mean depth, 1.8 m) South American and European lakes, while it did not improve predictions for chlorophyll *a* and only slightly those for total phytoplankton biovolume (Kosten et al. 2012).

Many cyanobacteria have higher temperature maxima compared to eukaryotic phytoplankton species, and their growth rates rise faster with temperature increases (Huisman et al. 2018). For example, cyanobacteria (*Dolichospermum cylindrica, D. flos-aquae, Merismopedia sp., Microcystis wesenbergii*) had the highest thermal optimum (30.6 ± 2.3°C) compared to 13 species and strains of chlorophytes (25.7 ± 0.1°C) and 4 species of diatoms (24.0 ± 0.4°C) as determined by laboratory incubation experiments (Nalley et al. 2018). Other phytoplankton taxonomic groups had even smaller temperature optima, so that dinoflagellate had the lowest optimum, with increasing optima by cryptomonads, chrysophytes, bacillariophyceae (i.e., diatoms), chlorophytes, and, finally, cyanobacteria.

Similarly, in another study, optimum temperatures were above 30°C for three widespread bloom-forming cyanobacteria taxa *Dolichospermum, Microcystis*, and *Cylindrospermopsis*

(Thomas and Litchman 2016). Optimum temperatures of naturally occurring cyanobacteria taxa ranged from 12°C (metalimnetic low-light and cold-adapted trait R, *Planktothrix rubescens*, in Mond See, Austria) to 22°C (*Cylindrospermopsis raciborskii* in Alte Donau, Austria) (Dokulil and Teubner 2000). But individual strains isolated from different natural waters of the same species (in this case, *Microcystis aeruginosa* and *Cylindrospermopsis raciborskii*, isolated from Australian reservoirs) varied widely in temperature and light dependency (Xiao et al. 2017), which helps explain the diverse flexibility and adaptability by various cyanobacteria species and their strains.

In contrast, another experimental study and a literature review did not find significant growth differences at elevated temperatures between specific cyanobacteria strains (including *Dolichospermum, Aphanizomenon gracile, Cylindrospermopsis, Synechococcus and two strains each of Microcystis aeruginosa and Planktothrix agardhii*) and specific chlorophytes (*Ankistrodesmus falcatus, Chlamydomonas reinhardtii, Desmodesmus bicellularis, D. quadricauda, Monoraphidium minutum, Scenedesmus acuminatus,* and *S. obliquus*) (Lürling et al. 2013). Cyanobacteria growth rates were significantly smaller only at the lowest tested temperature of 20°C. Cyanobacteria optimal growth temperature and growth rates at their optimum temperature were not significantly different from those of chlorophytes in a lab study for temperatures at 25–35°C. A literature analysis involving 130 records of the same species of cyanobacteria and chlorophytes supported this result "contradicting the suggestion that cyanobacteria dominate the phytoplankton community under warm circumstances owing to their relatively high growth rates compared with eukaryotic phytoplankton such as chlorophytes" (Lürling et al. 2013). Different responses to temperature changes may be caused by pre-conditioning and acclimatization effects explored below, as well as unconsidered environmental conditions.

A meta-analysis shows that in the distribution of cyanobacteria, lake temperature may be less important than nutrient and local variables. A study of 464 lakes covering a 14 000 km north–south gradient in the Americas (North, Centre, and South America) concluded in its title that "Nutrients and not temperature are the key drivers for cyanobacterial biomass in the Americas" (Bonilla et al. 2023). The authors determined TP as the primary resource explaining cyanobacterial biomass, while TN was most relevant in very shallow lakes (<3 m).

Considering human health aspects, there is the concern that increased temperature directly increases the abundance of cyanotoxins, so that blooms could become more toxic in a warming climate. Experimentally enhanced water temperatures yielded significantly increased growth rates of toxin-producing *Microcystis* as compared to non-toxic strains (Davis et al. 2009). Concurrent increases in temperature and P and N concentrations yielded the highest growth rates of toxigenic *Microcystis*, indicating the synergistic influence of nutrients and temperature on cyanotoxin production. A survey of 137 European lakes in the summer of 2015 found that "direct and indirect effects of temperature were the main drivers of the spatial distribution in the toxins produced by the cyanobacterial community, the toxin concentrations and toxin quota" (i.e., toxin concentration per unit algal biomass) (Mantzouki et al. 2018).

Despite the apparent preference for warmer temperatures, cyanobacteria can also be found at colder temperatures (below 15°C) in north-temperate lakes (Reinl et al. 2023). The different ways of cold bloom development were grouped into (1) initiation in cold water, (2) originating from the cold metalimnion, and (3) warm-water blooms prevailing at colder temperatures. Mode (1) involves concurrent upwelling of cold and potentially P-rich water from the hypolimnion, while Modes (2) and (3) are explained by the temperature preferences examined above. All modes of cold-water bloom development can be explained by access to nutrients from internal P loading and are discussed in Section 5.1.7.

The influence of temperature on cyanobacteria is further complicated by physiological characteristics such as acclimatization. Thermal acclimatization interacted with temperature changes influencing cyanobacterial growth and toxin production in laboratory experiments using four strains of common cyanobacteria (*Dolichospermum flos-aquae*, *Phormidium foveolarum*, and two strains of *Microcystis aeruginosa*) (Layden et al. 2022). Cold-acclimated cyanobacteria grew faster and produced more toxins at warming temperatures than ambient temperature-adjusted and hot-acclimated organisms at low and high nutrient concentrations. Such physiological pre-conditions could explain the recent increase in cyanobacteria and their toxicity in northern, often low-nutrient, lakes.

3.5.2 Light preference and related depth distribution

The effect of the light environment on phytoplankton in specific lakes is based in part on the length of time the phytoplankton are exposed to sufficient light. To quantify this light effect in a lake, a metric called light ratio, z_{eu}/z_{mix}, is often used. It compares the depth of the euphotic layer (z_{eu}, determined as 1% of surface penetration of photosynthetically active radiation) to the mixing depth (i.e., upper depth of the metalimnion, z_{mix}) as ratio z_{eu}/z_{mix}. Ratios greater than or equal to 1 indicates that algal communities are exposed to light regardless of position in the mixed layer, while ratios less than 1 indicate that position within the mixed layer is important and that algae likely spend some portion of the time under aphotic conditions.

Light ratios are related to water column stability, so that at z_{eu}/z_{mix} of 1, stratification can be strong with a shallow euphotic zone, or stratification is weak and the entire water column is illuminated as for mixed shallow conditions. At ratio less than 1, strongly stratified lakes may have a deep mixed layer as occurs in reservoirs with a deep-water outlet. The interplay of light and mixing affects nutrient availability from potentially P-enriched bottom waters (by internal loading) (Donis et al. 2021).

Cyanobacteria differ from most eukaryotic phytoplankton (except for eukaryotic cryptomonads that also have phycoerythrin) by containing the two pigments, phycocyanin and phycoerythrin, in addition to the green photosynthetic pigment, chlorophyll *a*, that is present in all phytoplankton cells. These specific pigments enable certain cyanobacteria to harvest light at low intensity and deeper in the water column. There are basically two types of cyanobacteria with respect to light dependency, and they respond differently to the light ratio. There are (1) surface bloomers (e.g., functional groups or traits H1, H2, M, Table 3.1, including *Microcystis*, *Dolichospermum*, and *Aphanizomenon*) with pigments of chlorophyll *a* and phycocyanin and (2) species adapted to low-light conditions at mixed conditions (S1, e.g., *Planktothrix agardii*) or in the deeper metalimnion (R, *Planktothrix rubescens*) with the red additional pigment of phycoerythrin. While mixed conditions are favoured by the low-light-adapted cyanobacteria (traits R and S1), they are growth-limiting to surface bloomers (traits H1, H2, M). A generalized schematic presents preferred locations, migration, and translocation paths and exchanges between the pelagic and sediment cyanobacteria pool (see Figure 3.11).

Surface-blooming cyanobacteria proliferate at z_{eu}/z_{mix} values of 1 or higher, while low-light adapted cyanobacteria are typically found at z_{eu}/z_{mix} values less than 1. Under typical summer conditions when the whole mixed layer (epilimnion) is illuminated (light ratio >1), *P. rubescens* inhabits the metalimnion where it forms deep phytoplankton maxima. Only when the mixed layer becomes deep compared to the illuminated water layer (light ratio $\ll 1$), *P. rubescens* migrates to the surface as in winter and spring blooms (Nürnberg et al. 2003). The adaptation to low light and cool temperatures enables *P. rubescens* to be located at the upper metalimnetic

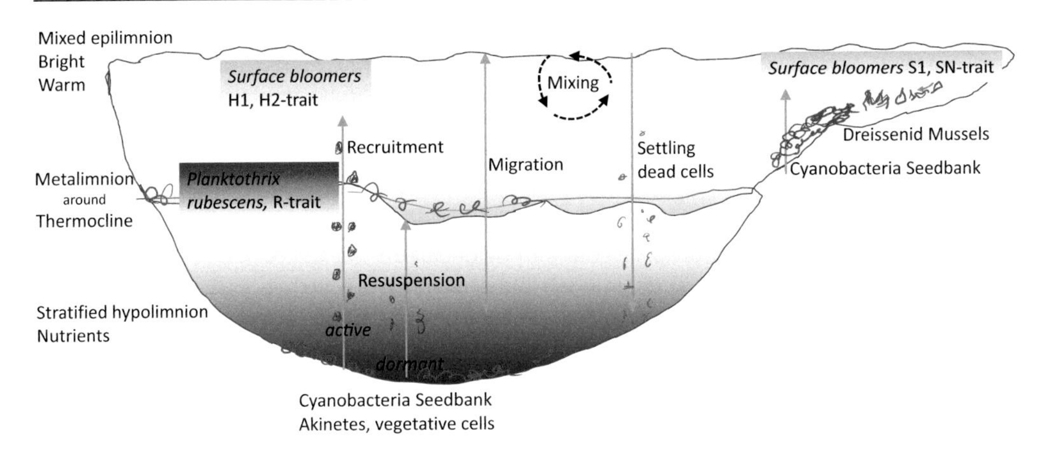

Figure 3.11 Schematic for typical summer cyanobacterial distribution in the water column of the stratified (left) and mixed part of a lake; details vary among cyanobacteria taxa.

border, just above the thermocline in deep, stratified, relatively clear, P-limited lakes that are oligo- to mesotrophic (TP concentration 10–30 µg/L) (Nürnberg et al. 2003). While their biomass increases with TP, it appears that *P. rubescens* can survive at relatively low metalimnetic TP concentrations because of the proximity to the P-rich hypolimnia. Luxury uptake in the P-rich hypolimnion has been suggested to sustain *P. rubescens* during upward migration.

Besides surface bloomers and low-light-adapted metalimnetic species, Chorus and Welker (2021) distinguished several more types of cyanobacteria distribution within the water column with possible transitions between them. In shallow lakes, cyanobacteria including *Planktothrix agardhii, Limnothrix*, and *Pseudanabaena* (Table 3.1) are typically homogenously distributed throughout the mixed water column with access to nutrients in the whole water column and the sediment surfaces. Further, there can be decaying, settled cyanobacteria at the bottom of lakes and live benthic cyanobacteria in streams (Section 5.7.2).

The distribution of surface blooms is highly dependent on other physical influences, especially thermal stratification, and wind. Surface blooms rarely generate at the location where they are detected and are usually controlled by wind strength and direction as well as lake depth, which influences strength of stratification (Chorus and Welker 2021). Typical surface bloomers (*Aphanizomenon* and *Dolichospermum*) can also occasionally inhabit deep, nutrient-rich layers with low light intensity and contribute to deep layer chlorophyll maxima, depending on their life cycle (e.g., for *Dolichospermum* (Zastepa et al. 2023a)).

Using light ratios, Donis et al. (2021) determined that light availability and stratification were more important for chlorophyll concentration than nutrients in 230 nutrient-rich European lakes. In a subset of shallow lakes, N was an additional predictor for chlorophyll, which could be explained by elevated internal P load from shallow lake sediments providing P replete conditions.

Underwater light was considered partially responsible for the prevalence of cyanobacteria in 143 shallow lakes mentioned in Section 3.5.1 (Kosten et al. 2012), leading to a conceptual model that includes a shade index as well as nutrients, temperature, and biological forcing via grazing. The shade index (ratio of average lake depth to Secchi depth transparency, *z/z_sec*) is informative for shallow lakes, where typically the whole water column is mixed with the phytoplankton dispersed throughout. Using regression models and correcting for temperature and pH effects, Kosten et al.

(2012) found that cyanobacteria out compete other phytoplankton at high shade indices, likely due to the widely distributed shade-tolerant Oscillatoriales (Table 3.1). Further, the shade index was inferred to be associated with thermal stratification and oxygen consumption that are favourable for sediment P release, enhancing the bottom-up stimulation of cyanobacteria by nutrients.

The impact of light can differ between different strains even for the same cyanobacteria species as was also found with respect to temperature dependencies (Xiao et al. 2017). There is a possible adaptation of cyanobacteria strains to protect against too much harmful light. Especially, ultraviolet (UV) radiation is harmful for phytoplankton but can be less so in toxic cyanobacteria strains. Lab experiments determined that at elevated UV and photosynthetically active radiation (PAR), a toxic strain of bloom-forming *Microcystis aeruginosa* had a competitive advantage over the non-toxic strain, because of more efficient antioxidant systems and stronger recovery capacity (Xu et al. 2019).

3.5.3 Preferred mixing state, depth, hydrology, and physiological adaptations

Cyanobacteria can control their buoyancy which enables vertical migration (Figure 3.11). Thus, cyanobacteria can take advantage of optimal nutrients (Sections 3.3 and 3.4), temperature (Section 3.5.1), and light conditions (Section 3.5.2) so that they can successfully compete against eukaryotic phytoplankton (reviewed in Watson et al. 2015). In general, buoyancy decreases at high irradiation and increases under sub-optimal light conditions that trigger cyanobacterial migration in the water column for optimal photosynthetic activity.

Cyanobacteria *buoyancy* is controlled in several ways:

(1) Many genera can control their buoyancy by the formation of so-called "gas vacuoles" (e.g., *Dolichospermum, Planktothrix, Microcystis*). These vesicles (they are not vacuoles in the cytological sense) are filled with air by diffusion, rendering the cyanobacterial cells less dense than water and therefore buoyant (Walsby 1994). Gas vesicles increased with light limitation, for example, in Lake Taihu *Microcystis* in situ and in batch culture (Duan et al. 2021).
(2) Some taxa use carbohydrate ballast to regulate buoyancy. This mechanism is based on the difference between carbohydrate weight from increased storage during the day by photosynthesis and reduced storage during the night by respiration (e.g., *Oscillatoria, Microcystis, Aphanizomenon*, and *Dolichospermum*) (Visser et al. 2016).
(3) Surface-blooming, scum-forming taxa can have additional uplift from trapped supersaturated oxygen gas produced by photosynthesis at high light and temperature (Medrano et al. 2016).

In these ways, surface bloomers can move up, while low-light adjusted taxa can remain at or move to specific depth layers. Some species are capable of diurnal migration depending on cell size, water temperature, and gas accumulation. Migration provides access to light at the surface during the day and access to nutrients at depth from sediment P release and the decomposition of seston and detritus (Oliver et al. 2012). Diurnal changes in lake stability, e.g., when morning quiescent conditions support buoyancy-controlled upward migration, can temporarily facilitate the entrainment of cyanobacteria near the surface.

The influence of nutrients, rather than light, on buoyancy control is less clear, but internal P sediment loading has been implied as a partial driver of buoyancy-controlled diurnal vertical migration by cyanobacteria such as *Microcystis* (Gobler et al. 2016). A decrease in buoyancy under N and P limitation and an increase under N and P replenishment have been hypothesized, but support in natural systems is rare as reviewed by Visser et al. (2016).

Besides migration based on buoyancy, cyanobacteria can recruit from benthic resting stages. Hence, the vertical migration of cyanobacteria is also possible by way of translocation of specialized cells from the sediment into the open water, because the bottom sediments support the continuation of cyanobacteria populations and conservation by "seed banks" (Figure 3.11). Sediments can harbour thick-walled resting cells of cyanobacteria, called akinetes, when water conditions are unfavourable as during drought periods, extreme temperature, and low-light periods, for example, in the winter. The overwintering akinetes are viable for decades and can inoculate summer blooms upon their germination. Akinete production thus provides a competitive advantage over genera that do not form akinetes, at least in the early stages of bloom formation. Akinetes are formed by several taxa, including *Dolichospermum*, *Aphanizomenon*, *Cylindrospermopsis* (*Raphidiopsis*), *Gloeotrichia*, and *Nodularia* (Table 3.1).

Other taxa, for example, the ubiquitous *Microcystis*, do not form akinetes but overwinter in a vegetative state from which they transfer and inoculate the water column in early summer. Some of the akinete-forming taxa can also grow vegetative resting forms as an additional means of survival during inopportune periods. A review of studies that include benthic cyanobacteria stages in addition to pelagic stages illuminates the range of habitats utilized by different life stages of cyanobacteria (Cottingham et al. 2021). Cyanobacteria recruitment from akinetes has been proposed to contribute to internal P loading especially from shallow areas of lakes. Estimates on the extent of P translocation from the sediment to the water are variable and further discussed in Section 5.4.

Variations of *water stability and thermal stratification* have a large effect on cyanobacterial distribution and prevalence. For example, vertical mixing produces sub-saturating light conditions thus decreasing growth efficiency and impacting phytoplankton prevalence (Köhler et al. 2018). Because water column stability and mixing are much controlled by long- and short-term lake hydrology and morphometry (e.g., the morphometric ratio, Equation 2.1), they are important drivers of cyanobacteria location, proliferation, and variability.

Lake-mixing events affect phytoplankton differently depending on whether they occur in deep or shallow lake sections (Figure 3.11) (Cottingham et al. 2021). Deep mixing events can have an adverse effect on cyanobacteria because of diminished light and stability. Conversely, mixing events in shallow areas are expected to help in recruitment from the bottom sediments and nutrient entrainment as internal loading, which would support cyanobacteria proliferation.

In a meta-analysis of deep chlorophyll maxima in 51 mostly stratified, north-temperate lakes (southern Quebec, Canada), thermal stability predicted the vertical distribution of buoyancy-regulating cyanobacteria (but not of most eukaryotic algae taxa) better than nutrients, light, and biological variables (such as grazing pressure, Section 3.6) (Lofton et al. 2020). Using pigment profiles (rather than plankton counts), the authors established that three abundant buoyancy-controlling taxa could be responsible for this pattern: *Coelosphaerium, Dolichospermum*, and *Microcystis*. Similarly, the cyanobacteria community shifted from *Planktothrix agardhii* during mixed conditions, to *Microcystis* during calmer and warmer episodes, and then to *Dolichospermum* in very calm and dry periods in Lac au Duc, France, a shallow eutrophic reservoir (Le Moal et al. 2021).

The importance of thermal stratification in contrast to air temperature was established by observing cyanobacteria proliferation during two hot Central European summer heatwaves in shallow, eutrophic Müggelsee (Huber et al. 2012). Cyanobacteria thrived only in the summer (2006) with prolonged quiescent and thermally stable conditions combined with elevated air temperature. An equally hot summer (2003) failed to produce blooms, which was attributed to the prevention of thermal stratification by occasional mixing during episodes of high wind. In quiescent 2006, surface-blooming *Dolichospermum* (trait H1, Table 3.1) benefitted directly from the stable water column, while the authors explained that the low-light-adapted metalimnetic *Planktothrix rubescens* (trait R, Table 3.1) also proliferated by taking advantage of the stratification-induced internal nutrient loading.

Also, changes in *Microcystis* were explained mainly by stability and water temperature related to climate in shallow Nieuwe Meer, as long as nutrients were replete (Section 4.4.1) (Jöhnk et al. 2008). In this whole-lake experiment, bi-weekly artificial destratification decreased *Microcystin* biomass despite saturated nutrients and high air temperature, when mixing decreased the water temperature and available light (see also destratification as restoration treatment as described in Section 6.3.3).

Water column stability and mixing are much affected by the water flushing rate (or its inverse, the water residence time). Explicit enhancements of cyanobacteria proliferation with increased water residence time and decreased flushing creating more stagnant conditions have been reported in many lakes and reservoirs as reviewed by Carey et al. (2012). For example, in 28 shallow Chinese lakes, water retention time controlled whether spring and summer cyanobacterial biovolume were dependent on the nutrients TN and TP (see also Section 3.3.1). Significant nutrient dependencies were only found in lakes with water retention time longer than 100 days (and no macrophyte cover), while cyanobacteria were absent in rapidly flushed lakes (Zou et al. 2020). This dependency is especially relevant in reservoirs, where water flow is variable and flushing can be more than three times per year, which complicates simple nutrient dependencies.

Similarly, cyanobacteria were directly dependent on *flushing rate* in a chain of six shallow, subtropical lakes (Havens et al. 2019). Biomass was highest at low flushing rates and consistently lower at higher rates during the examined 18-year period (1999–2016). Since elevated flushing only occurred for above-average water levels, high cyanobacteria biomass did not occur at high water levels or elevated lake volumes. In these lakes, lake level (and flushing rate) were significantly correlated with 25-month antecedent cumulative rainfall. Rainfall in turn was correlated to regional weather patterns like the ENSO (El Niño – Southern Oscillation), pointing to the overall relationship between climate and climate change with cyanobacteria biomass (further discussed in Section 4.4).

Likewise, drought-induced longer hydraulic residence in a run-of-river Texan reservoir was associated with earlier stratification and hypoxia (Bellinger et al. 2018). These physicochemical changes were coupled with a quadrupled probability of cyanobacterial bloom occurrence with earlier beginnings and longer durations.

An increasing number of studies on cyclical water level changes describe low P concentration and low cyanobacteria biomass at high water levels and high P concentration and cyanobacteria biomass at low water levels in lakes and reservoirs worldwide. For example, large Tonle Sap Lake, Cambodia, derives its major nutrient supply from the Mekong River basin by monotonal flood pulses. The lake fluctuates from below 2 m to 10 m in its floodplain (85 800 km^2) (Uk et al. 2022). Internal P load was determined as the cause of high bioavailable P concentration and associated cyanobacteria (*Dolichospermum*, *Lyngbya*, and *Microcystis*) during the turbid low-water period, while P concentration is low and the phytoplankton mainly consist of diatoms during the relatively transparent high-water period.

Cyanobacteria blooms as quantified by the period of Low-Nitrate-Days (LND, Box 3.3) in a river and reservoir system of the Lake Erie catchment were significantly negatively correlated with summer inflow and also with the winter-NAO climate index, indicating the negative effect of summer inflow and of precipitation during the December–March period on cyanobacteria biomass (Figure 3.12). In this case, LND decreased over the period of 38 years, reflecting catchment management efforts to decrease TP loads and the invasion by dreissenid mussels in 2003, which obscures the long-term climate affects. However, climate warming (air temperature

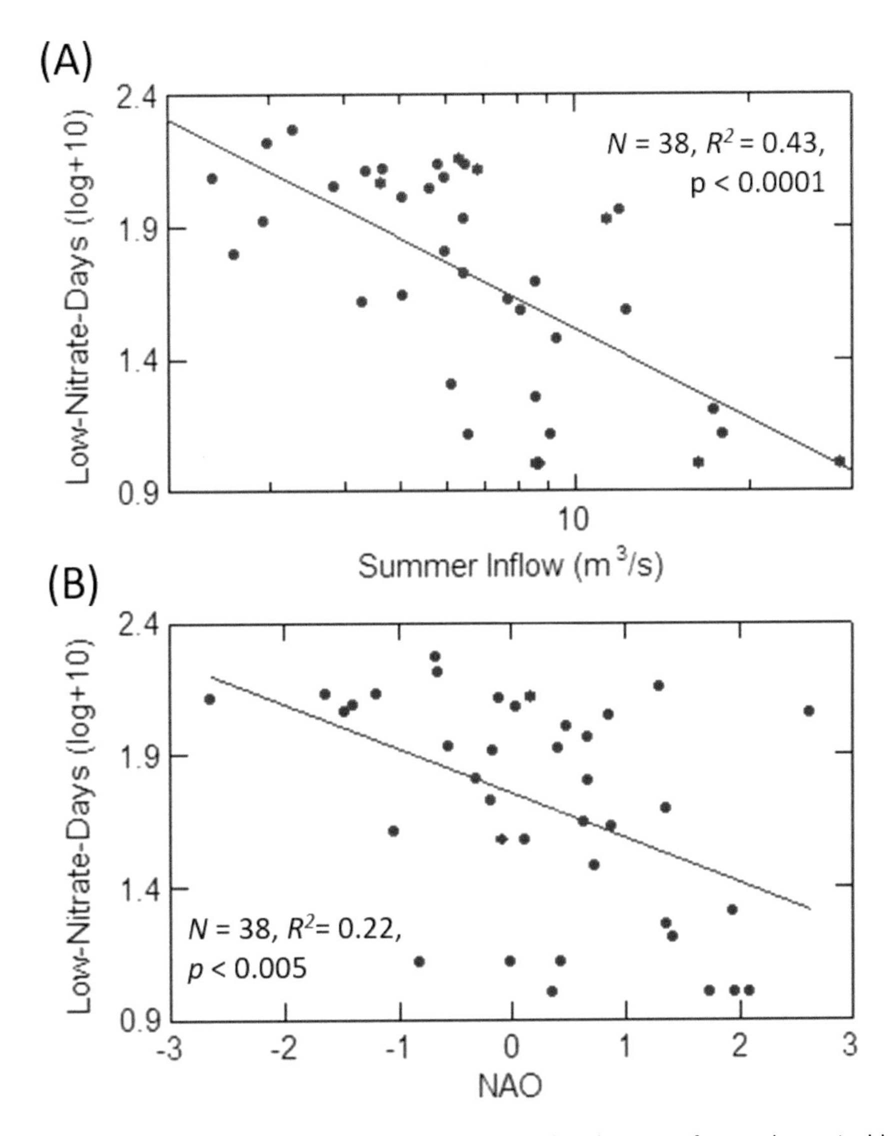

Figure 3.12 Relationships between the log-transformed indicator of cyanobacteria blooms (Low-Nitrate-Days, Box 3.3, 1966–2005) and (A) the main summer inflow of Fanshawe Lake, Upper Thames River, Ontario (1954–2005), and (B) the winter North Atlantic Oscillation Index (NAO), *redrawn from Nürnberg (2007a).*

increases that may decrease flushing) since 1991 and an increase in the ice-free period since 1982 observed in an adjacent catchment basin (Erratt et al. 2022a) will possibly increase cyanobacteria blooms into the future.

3.6 Influences from biota

Biological effects on cyanobacteria proliferation include influences by other organisms. Such influences, including grazing, competition, and exuded chemical compounds that negatively affect surrounding competitors (allelopathy), are here shortly described. Biological conditions related to the food web affect cyanobacteria distribution and are "top-down" controls as opposed to "bottom-up" controls that include nutrient and physical controls discussed in Sections 3.3, 3.4, and 3.5. Internal P loading is less likely to directly influence biological controls but, in turn, could influence interactions, although they are seldom quantified in a larger context. The invasions of the predatory zooplankter spiny water flea (*Bythotrephes cederströmii*) and dreissenid mussels were related to an increase in phytoplankton and cyanobacteria biomass that was modelled to cause a significant increase of the anoxic factor in Lake Mendota, Wisconsin, USA (Rohwer et al. 2024, 2023). But observed increased lake stability caused by climate change could also have increased anoxia and perhaps internal P loading. The poorly understood correlation of invasive dreissenid mussels with cyanobacteria is investigated in more detail in Section 5.1.11, where it is explained by enhanced anoxia and sediment P release.

As reviewed by (Watson et al. 2015), defence and counter-defence mechanisms include morphological (related to digestibility, size, or mucilage), chemical (taste and olfactory cues, toxins), life cycle timing, and behavioural adaptations. For example, large colonial cyanobacteria avoid predation by size-selective *zooplankton* grazers such as *Daphnia* by their size, and the shape of filamentous cyanobacteria can interfere with the feeding apparatus of zooplankton and benthic invertebrates. Another review found that *Microcystis* had the largest negative effect on zooplankton population growth, independent of their microcystin content (Tillmanns et al. 2008).

A recent study in polymictic eutrophic Lake Taihu determined that two zooplankton groups differed in their co-occurrences with cyanobacteria. Cladocerans (includes *Daphnia* species) were more abundant with high cyanobacteria biomass than rotifers, which were most dominant when cyanobacteria were low or absent (Zhao et al. 2022). Also, cyanobacterial bloom intensity was negatively associated with phytoplankton species and functional richness, as expected, but was positively associated with zooplankton functional richness.

Grazing pressure, in addition to nutrients, water temperature, and light, was included in the conceptual model for the phytoplankton community composition constructed for 143 lakes in South America and Europe ((Kosten et al. 2012), Section 3.5.2). These interactions became evident when added nutrients led to chlorophyte dominance, but added nutrients in combination with zooplankton led to cyanobacteria dominance (Wang et al. 2010). Similarly, zooplankton groups were some of the significant predictors (after TP concentration) of cyanobacteria biomass increases in a survey on 640 Canadian lakes (MacKeigan et al. 2023).

Many cyanobacterial genera (*Aphanizomenon, Calothrix, Cylindrospermopsis, Dolichospermum, Fischerella, Gomphosphaeria, Hapalosiphon, Leptolyngbya, Microcystis, Nodularia, Nostoc, Oscillatoria, Phormidium, Scytonema*, and *Trichormus*) are potent producers of bioactive *allelochemicals* targeting other cyanobacteria, micro- and macro-algae, and angiosperm vegetation (Leão et al. 2009). Allelopathic compounds could play a role in the proliferation of cyanobacteria in eutrophic lakes constraining phytoplankton succession and avoiding predation. The identified allelochemicals include cyanotoxins and volatile organic compounds (VOCs) like geosmin (*trans*-1,10-dimethyl-*trans*-9-decalol) and 2-MIB (2-methylisoborneol) that negatively affect taste and odour characteristics. However, much of this information is based on laboratory studies, and the importance of allelopathy in ecosystems requires further study (Leão et al. 2009).

Extensive *macrophyte* cover seems to prevent the establishment of large cyanobacteria accumulations so that lakes have been classified into two groups: macrophyte-dominated or plankton-dominated (van Nes et al. 2007; Scheffer et al. 1993). The change from a macrophyte-dominated to a plankton-dominated shallow lake system is expressed in the paradigm of *alternative states* (Scheffer et al. (1993), critically discussed by Davidson et al. (2023)). The complex interactions between macrophytes and cyanobacteria include slowed water movement with decreased sediment resuspension, increased sediment accumulation, pH changes in macrophyte beds (James et al. 1996; James and Barko 1990; Madsen et al. 2001) (Section 2.4.2), and decreased light penetration and heat absorption by excessive growth of cyanobacteria (Kumagai et al. 2000) (Section 5.4.2).

The *dreissenid mussels Dreissena polymorpha* (zebra mussel) and *D. rostriformis bugensis* (quagga mussel) have invaded European and North American freshwaters since the 1990s (Karatayev et al. 2015). While zebra mussels are restricted to shallower lakes and lake regions, quagga mussels are distributed throughout all depths and form larger populations, filter larger water volumes, and may have greater system-wide effects (Figure 3.11). Recently, quagga have replaced zebra mussels in most nearshore areas and have successfully spread throughout the North American continent (Cha et al. 2013).

These dreissenid mussels seem to facilitate *Microcystis* blooms (Bahlai et al. 2021), but the various biotic and nutrient interactions are complex. Changes in the food web (top-down effects) including "selective rejection" of *Microcystis* by dreissenids were observed in Saginaw Bay, Lake Huron (Vanderploeg et al. 2001), and decreased competition and predation by preying on other phyto- and zooplankton were hypothesized. Similarly, a selective grazing effect that would diminish spring blooming phytoplankton during complete mixing but not affect phytoplankton during stratified conditions was confirmed by modelling quagga mussels in Lake Michigan (Rowe et al. 2015). Further, the uptake of mussel-excreted nutrients by *Microcystis* migrating to nutrient-rich bottom waters around zebra mussel colonies was proposed (Bahlai et al. 2021).

In addition, studies on dreissenid relationships with phytoplankton found changes in the nutrient cycle, especially of P, suggesting that bottom-up processes with P are involved. It is highly probable that short-lived redox potential changes from calming conditions surrounding the mussel habitat (Ackerman et al. 2001) and benthic enrichment in nutrients and organic matter facilitate sediment P release (Section 5.1.11). The impact of such P enrichment would be most obvious in low-P systems. Indeed, lakes with dreissenid mussels of 39 lakes with low TP ($\leq$ 20 µg/L) in Michigan, USA, had 3.3 times higher microcystin concentration and 3.6 times higher biomass of *Microcystis aeruginosa* than lakes without the mussel while *Dolichospermum spec.* biomass was higher in these lakes (Knoll et al. 2008). The enhancement of *Microcystis* biomass was corroborated by enclosure experiments.

A similar increase of dreissenids along with cyanobacteria blooms (cyanoHABs with unspecified species), as well as toxins, was documented by a volunteer-based survey of about 168 lakes in New York State, USA (Matthews et al. 2021). A dependence on trophic state determined by chlorophyll and TP concentrations highlights the bottom-up connections. In 15 "oligotrophic lakes (Chl-a < 4 μg/L), the presence of dreissenid mussels nearly doubled the occurrence of cyanoHABs and increased cyanotoxins nearly 20-fold"; similar increases occurred in 41 mesotrophic lakes (Figure 3.13). In contrast, "productive lakes with TP > 30 μg/L and Chl-a > 15 μg/L had a high probability (>75%) of cyanoHABs regardless of the presence of dreissenids".

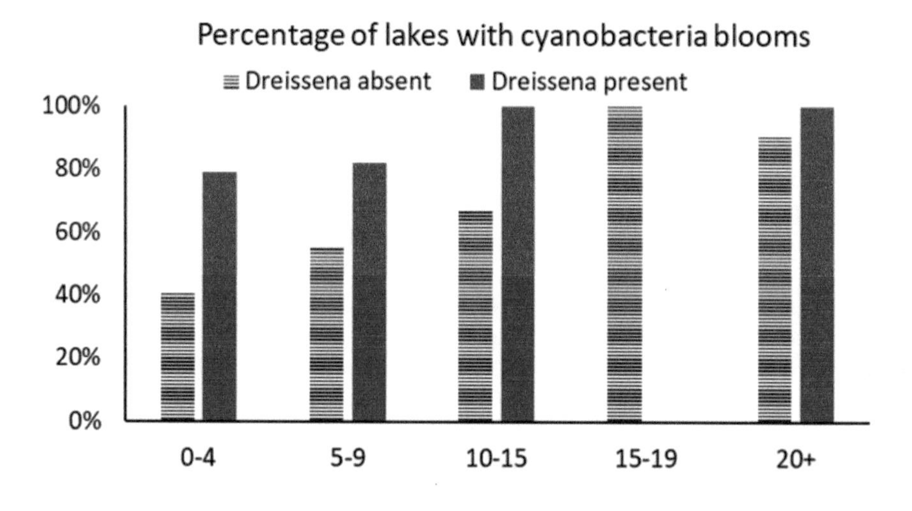

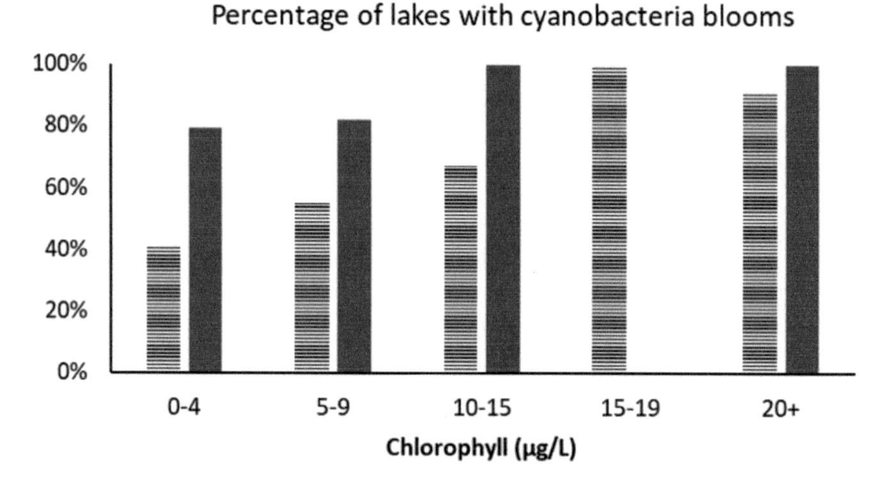

Figure 3.13 The effect of dreissenid mussels on the occurrence of cyanobacterial blooms as a function of chlorophyll (grouped by increasing concentration). Percentage of lakes with cyanobacterial HABs (top) and of those with cyanotoxins (microcystin > 20 μg/L, bottom) is shown. Data are from 168 New York lakes surveyed by the Citizens Statewide Lake Assessment Program (CSLAP) during 2012–2017. *Figure drawn with data provided by David A. Matthews (Matthews et al. 2021).*

3.7 Conclusion with respect to cyanobacteria

Cyanobacteria and their toxicity present an increasing global problem as determined from the distribution of cyanobacteria and cyanotoxins and their increased occurrence over time. Advances in genetic approaches have commenced to enhance understanding, particularly with respect to mechanisms and drivers of cyanotoxicity and the distribution of cyanobacteria strains and strain succession.

The large class of freshwater cyanobacteria responds to multiple nutrient, habitat, and climate conditions in various ways (Table 3.3). Overall, bottom-up effects of nutrients determine the impact of top-down effects from biotic interactions (including grazing, competition, and allelopathy) on cyanobacteria proliferation. Physiological temperature adaptation to high optima is unlikely the only reason for cyanobacteria proliferation in warming lakes leaving room for other causes, such as internal P loading.

Table 3.3 Cyanobacteria response (+ positive or – negative) to nutrients, biota, lake, catchment, and climate characteristics.

Characteristic		*Effects on cyanobacteria*
Phosphorus (especially from internal loading):		
	Amount	(+)
	Location of source	Direct source from sediments: + Via migration, benthic seedbanks
	Chemical form	Orthophosphate, oligo-phosphates
	Bioavailability	(+)
	Timing of input	(+) Summer-fall
Nitrogen:		
	Amount	(+)
	Chemical form	(+) at higher N:P ratio and for N compounds that are chemically reduced (NH_4, urea); organic N
Biota:		
	Dreissenid mussels	(+) in oligo-, mesotrophic lakes
Lake characteristics:		
	Lake stability	(+)
	Water temperature	(+)
	Trophic state	(+)
	Pollution history	(+)
	Light	(+)
Catchment characteristics:		
	Land use	Variable (e.g., agriculture > forest)
	Geochemistry, geology	Variable (e.g., hardwater > softwater)
Climate variables and climate change (see Section 4.6**for specifics):**		
	Air temperature	(+)
	Precipitation, flows	(–)
	Drought	(+)
	Lake level	(–/+)
	Wind/storm	(–/+)

Specific characteristics render cyanobacteria competitive in comparison to eukaryotic phytoplankton:

- Enhanced N uptake due to the potential for N fixation in some taxa;
- Migration by buoyancy control, which enables P translocation from bottom waters and sediments into the mixed surface water and access to redox-related metals;
- Preferences for light (high), temperature (high), and mixing state (quiescent) that are optimized in some predicted climate change scenarios (Chapter 4);
- Hibernation by vegetative cells (e.g., *Microcystis*) and the production of seed banks (akinetes) (e.g., *Dolichospermum*, *Aphanizomenon*, *Cylindrospermopsis* (*Raphidiopsis*), *Gloeotrichia*, and *Nodularia*) that are supported by nutrient replenishment from lake bed sediments;
- Morphological and chemical adaptations, and the production of allelochemicals and toxins that limit top-down pressure by zooplankton and benthic invertebrate grazers;
- A high general adaptivity to changing nutrient and environmental conditions because of the abundance of different strains.

Chapter 4

Climate and climate change effects on freshwater systems

Air temperature increases from climate change mean an increase in energy to the earth atmosphere. Such energy increase results in increased water temperature in lakes and oceans and other related changes such as enhanced atmospheric turbulence manifested as increased wind speed and tornado activity, increased variability in humidity and rainfall, and generally unsettled climate conditions.

The effects of climate variability on lakes are often documented by observing the relationships between climate indices and lake effects. Climate indices numerically describe global temperature and rainfall patterns; they include the El Niño – Southern Oscillation (ENSO) index that is based on temperature changes in the tropical Pacific Ocean, the North Atlantic Oscillation (NAO) index that is based on sea-level pressure differences, and the Pacific Decadal Oscillation (PDO) index based on North Pacific Sea surface temperature; in addition, there is the Pacific Northwest Index (PNI) that is based on three terrestrial climate variables in the northwestern United States (Abtew and Trimble 2010; Lee et al. 2014; Marcé et al. 2010; Nõges et al. 2010; Nürnberg 2007a, 2002).

Studies involving climate indices determine the impact of specific weather patterns and climate variability on the limnological conditions and water quality in lakes and reservoirs of specific regions. Where they indicate climate change, they help in the separation of changes in climate from other environmental stressors (e.g., urbanization, watershed usage changes) that may cause specific water quality conditions.

Several approaches can be used to forecast or predict general or specific responses to climate change in freshwater systems (Jeppesen et al. 2014; Nürnberg 2020a). They vary in strengths and weaknesses and include:

- Historic observations in time series (at least decadal times scales needed, usually available for specific lakes only).
- Paleolimnological records (long-term, i.e., millennial, centennial, or decadal timescales, specific methods needed).
- Space-for-time substitution (e.g., investigating the adaptation to temperature changes forecasted for climate change by comparing contemporal lakes at different locations naturally exhibiting these differing temperatures).
- Controlled experiments (a direct test, based on theoretical understanding, but could lack the consideration of important interactions).
- Empirical mathematical models (good testability and predictability but could include invalid assumptions and lack relevant variables).
- Mechanistic, process-based mathematical models (large required data input, low testability).

DOI: 10.1201/9781003301592-4

Historic time series: Historic observations based on monitoring data provide direct indications of temporal changes and may be intuitively conclusive (Ibelings et al. 2016a; Nelligan et al. 2019b; Pilla et al. 2021; Rastetter et al. 2021; Yan et al. 2008; Zinnert et al. 2021). The drawback is that unconsidered potential influences may vary (so that their variation is not controlled in the statistical sense), for example, changes in trophic state, nutrient loading from the catchment, or anthropogenic impacts.

Paleolimnology: Lake bed sediments keep a record of past conditions. The method of paleolimnological reconstruction and the investigation of specific physical, chemical, and biological indicators of specific conditions have been successfully used to record climate change effects in lakes. Numerous paleolimnological studies on sediments of global lakes compare climate data with lake data specific to previous nutrient status (trophic state), hypoxia, thermal stability, and mixing status, as well as phytoplankton records, including cyanobacterial records (Smol 2008). Some of these studies are listed in Table 4.1 and Table 4.2 and discussed throughout Chapters 4 and 5.

To extend the climate records beyond those available by human observations, paleo-temperature, for example, has been reconstructed from lake sediment archives of microbial membrane lipids and chironomids (Raberg et al. 2021). Besides warm-season temperature (for which a method involving microbial membrane lipids worked best), paleo-conductivity and pH trends were determined.

Diatoms are especially suited to provide long-term sedimentary records. Consistent changes in diatom species revealed recent climate change as the main driver of limnological variation (Rühland et al. 2023, 2015). For example, lake-mixing status was inferred from diatom species with separate mixing depth optima in two lakes in the US Rocky Mountain area since 1750 AD (Saros et al. 2012). The paleolimnological information correctly predicted shallower mixing depth in the alpine lake with advanced tree line (producing wind stilling) and deeper mixing depth (more mixed conditions) for a boreal lake where wind strength had increased.

Space-for-time substitution: To extend the range of conditions that may be expected with climate change, simultaneous observations from many lakes are studied and results transferred to temporal changes. For example, light availability and stratification dependencies of chlorophyll concentration in a summer with an extreme heatwave were simultaneously determined by 198 authors in 230 nutrient-rich European lakes and found to be more important than nutrients (Donis et al. 2021). This method also helped to distinguish climate change effects in 41 lakes on the western Anatolian plateau of Turkey (Beklioğlu et al. 2020).

Controlled experiments provide direct answers to specific questions. For example, mesocosm experiments with urban Dutch lake seston determined that nutrients were more important than temperature in increasing cyanobacteria, chlorophyll, and microcystins (Lürling et al. 2017).

Models extend available observations and conditions. The benefit of testability of empirical models is replaced by theoretical and mechanistic insights in process-based models. Model choice depends on data availability, financial and technical resources, and the application history of specific models, while testability and verifiability combined with the principle of parsimony are considered most important in any model (Nürnberg 2020a).

As the following chapters clarify, projected climate change (Section 4.1) causes limnological responses in physical (Section 4.2), chemical (Section 4.3), and biological effects specific to cyanobacteria (Section 4.4) and selected organisms (4.5) (Figure 4.1).

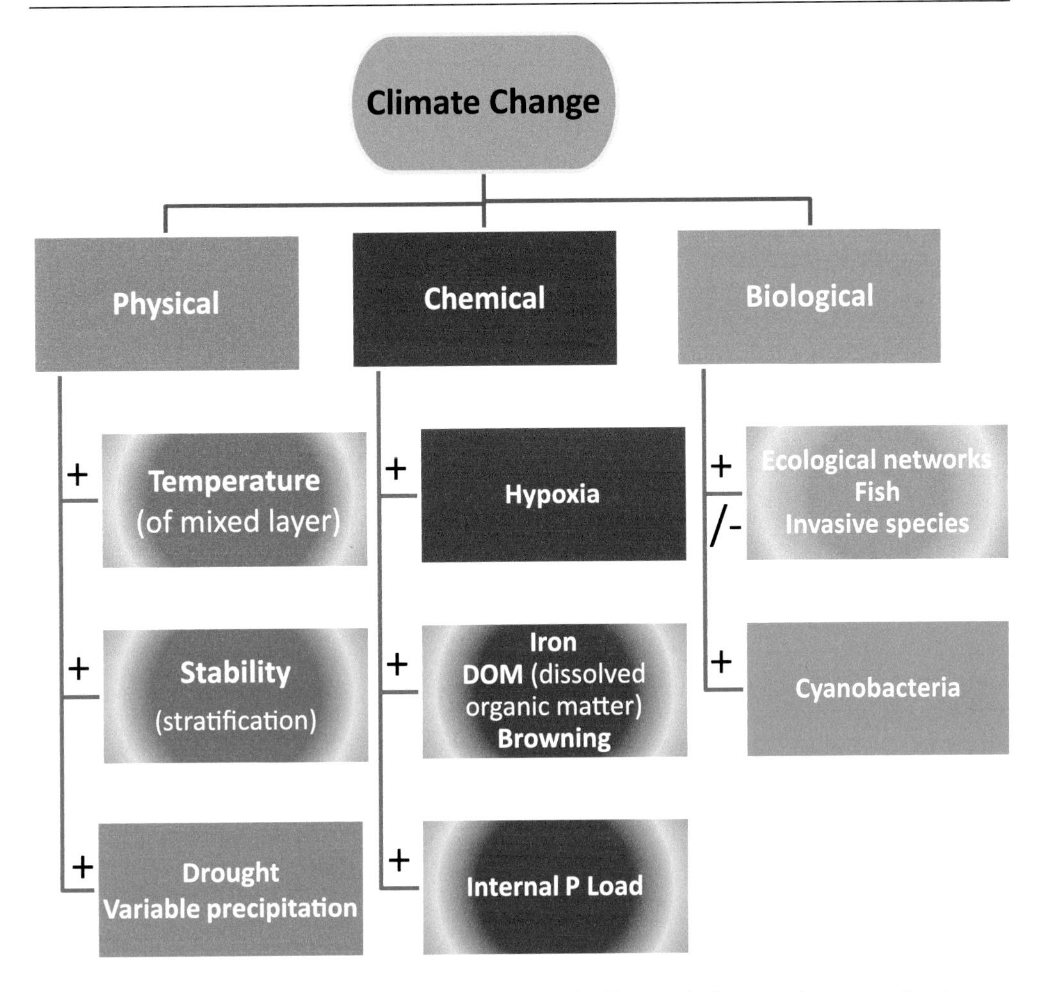

Figure 4.1 Typical physical, chemical, and biological effects of climate change on freshwater indicated as increasing (+), decreasing (–), or both (+/–) trends, as discussed in Sections 4.2 to 4.5. Direct increasing effects on cyanobacteria are circularly shaded and framed in green. This is a simplified generalization of often encountered effects, but exceptions have been recorded.

4.1 Experienced and projected climate change

Climate models have predicted climate change globally with regional forecasts being updated frequently (Bush and Lemmen 2019). For example, the annual average air temperature in Canada increased by 1.7°C between 1948 and 2016, almost twice the increase observed for the whole earth (Bush and Lemmen 2019). The year 2023 was confirmed as the world's hottest year on record (BBC, January 9, 2024. https://bbc.com/news/science-environment-67861954). The overall and general trend of the climate is its warming (Figure 4.2, Figure 4.3), with specific predictions dependent on the extent of carbon sequestration in the future (Figure 4.4). Low, medium, and high imbalances are used in typical climate scenarios. In this context, the imbalance between the solar radiation entering the climate

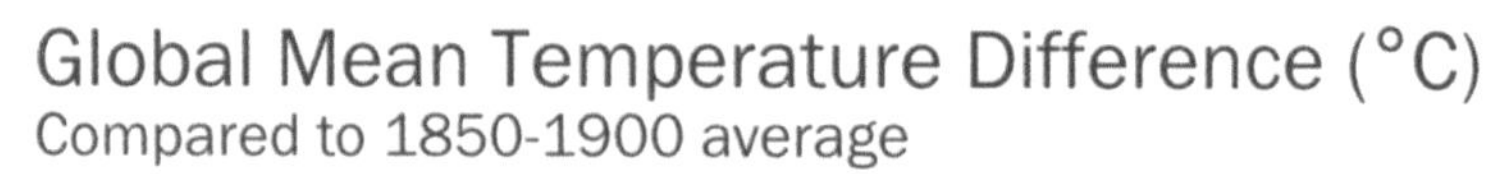

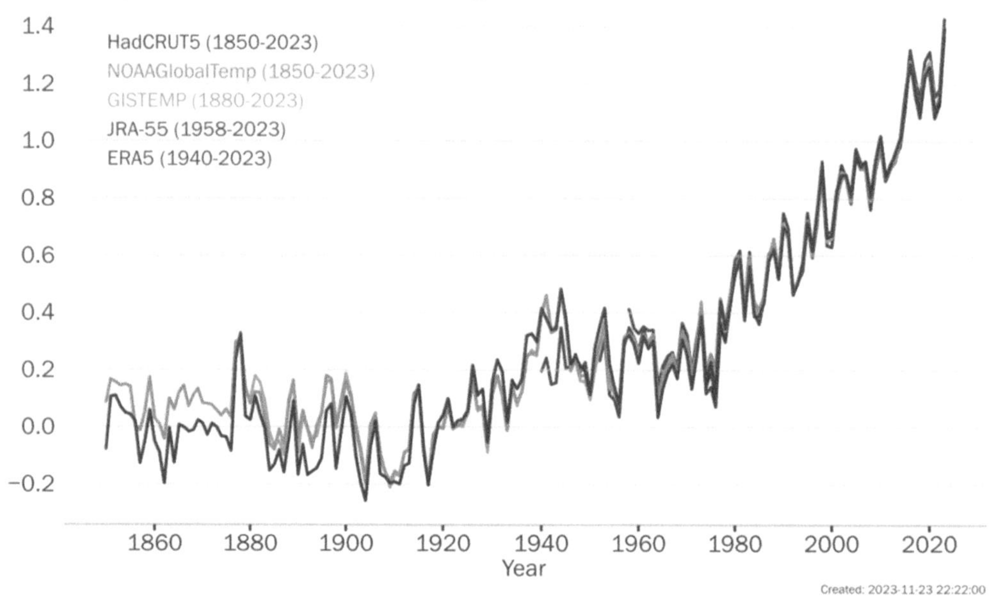

Figure 4.2 Annual global mean temperature anomalies (relative to 1850–1900) from 1850 to 2023. The 2023 average is based on data January to October. Data are from five data sets (WMO 2023). *Figure 2 of WMO [World Meteorological Organization] (2023): WMO Provisional state of the global climate 2023.*

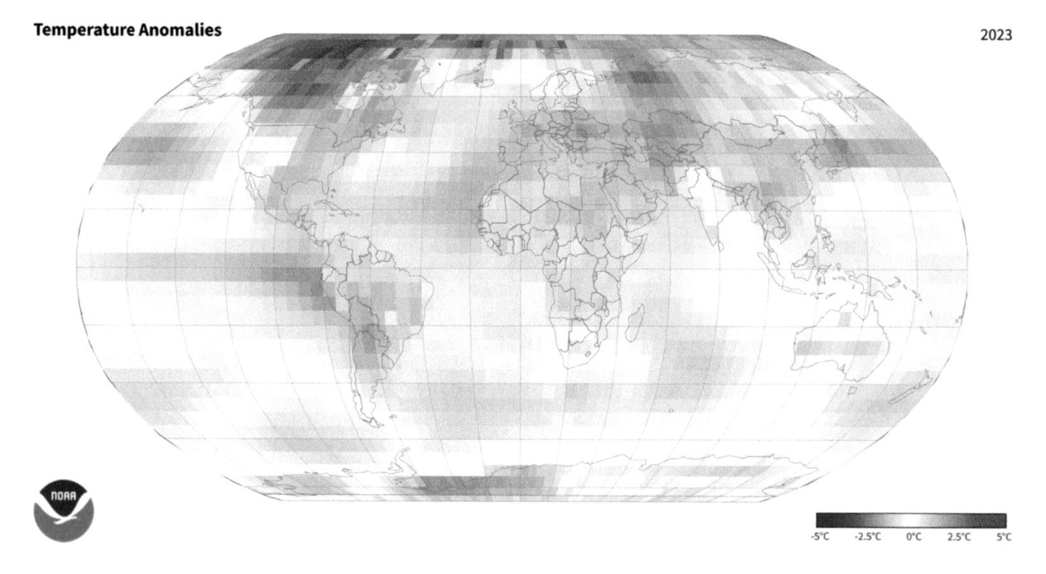

Figure 4.3 Map of global 2023 temperature anomalies based on the 1991–2020 mean. *Figure source: www.ncei.noaa.gov/access/monitoring/climate-at-a-glance/global/mapping/2023 Downloaded March 2, 2024.*

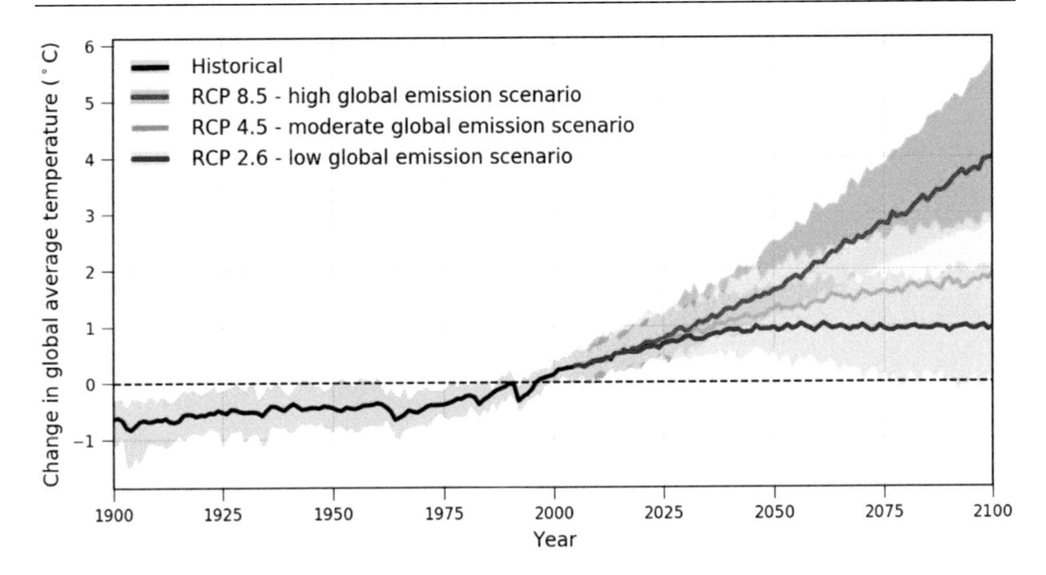

Figure 4.4 Predicted annual global surface temperature deviations from the average over 1986–2005 reference period for different climate scenarios. Representative Concentration Pathways (RCPs) represent the imbalance, caused by greenhouse gases, between the solar radiation entering the climate system and the infrared (longwave) radiation leaving it, where 2.6 is low, 4.5 is moderate, and 8.5 is high. *Figure 1. www.canada.ca/en/environment-climate-change/services/climate-change/canadian-centre-climate-services/basics/scenario-models.html Downloaded March 2, 2004.*

system and the infrared (longwave) radiation leaving it caused by greenhouse gases and other external drivers is called Representative Concentration Pathway (RCP), where 2.6 is low, 4.5 and 6 are intermediate, and 8.5 is high.

According to the Intergovernmental Panel on Climate Change (IPCC) report for 2021 (Intergovernmental Panel On Climate Change 2023) worldwide unprecedented changes in temperature and precipitation patterns have been recorded globally in recent decades, and further change is predicted to occur in the near future, mainly as the result of human activity. Similar observations were made in previous IPCC reports (Jeppesen 2015). The 2021 IPCC report described and investigated CO_2 concentration, precipitation, global surface temperature (that translates into air temperature), sea level, and ocean heat content (Figure 4.5) but does not adequately address the impacts on freshwater systems (Douville et al. 2022).

Besides increased air temperature and changes in precipitation patterns, wind patterns have been affected by climate change with increases in extreme storm events (tornado, hurricane) (Douville et al. 2023), as well as decreases associated with atmospheric stilling in China, the United States, Australia, the Netherlands, and the Czech Republic (Vautard et al. 2010; Woolway et al. 2019). Decreased wind speeds have also been caused by more indirect effects of climate change, including the forestation in alpine regions that were formerly above the treeline, where increased tree growth invokes stratification and hypoxia in alpine, especially coloured lakes (Klaus et al. 2021), Section 4.2.3.

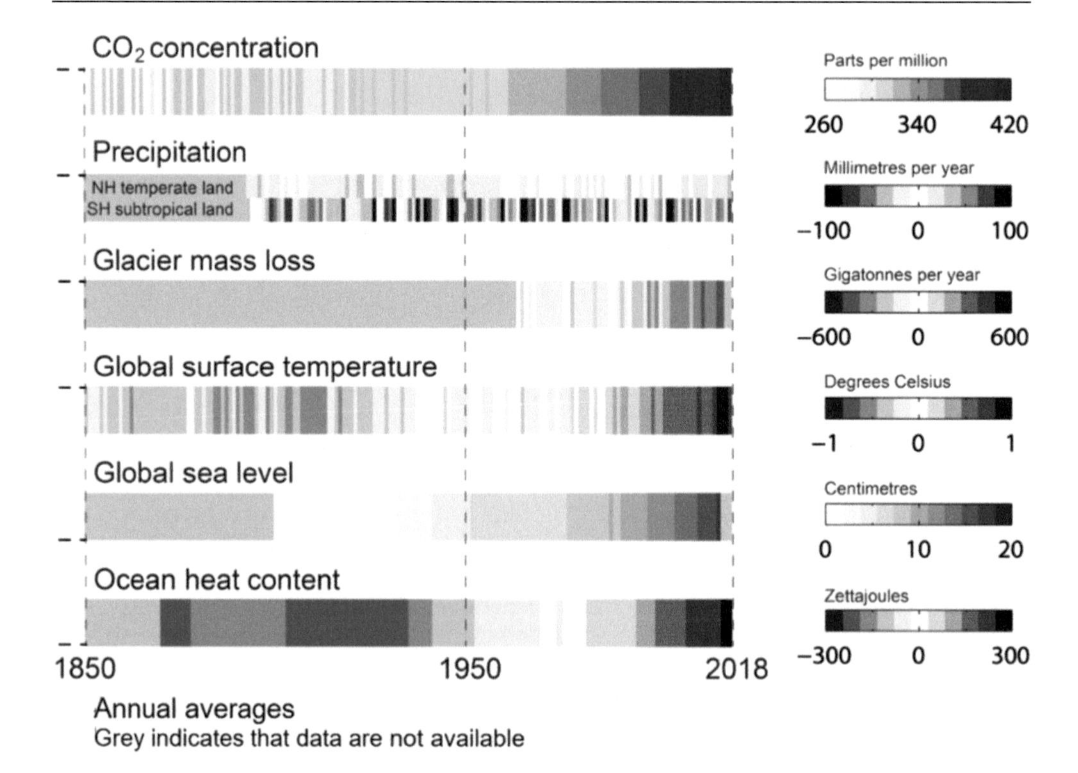

Figure 4.5 Observed climate change between 1850 and 2018 based on six indicators: CO_2 concentration, glacier mass loss, and annual mean anomalies relative to a multi-year baseline for land area precipitation (NH, in the northern temperate hemisphere and SH, in the subtropical hemisphere), global surface temperature, global sea level, and ocean heat content; *Figure 1.4, p. 158 of Intergovernmental Panel On Climate Change (2023), open access.*

4.2 Physical changes in freshwater from climate change

Climate warming and related atmospheric changes over the last several decades have resulted in warmer epilimnetic temperatures and stronger thermal stratification that has led to longer growing periods during enhanced ice-free periods in many freshwater lakes. Many of these effects were summarized for lakes in different climate regions by the influential study of Adrian et al. (2009), "Lakes as sentinels of climate change". Adrian's et al. (2009) observations have been supported by many more studies since then, some of which are described in the following sections. Reported here are more recent studies (mainly 2016–2023). Studies particularly relevant in the description of any physical effects on freshwater by climate change are listed in Table 4.1 and discussed in the following subsections. They include 31 studies with epilimnetic temperature increases, 29 with stratification lengthening, 17 with increases in ice-free days, 3 related to snowpack changes, 13 with increased hypoxia, 6 related to light availability, and 12 studies with additional physical changes including changes related to hydrology, water level, and hypolimnetic characteristics.

4.2.1 *Surface and bottom water temperature as affected by air temperature increases and heatwaves*

Lake water temperature increases have been demonstrated as a consequence of growing air temperature with in-situ and satellite-sampled records (Table 4.1) and in model approaches (Piccolroaz et al. 2024). Some of the global summaries are described here.

Air temperature rose by 0.24°C per decade, on average, from 1985 to 2009 and was correlated, often significantly, with lake summer surface temperature in 193 global lakes (O'Reilly et al. 2015) (Figure 4.6). Lake morphometry, radiation, and cloud cover also influenced lake temperature. Lake temperature increases were demonstrated globally by partitioning lakes into nine thermal regions dependent on latitude (Maberly et al. 2020). Climate models predicted that 12%, 27%, and 66% of lakes will change to a lower latitude thermal region by 2080–2099 for low, medium, and high greenhouse gas concentration trajectories defined above (RCPs of 2.6, 6.0, and 8.5).

Long-term data sets established significant changes between thermal habitats in lakes. Thirty-two million temperature measurements from 139 lakes determined the effects of long-term air temperature change on thermal environments (Kraemer et al. 2021). On average, 6.2% thermal habitats had changed without overlap in recent (1996–2013) time compared to the baseline period (1978–1995). Non-overlap was (even) higher (19.4%) when habitats were examined separately for season and depth. Tropical lakes exhibited substantially higher thermal non-overlap compared to temperate lakes. Many Arctic lakes and ponds have already passed critical ecological thresholds due to recent climate warming as determined from paleolimnological studies (Smol and Douglas 2007).

Lake bottom temperature is not consistently affected by climate warming. Half of 231 lakes across northeastern North America experienced warming and half cooling deep-water temperature (Richardson et al. 2017). Similarly, the bottom temperature change was on average very low (+0.06°C/decade) in data of 102 global lakes (Pilla et al. 2020). Cooling bottom water can be related to increased stratification and early ice-out (Section 4.2.3).

Heatwaves in freshwater environments can be defined as a period of at least five days in which lake or river water surface temperatures exceed a local and seasonally varying 90th percentile threshold, relative to a baseline mean (Woolway et al. 2021a). Heatwaves have been dramatically increasing in intensity and duration since about 2006 as determined with historic data and are projected to keep increasing, depending on the climate scenario (Figure 4.7, Woolway et al. 2021a). Heatwaves have different effects on shallow lakes compared to deep lakes. Their intensity decreases with depth while their duration increases, suggesting differing effects on chemical and biological reactions, including stratification, anoxia, and internal loading (Zhan et al. 2021).

Riverine heatwave frequency increased with a doubling of mean heatwave days per year from 11 in 1996 to 25 in 2021 in a study involving 70 US rivers. Heatwaves occurred mainly in the summer and fall, in mid- to high-order streams, and at low discharge conditions; they were also affected by the position relative to a reservoir (Tassone et al. 2023). Riverine heatwaves were often associated with normal or below-normal discharge conditions, increasing the occurrence of droughts.

Besides increases and extremes, the *seasonality* of climate-related temperature changes affect lake functioning. For example, water temperature has risen preferentially in late summer and fall, potentially enhancing sediment P release. A meta-analysis determined that lake warming throughout the United States occurs mostly at the end of the stratification period in the fall

Table 4.1 Studies indicating or predicting climate change effects on physical characteristics and hypoxia in lakes (+, increase; −, decrease).

	Reported limnological trends						
	Epilimnetic temperature	*Stratification*	*Ice-free days*	*Snow-pack*	*Hypoxia*	*Light availability*	*Others*
Count of studies:	**34**	**28**	**17**	**3**	**13**	**6**	**12**
	Increased air temperature						
	+ (n = 16)	+	+	+			Water level
	−0.06 to 0.66°C/decade; hypolimnion: −0.44 to 0.31°C/decade	+			−55 to 0.7 µg/L per year hypolimnetic DO change		
	+						Decrease in flushing rate
	+	+ 28–37 days, increased stability (Schmidt)	+		+		Bottom temperature: none in 2 lakes, slight + in HA
	+0.32–0.47°C/decade		+ 11.7–14.8 days/decade				

Study Characteristics				
Number of lakes, characteristics	*Location*	*Period*	*Result summary*	*Source*
Review of many global lakes			Classic review. "Lakes as sentinels of climate change". Shows climate change effects in lakes.	Adrian et al. 2009
Oligotrophic lakes	Adirondack Mountain region, New York State, the US	1994–2021	Warming and thinning of epilimnia, cooling and expansion of hypolimnia, and declining hypolimnetic dissolved oxygen	Bukaveckas et al. 2024
Esthwaite	The UK	2003 and model	*Anabaena* and *Aphanizomenon* were more abundant, dominant, and showed increased bloom duration with increasing water temperature and decreasing flushing rate.	Elliott 2010
3 alpine lakes: Irrsee, Mondsee, Hallstätter See	Austria	1975–2015	Expanding seasonal hypoxia and anoxia were correlated to prolonged seasonal stratifications, and to increased TP concentrations in bottom water, but not in whole water column (volume-weighted TP). Expected transition from di- to monomixis.	Ficker et al. 2017
2 Bohemian Forest lakes	Czech Republic	1998–2022	Increasing air temperatures during most months and increasing solar radiation especially in March and November (the months preceding ice-on/off). Decreasing snow cover in winter (by 3.8 cm/decade)	Kopáček et al. 2024

(Continued)

Table 4.1 (Continued)

Reported limnological trends						
Epilimnetic temperature	*Stratification*	*Ice-free days*	*Snow-pack*	*Hypoxia*	*Light availability*	*Others*
	+ by mean depth, average lake temperature, difference between surface and bottom temperature					
+						
+	+ 0.2 days/yr	Earlier ice off: −8.1 ± 6.4 days				
+ (Thermal region)						
+ (bottom: −)	+ (Schmidt stability: +)					
	+			+		Bottom temperature: − in coloured stratified lake
+	+					

Study Characteristics				
Number of lakes, characteristics	*Location*	*Period*	*Result summary*	*Source*
26		1970–2010	Lake stratification is becoming more stable with deeper and steeper thermoclines	Kraemer et al. 2015
139	Global	1978–1995 versus 1996–2013	Lake thermal habitat changed (non-overlapping) by 19.4% for similar seasons and depths and is more severe in tropical lakes and for endemic species.	Kraemer et al. 2021
963 lakes satellite-sampled	North of 30 °N	1979–2020	8-day advancement in the average timing of ice break-up.	Li et al. 2022
732 satellite-sampled	Global	16 yrs (1996–2011)	Partitioning lakes globally into 9 thermal regions predicts a change into a lower latitude (warmer) thermal region of 12–66% lakes for 3 different greenhouse gas trajectories.	Maberly et al. 2020
3: Mendota, Fish, Wingra, simulation	Wisconsin, the US	1911–2014	Wind speed has a large effect on temperature and stratification	Magee and Wu 2017
Kuivajärvi, Finland, dimictic, brown (DOC 13 mg/L); Vendyurskoe, Russia, polymictic, clear (DOC 5.5. mg/L)	Boreal lakes in northern Europe	2010, 2013, 2015	Warm summers turned a clear, typically polymictic lake into a dimictic lake, leading to anoxia. A hot spring resulted in an early onset of stratification with low hypolimnetic temperatures and persistent oxygen depletion in a dimictic coloured lake.	Mammarella et al. 2018
46			Lake forecasting frameworks (models for short and long term) can be set up using open-access software and data, for sites with low-frequency monitoring data, forced with freely available meteorological data and produce high-quality forecasts.	Moore 2020

(Continued)

Table 4.1 (Continued)

Reported limnological trends						
Epilimnetic temperature	*Stratification*	*Ice-free days*	*Snow-pack*	*Hypoxia*	*Light availability*	*Others*
+ 0.34°C/ decade					+ (solar irradiation)	
+ in summer only	no trend					
+	+					
				+		
+ 0.37°C per decade						
+, whole-lake warming of polymictic lakes, bottom 50% + and 50% −.	+					

Study Characteristics				
Number of lakes, characteristics	*Location*	*Period*	*Result summary*	*Source*
118 in situ sampled; 128 satellite-sampled;	Global	1985–2009, <=13 yrs	Ice-covered lakes are warming faster than air temperatures; warming influenced more by lake characteristics than geography.	O'Reilly et al. 2015
17 lakes undergoing restoration	Denmark	1989–2008	Nutrient management during restoration partially overrides warming effects, but internal load counteracts restoration attempts.	Özkan et al. 2016
7, 5	South-central Ontario, Canada (7); North-central Wisconsin, the US (5)	25 yrs	Increased epi and metalimnetic temperature lead to increased stratification and delayed fall turnover, also influenced by DOC.	Palmer et al. 2014
14 oligotrophic with winter ice cover, 1–16 ha, 7–50 m max depth	French Alps	2015–2019	Deep-water warming increased DO depletion under ice.	Perga et al. 2023
102	Global	1970–2009	Deep-water temperature trends (+ 0.06 C/decade on average), with trends in individual lakes ranging from −0.68 to + 0.65 C/decade. Deep-water temperature trends were not explained by surface water temperature or thermal stability.	Pilla et al. 2020
231	Northeastern North America	1975–2012	Half of the lakes experienced warming and half cooling deep-water temperature. More transparent lakes (Secchi transparency >5 m) tended to have higher near-surface warming and greater increases in strength of thermal stratification than less transparent lakes. Whole-lake warming was greatest in polymictic lakes.	Richardson et al. 2017

(Continued)

Table 4.1 (Continued)

Reported limnological trends						
Epilimnetic temperature	*Stratification*	*Ice-free days*	*Snow-pack*	*Hypoxia*	*Light availability*	*Others*
+ Jul-Aug-Sep						
		+17 (Earlier ice-off: −6.8 days; later ice-on: 11.0 days)				
	Earlier onset and longer duration					
+						
Whole-lake temperature + 0.042°C/yr						
+ Days over 20°C increased at a rate of 0.72 d/yr, 75% occurred in the fall	+					Bottom temperature: + (slight)
Median: +2.5°C; maximum: 5.5°C	16% increased stability	+25%				

Study Characteristics				
Number of lakes, characteristics	*Location*	*Period*	*Result summary*	*Source*
2: Crystal & Sparkling; Simulation: 1894 lakes	Wisconsin, the US	1982–2011; simulation: 1979–2012	Surface and bottom temperatures increases in clear lakes; surface temperature increases only in lakes with decreasing Secchi transparency (~1%/year).	Rose et al. 2016
60	36°N to 66°N: North America (40), Europe (18), Asia (2)	107 to 204 lake-yrs before 2016	Trends in ice-on and ice duration were six times faster in the last 25-year period (1992–2016) than previous quarter centuries.	Sharma et al. 2021
Lake Simcoe, 3 basins of different stability	Ontario, Canada	1980–2008	All basins were stratified earlier in the spring and mixed later in the fall.	Stainsby et al. 2011
Modelled on data from 168 lakes; verified on 9 lakes	China	1980–2100	Surface temperature increased by 0.11°C and heatwaves by 7.7 d per decade during 1980–2021. Projection for the high-emission scenario in 2100: 2.2°C and 197 d.	Wang et al. 2023a
142	Wisconsin, the US	1990–2012	Temperature increase was similar across all depths in lakes > 0.5 km^2, but was restricted to shallow waters (< half of maximum depth) in smaller lakes.	Winslow et al. 2015
6	Northern Wisconsin, the US	1981–2015	Lake temperature increased more than air temperature. Monthly temperature and stability increased most in September during June to October period.	Winslow et al. 2017
635	Global	2080–2100 (RCP 6.0), using 4 climate models	Lakes will become warmer, have shorter ice-cover periods, and will become more thermally stable for scenarios where the values of RCP are 2.6 and 6.0 (values for RCP 6.0 are presented here).	Woolway and Merchant 2019

(Continued)

Table 4.1 (Continued)

	Reported limnological trends						
	Epilimnetic temperature	*Stratification*	*Ice-free days*	*Snow-pack*	*Hypoxia*	*Light availability*	*Others*
	Lake heatwaves, increase of intensity: 3.7 ± 0.1 to 5.4 ± 0.8; of duration: 7.7 ± 0.4 to 95.5 ± 35.3 days						
		Stratification sets up 22.0 ± 7.0 days earlier and end 11.3 ± 4.7 days later by the end of this century			+		
	+			-			Increased evaporation; changes in lake volume
		+			+		Bottom temperature: –
		Earlier onset	-				Earlier warming of shallower regions
	+	+					

Study Characteristics				
Number of lakes, characteristics	*Location*	*Period*	*Result summary*	*Source*
>100	Global	Since 1901–2099	Lake heatwaves have and will become hotter and longer in 2006–2099, compared to 1970–1999, based on historic data and simulation and will keep increasing in response to three climate scenarios. Surface heatwaves are longer-lasting but less intense in deeper lakes (<60 m) than in shallower lakes.	Woolway et al. 2021a
>100	Global	Since 1901–2099	"It is very likely that this 33.3 ± 11.7-day prolongation in stratification will accelerate lake deoxygenation with subsequent effects on nutrient mineralization and phosphorus release from lake sediments"	Woolway et al. 2021b
Four specific lakes; global summary and projections			Rapid changes after tipping points produced by climate change are more likely than gradual changes.	Woolway et al. 2022b
Lake Zurich, warm monomictic (Jan to Apr)	Switzerland	1977 (2009 for some variables) – 2016	Indirect effect: Re-oligotrophication by lake warming, increasing thermal stability, preventing holomixis and upwelling of phosphorus (SRP) and down-welling of aerated water.	Yankova et al. 2017
Laurentian Great Lakes	The US, Canada	1982–2012 (extreme warming during 1997–1998 El Niño event)	Regime shift after warming in winter and spring and increased solar radiation in spring.	Zhong et al. 2016
Laurentian Great Lakes	The US, Canada	1982–2012	Superior: 1.2/decade, central-northern Lake Michigan, Huron, Ontario: 0.5–0.6:1.2°C/decade; Erie: 0.3°C/decade	Zhong et al. 2018

(*Continued*)

Table 4.1 (Continued)

Reported limnological trends						
Epilimnetic temperature	*Stratification*	*Ice-free days*	*Snow-pack*	*Hypoxia*	*Light availability*	*Others*
Ice cover decrease						
		+13 days				
		+24 d (9 days earlier + 15 days later) Ice-out date				
		Increasing with time and winter temperature				
	0.2 d/yr earlier	8 d later ice-out				
		+ 5.8 d/100 yrs later ice-in, 6.5 d/100 yrs earlier ice-out				
		+90 days				
		+ Intermittent ice cover (i.e., no consistent annual ice cover)				

Study Characteristics				
Number of lakes, characteristics	*Location*	*Period*	*Result summary*	*Source*
215 temperate (5–1768 m ASL)	Northern Hemisphere	1983–2018	Ice-cover duration is significantly positively correlated with lake latitude and elevation.	Christianson et al. 2021
7 Green Lakes (alpine 3126–3620 m ASL)	Colorado	1983–2018	Ice cover decreases 50% faster in alpine than in non-alpine lakes.	Christianson et al. 2021
7	New Hampshire, Maine, the US	Paleolimnology, since early 1800s	Ice-out date is related to *Gloeotrichia* biomass (pigments) when watershed to lake area ratio is small (indicating importance of internal P load).	Ewing et al. 2020
122	Global	1939–2016	Increasing ice-free days are related to increasing winter temperature.	Filazzola et al. 2020
963	North of 30 °N	1979–2020	Satellite observations	Li et al. 2022
39 lakes and rivers	Northern Hemisphere (Russia, Finland, Switzerland, Japan, the US, Canada)	1846–1995, some since 16th century	Freeze dates were delayed by 5.8 days per 100 years later, and break-up dates averaged 6.5 days per 100 years earlier, leading to increasing air temperatures of about 1.20°C per 100 year with increased rates since 1950.	Magnuson et al. 2000
Mendota and other US lakes	Wisconsin, the US	Since mid-1800 to ~2006	Ice cover declined by 40%, from 4 months to 2.5 months per winter on average, with an increasing rate since 1970.	Magnuson 2007
513	Northern Hemisphere		Intermittent ice cover (no consistent annual ice cover, presently in 14 800 lakes) to increase by 2.4× for projected 2°C and by 15.6× for an 8°C increase of air temperature. Drivers: air temperature, lake depth, elevation, and shoreline complexity.	Sharma et al. 2019

(Continued)

Table 4.1 (Continued)

Reported limnological trends						
Epilimnetic temperature	*Stratification*	*Ice-free days*	*Snow-pack*	*Hypoxia*	*Light availability*	*Others*
+		Ice-out date	-			
Atmospheric Stilling						
				+		
+ (no change in bottom)	+38 days			+16% of hypo-limnion		Chlorophyll: +37%
	+				+ (from decreased resus-pension)	
	+			+		
Sign increase of 3.6°C in 106 yrs		21 d in 53 yrs				
+ (bottom: −)	+					

Study Characteristics				
Number of lakes, characteristics	*Location*	*Period*	*Result summary*	*Source*
Emerald Lake (alpine)	Sierra Nevada, California, the US		Lake temperature affects snowpack via ice-off timing.	Smits et al. 2020
Lake Taihu (shallow)	China, Yangtze Delta	1996–2017	Decreasing wind coincides with increasing hypoxia and nutrient concentrations	Deng et al. 2018
Blelham Tarn	Lake District in the UK	1968–2008	Anoxia dependent on June to September monthly maximum windspeed, which decreased by 1.9 m/s	Foley et al. 2012
Oligotrophic Harp Lake	On Precambrian Shield, Ontario, Canada	1979–2009	Increased air temperature and decreased wind drive enhanced thermal stability since 1979 and at an increasing rate after 1996.	Hadley et al. 2014
Võrtsjärv (shallow)	Estonia	1964–2017	30% decrease in average wind speed; decreased resuspension, increased light	Janatian et al. 2020
40 arctic-alpine (367–998 m a.s.l.)	Sweden		Arctic-alpine lakes with taller shoreline vegetation relative to lake surface area stratify more strongly and have lower near-bottom oxygen concentrations. Primarily because of higher dissolved organic carbon concentrations and secondarily because of lower wind speeds.	Klaus et al. 2021
Greater Slave	Canadian Arctic		Decrease in wind speed by 2.5 km/h in 62 years, detection of scaled chrysophytes (golden algae) since 2010.	Rühland et al. 2023
650	Northern Hemisphere	1980–2016	Atmospheric stilling contributes 15%, 27%, and 23% of the calculated increases in lake surface temperatures; thermal stability; and lengthening of stratification.	Woolway et al. 2019
Vesijärvi				

(*Continued*)

Table 4.1 (Continued)

Reported limnological trends						
Epilimnetic temperature	*Stratification*	*Ice-free days*	*Snow-pack*	*Hypoxia*	*Light availability*	*Others*
Wind, storm, precipitation						
	+			+		
– after mixing	– after mixing				– after mixing	
						53% volume decrease
Light regime, browning						
	+ 5–50%			+ increased methane	Decreased light penetration due to browning and eutro-phication	Bottom water cooling in shallow (< 5 m mean depth), browning lakes, temperature increase in deeper, clear lakes
+	Earlier onset, stronger			+	–, + DOC and colour	
+	+ decreased epilimnion depth			+	Decreased light penetration from brow-nification, increased DOC and pH.	Bottom tem-perature: – DOC: +

Study Characteristics				
Number of lakes, characteristics	*Location*	*Period*	*Result summary*	*Source*
Lake Giles, 24 m maximum, clear	Pennsylvania, the US	1988–1997, 2007–2014		Knoll et al. 2018
Meta-analysis of 28 studies, mostly shallow lakes	Global	1976–2012	The larger the shallow lakes, the more fetch the more internal P load	Stockwell et al. 2020
1 972 large lakes	Global	1992–2020	Volume decrease in natural lakes is attributed to climate change (decreased precipitation) and human water consumption; in reservoirs, it is due to storage losses, based on satellite observations, climate data, and hydrologic models.	Yao et al. 2023
205 from literature review	Europe, North America, Russia, Japan, Tanzania		Warming and transparency loss due to eutrophication or browning overrides atmospheric warming alone, possible bottom water cooling, extended stratification, and methane increase.	Bartosiewicz et al. 2019
17 oligotrophic, recovering from acidification	Adirondack Mountains	1994–2021	Lakes with thin hypolimnia exhibited low DO concentrations, which were exacerbated by the strengthening of vertical temperature gradients and earlier onset of stratification.	Bukaveckas et al. 2024
2: Giles and Lacara	Pennsylvania, the US	1988–2014	Increased precipitation (climate change) and pH (recovery from acidification) decrease water transparency, increase heat and light adsorption in the surface layer leading to an overall increase in DOM (brownification).	Pilla et al. 2018; Knoll et al. 2018

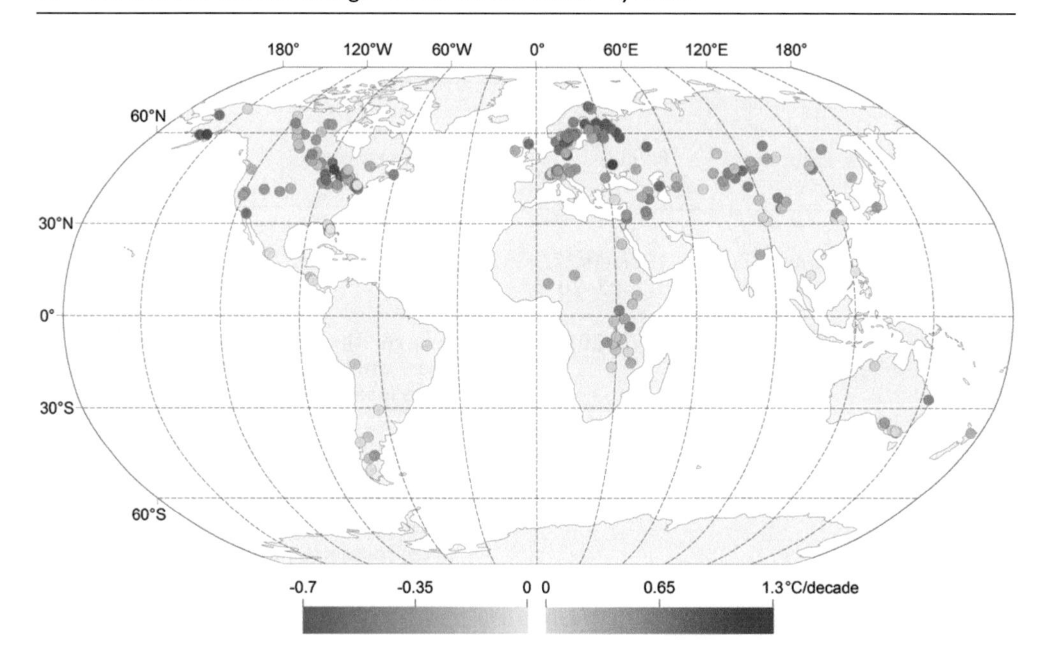

Figure 4.6 Map of trends in lake summer surface temperatures from 1985 to 2009. *Figure 1 of O'Reilly et al. (2015), open access.*

(Figure 4.8). Increases in air temperature in the fall are also reported in individual studies (e.g., Isles et al. (2017b) and Table 4.1). Such increases in air temperature were highly correlated (R^2 = 0.73) to water temperature in six northern Wisconsin lakes from 1981 to 2015 (North Temperate Lakes Long-Term Ecological Research Network, NTL-LTER lake study, Winslow et al. (2017)). Most water temperature increases occurred in the summer and fall, when 75% of the increase in the number of days-above-20°C occurred. September surface water temperatures warmed almost twice as fast (0.073°C/yr), as summer average temperature (0.039°C/yr July–September). Surface water for the June to October growing period warmed up 35% more (0.042°C/yr) than air temperature (0.027°C/yr). Importantly and supporting internal P loading, bottom water temperature and water column stability (expressed as Schmidt stability) increased in five of the six study lakes.

4.2.2 *Winter temperature, ice cover, and snowpack*

The shortening of ice-cover duration due to warming winters has been one of the earliest quantitative indications of climate change impacts on lakes. The early influential paper by Magnuson et al. (2000), "Historical trends in lake and river ice cover in the northern hemisphere" (1500 citations by 2023, Google Scholar search, 8 August, 2023), has been succeeded by numerous studies (Table 4.1). By recording an increase in air temperature of 1.2°C per 100 years with accelerated warming in recent decades, an undeniable link between climate change and lake response across the northern hemisphere was demonstrated for the first time.

Multiple recent studies listed in Table 4.1 support the early findings. More than half of the world's 117 million lakes experience intermittent ice cover (when winters without complete ice

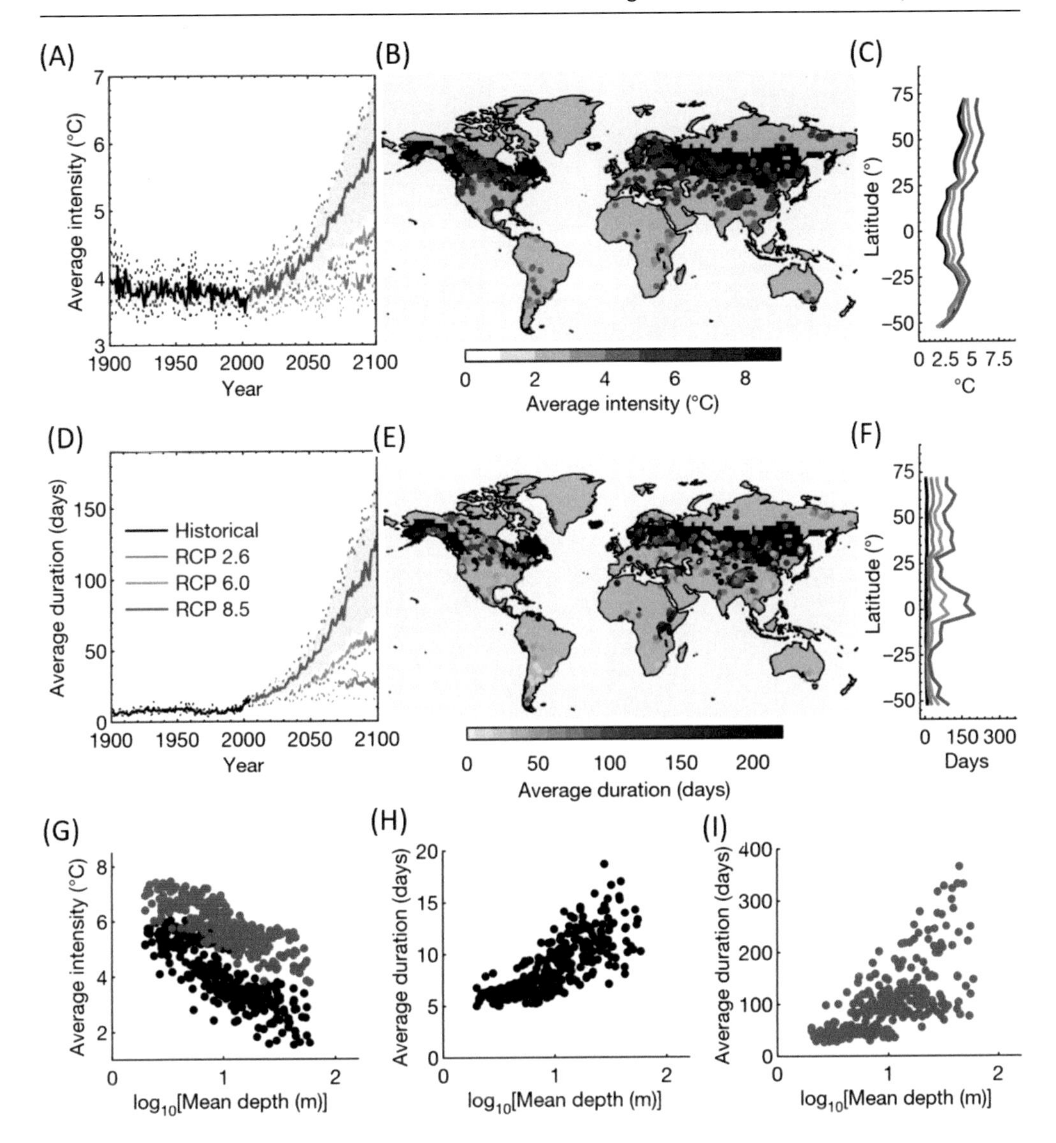

Figure 4.7 Increasing heatwave intensity (A to C) and duration (D to F) since 1900 and for lakes with different depths (G to I). Historic values (black), predicted future values (red). *Figure 1 of Woolway et al. (2021a) with permission.*

cover alternate with winters with full cover), and an increase in such lakes is projected (Sharma et al. 2019) (Figure 4.9). Similarly, a meta-analysis using multiple generational (78-year) ice records from 122 lakes (Filazzola et al. 2020) reported that:

> (1) extreme ice-free years are becoming more frequent and severe, (2) winter air temperature is a significant predictor of ice cover that is driven by large-scale climate oscillations, (3) extremes in temperature are closely related to extremes in ice cover, and (4) ice-free years are forecasted to result in significant loss of ice-cover in the future.

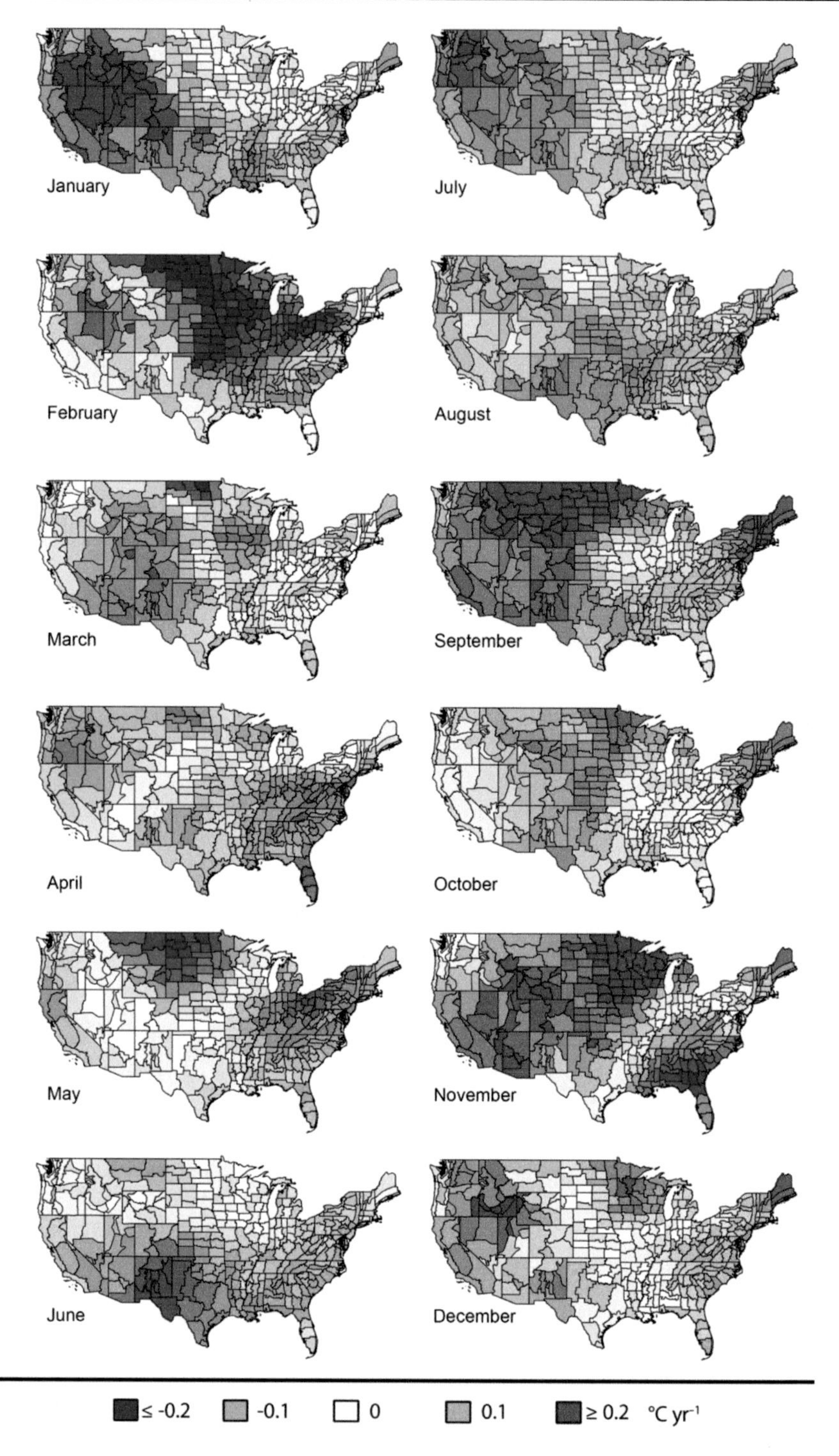

Figure 4.8 Map of air temperature trends by month across the contiguous United States from 1981 to 2015. *Figure 4 of Winslow et al. (2017), open access.*

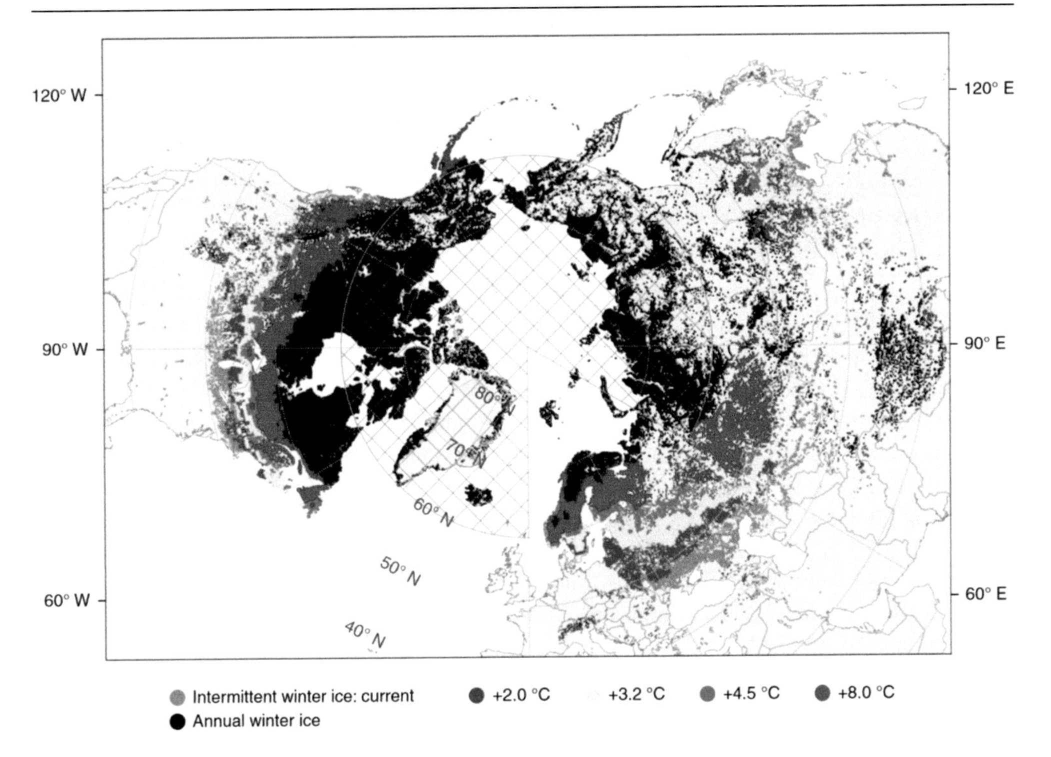

Figure 4.9 Spatial distribution map of current and future Northern Hemisphere lakes (larger than 10 ha) that could experience decreased ice cover with climate warming. Intermittent winter ice-cover projections were based on current conditions (orange) and established air temperature projections of +2.0°C (purple), +3.2°C (yellow), + 4.5°C (blue) and +8.0°C (red). All other lakes are shown in black. *Figure 2 of Sharma et al. (2019), with permission.*

The variability in ice-cover duration is driven by two independent factors: (1) directional shifts to earlier ice-off and later ice-on dates due to climate change as described above and (2) high-frequency climate oscillations (Schmidt et al. 2019). The changes in ice-cover duration influence ecosystem function in both the current winter and the subsequent ice-free season (Wilkinson et al. 2020). In particular, decreased ice duration typically prolongs the summer thermal stratification period (Pilla and Williamson 2022), which increases the period of cyanobacteria proliferation and possible sediment P release.

Winter internal P and redox-controlled metal (Fe, Mn) loading has been observed from nutrient-rich sediments of Lake Champlain's shallow eutrophic Missisquoi Bay and is also affected by ice-cover duration (Schroth et al. 2015). When the duration of ice cover decreases, the period of stratification under ice decreases (Woolway et al. 2022a). This could decrease the length of winter anoxia and sediment P release in favour of summer anoxia and summer P release. Such a switch of release periods would increase the overall annual internal P loading and the eutrophication effect. P release rates are lower in the winter than in the summer (because of lower temperature), so that sedimentary accumulation of mobile P would increase in the winter and lead to more P release in the summer.

Increased light from snow removal increased photoautotrophic growth in a Minnesota, USA, lake and growth was further stimulated by nutrient additions (Knoll et al. 2024). Therefore, declines in snowpack and ice cover likely enhance the effect of climate warming on eutrophication, especially when combined with increased nutrients that further increase phytoplankton biomass.

4.2.3 Water column stability, stratification, and mixing as affected by temperature increases, atmospheric stilling, and storms

Increased *stratification* duration and strength have been observed and projected in lakes worldwide (Table 4.1). A study on 635 lakes determined that at least 100 of these lakes would experience changes in their mixing regimes under two emission scenarios (RCP 2.6 and RCP 6.0) due to modelled changes of temperature and ice cover (Woolway and Merchant 2019). Some of the currently monomictic lakes are projected to become meromictic (no annual mixing of the stagnant bottom layer), and some of the dimictic lakes are to become monomictic. In addition, and as discussed in previous sections, the study also projected many lakes to become permanently ice free by the end of this century, while surface temperatures are projected to be warmer by 2.5°C and 5.5°C for two climate scenarios (RCP 2.6 and RCP 6.0). In other studies, shallow, historically polymictic lakes have shown increasing stability displayed as increased periods of stratification that are often associated with hypoxia (Favot et al. 2024; Mammarella et al. 2018; Wilhelm and Adrian 2008).

Climate change effects also depend on lake transparency besides lake depth (Section 4.4.4). A meta-analysis of 231 lakes across northeastern North America found that more transparent lakes (Secchi transparency >5 m) had higher near-surface warming and greater increases in strength of thermal stratification than less transparent lakes (Richardson et al. 2017). Using long-term observations, lake stratification changes were modelled across the Northern Hemisphere from 1901 to 2099. Under the high greenhouse-gas-emission scenario, stratification was predicted to begin 22.0 ± 7.0 days earlier and end 11.3 ± 4.7 days later by the end of this century, resulting in a 33.3 ± 11.7-day prolongation in stratification (Woolway et al. 2021b).

Atmospheric stilling is another stabilizing effect on lake thermal structure due to climate change and has been studied frequently (Table 4.1). Surface wind records at 822 weather stations indicate declining windspeeds by 5–15% in the northern mid-latitudes between 1979 and 2008 (Vautard et al. 2010). Likewise, stability increased with decreased wind and increased air temperature in a small Canadian lake on the Precambrian Shield, while the duration of stratification did not change (Hadley et al. 2014).

The effects of increased stratification and wind stilling on lake ecosystems are numerous and differ with lake depth and season (Magee and Wu 2017). In shallow, historically polymictic lakes, increased periods of stratification are often associated with increased hypoxia (Favot et al. 2024; Wilhelm and Adrian 2008). As modelled (Bartosiewicz et al. 2019) and observed in 650 lakes (Woolway et al. 2019), atmospheric stilling can have a cooling influence on the sediment–water interface and bottom water. Polymictic lakes with increasing eutrophication and browning can experience cooling in the water column. Early ice-out with subsequent rapid stratification can decrease hypolimnetic summer temperature in stratified lakes.

In large shallow lakes, wind stilling results in lower bottom shear stress (which is a function of ground wind speed, wind fetch, and water depth), thus decreasing sediment erosion and resuspension. Such events resulted in increasing low-light adapted Oscillatoriales biomass in the Estonian lake Võrtsjärv (Janatian et al. 2020). Stilling has also been observed in alpine lakes, where the treeline moved to higher elevations due to climate warming, thus reducing lake exposure to wind (Klaus et al. 2021).

Storm events, including extreme wind storms with possible heavy precipitation, are predicted to increase in frequency and intensity (Intergovernmental Panel On Climate Change 2023). Severe storms have immediate effects on physical lake processes by mixing of the water column, causing bottom sediment resuspension, and changing surface water temperature and indirect effects through runoff of terrestrial nutrients and sediments amplified by increased precipitation (Thayne et al. 2022).

The effects of individual storms depended on antecedent lake conditions, including turbidity, dissolved oxygen saturation, and stability in the eutrophic, polymictic German lake Müggelsee (Thayne et al. 2022). Pre-conditions were also important in mixed Swedish Lake Erken in a study comparing the present with simulated future (RCP8.5) wind conditions (Mesman et al. 2022). High wind speeds decreased phytoplankton concentration compared to medium speeds. But wind had little effect on phytoplankton concentration if the mixed layer was already deep prior to the wind exposure. In this simulation, atmospheric warming did not strongly influence the effect of wind events on phytoplankton. In general, models predicted decreased stability, deepened mixed layer associated with interrupted bottom water hypoxia, and increased phytoplankton concentration after the wind events due to increased nutrients from the bottom water (internal P load, further discussed in Section 4.3.2).

A meta-analysis of 28 studies on worldwide, mostly shallow lakes determined the physical, chemical, and biological effects on three possible storm manifestations: (a) dry storm (wind), (b) rain only (rain), and (c) wet storm (wind and rain) (Stockwell et al. 2020) (Table 4.1). Whereas dry storms were recorded to preferentially affect thermocline depth, mixing, and light conditions, rain affected nutrients and flows, and wet storms affected mixing, nutrient, and light conditions. Consequently, the different types of storms affected cyanobacteria differently (Section 4.4.2).

4.2.4 Lake hydrology as affected by precipitation variability and drought

Lake hydrology, including flushing rate, annual water load, and water level, controls nutrient mass balances including nutrient retention and fluxes (Section 2.4) and affects lake productivity and cyanobacteria biomass (Sections 3.5.3 and 4.4.2). The multiple effects of individual rain events include increased nutrient and water flow and are described in connection with storm events (Section 4.2.3).

Intensity of rainfall and length of dry periods affected at least seven conditions in fresh and saline water systems in a literature review (Reichwaldt and Ghadouani 2012). Water residence time, anoxia, turbidity, water temperature, conductivity, water stability, and the biomass of phytoplankton all increased with lengthening dry periods and decreasing precipitation. The authors concluded that these changes tend to accelerate eutrophication processes including cyanobacterial proliferation, except when heavy rain events can lead to a temporary disruption of cyanobacteria blooms by flushing and destratification (further discussed in Section 4.4.3).

Changes in precipitation patterns due to climate change (Figure 4.5) have affected lake conditions and are expected to continue (Douville et al. 2023). Globally, the areas with decreased runoff and increased evaporation will expand with climate change (Mishra and Singh 2010), causing decreased future flushing. As a result, decreased lake and reservoir volume has become realized in recent years (Woolway et al. 2022b). Satellite observations, climate data, and hydrologic models determined a 53% volume decrease in 1 972 global large lakes (Yao et al. 2023) (Table 4.1).

But the severity of these effects differs with location and past and current conditions. For example, a decrease in precipitation is especially detrimental in already drought-stressed areas, where decreasing water levels impact water quality along with quantity, as in reservoirs of semi-arid regions of western North America (Yasarer and Sturm 2016), Brazil (Brasil et al. 2016), Australia (Baldwin et al. 2008), the Mediterranean (Coppens et al. 2016; Jeppesen et al. 2015; Marcé and Armengol 2010), China (Sun et al. 2019), and Africa (Dibi-Anoh et al. 2023).

Climate-related decreases in precipitation and increases in droughts have also been reported in relatively wet climates of subtropical lakes in the southern United States. In these lakes, low flows since 1999 were linked to increases in SRP, TP, chlorophyll concentration, and cyanobacteria (Havens et al. 2019). On the other hand, there were slight long-term increases in precipitation in temperate and arctic North American lakes since the 1930s (Favot et al. 2019; Sivarajah et al. 2021).

Precipitation changes can also affect underwater light regime by changing water transparency and colour associated with lake browning (Table 4.1). The underlying mechanism is based on changes in iron and DOC; therefore, this effect is discussed together with the chemical effects from climate change on internal processes in Section 4.3.2.

Different from observations on decreasing lake levels in lakes and reservoirs of all sizes, the long-term (1992–2019) trends in water levels in 200 large global lakes based on satellite altimetry data were slightly positive (+0.8 cm/yr, p value 0.02, after statistically removing the effects of background climate variation) (Kraemer et al. 2020). However, the authors conceded that the 200 large lakes (total volume of 80 241 km^3) are not representative of most other lakes, which tend to have smaller surface area and shallower maximum depths.

Projected changes in precipitation patterns are further presented in the sections that discuss their effects on lake functioning with respect to cyanobacteria (Sections 4.4.2 and 5.1.9).

4.3 Chemical changes in freshwater from climate change

Climate change effects on freshwater chemistry, relevant to water quality in general and to cyanobacteria proliferation in particular, are presented in Table 4.2. Studies are primarily related to changes in oxygen state (Section 4.3.1) and related processes like redox-dependent metal release and browning, all of which can influence nutrient availability and internal P loading (Section 4.3.2). Some oxygen-related effects are related to physical factors and were studied together with changes in lake-mixing states and thermal stratification; they are listed in Table 4.1.

Studies with chemical and biological effects related to climate change include 41 reports related to cyanobacteria proliferation, 8 on water transparency (Secchi), 16 considering lake water TN concentration and 24 considering TP concentration. 14 studies report increased hypoxia (in addition to 13 listed in Table 4.1), 27 studies determined internal P loading as a contributing factor to the cyanobacteria increase, and 11 provide related information that implicates internal P loading.

4.3.1 Decreased dissolved oxygen concentration and redox potential

Dissolved oxygen (DO) is essential to the respiratory metabolism of aquatic organisms, ecosystem functioning, and water quality of freshwater systems. DO supply affects the abundance and distribution of freshwater organisms and the redox-related geochemical cycles that help control internal nutrient and metal cycling.

Many established climate indices are related to lake hypoxia. ENSO affects reservoir flow and anoxia in the subtropical United States in Florida (Abtew and Trimble 2010) and in southern Europe in Spain (Marcé et al. 2010). PNI is correlated with the hypoxic factor, a measurement of spatial and temporal extent of hypoxia (Section 2.2, Box 2.1) in a western US reservoir (Brownlee Reservoir, Figure 4.10A, B). Records of a related PNI (1891–2020) indicate much hotter and dryer conditions since the late seventies (Figure 4.10C), consistent with climate change. Other studies also described the directional changes of ENSO indicating climate change trajectories on lake and reservoir anoxia (Intergovernmental Panel On Climate Change 2023).

Increased hypoxia (low oxygen concentration e.g., < 3–5 mg/L (Nürnberg 2019)) and anoxia (< 1–2 mg/L) in the bottom water due to climate change are well-established, and supporting evidence includes direct observations in lakes (in 27 studies, Table 4.1, Table 4.2), estuaries, and saline waters (Diaz 2001; Diaz and Rosenberg 2008; Gilbert 2017; Jane et al. 2021; Lewis et al. 2024b).

A comprehensive meta-analysis on almost 400 lakes with the informative title "Widespread deoxygenation of temperate lakes" distinguishes DO trends in surface water of lakes from their bottom waters (Jane et al. 2021). DO decline in surface waters relates to reduced solubility under warmer water temperatures, except for DO increases from enhanced phytoplankton photosynthesis in highly productive warming lakes (Figure 4.11A). DO decline in deep waters (Figure 4.11B, C) is associated with stronger and longer thermal stratification and loss of water clarity associated with photosynthesis (Table 4.2). A separate study on more than 400 widely distributed lakes (Jane et al. 2023) determined that the increases of anoxia (0.5 mg/L DO) and hypoxia (5.0 mg/L DO) were not attributed to changing DO depletion rates (potentially from increased browning or eutrophication) but to climate-induced lengthening of seasonal stratification. DO decline in freshwater is 2.7–9.3 times that of marine systems, where DO depletion has been recognized for several decades earlier (Diaz 2001).

The most important effect on the rise of summer hypoxia is probably increased water column stability due to longer and more widespread stagnant conditions in all lakes and warmer bottom temperatures in shallow lakes (Section 4.2.3).

An opposite effect, potentially decreasing open-water hypoxia, is possible when ice-out and the formation of thermal stratification occur earlier and more rapidly in the spring so that colder water is trapped in the bottom layer. Early ice-out with subsequent rapid stratification can decrease hypolimnetic summer temperature in stratified lakes, hence decreasing the rates of oxygen depletion and sediment P release (Bartosiewicz et al. 2019; Woolway et al. 2019). However, the lengthening of the stratified growing period could offset these decreases because the periods of anoxia and sediment P release are lengthened.

Relationships between hypoxia, stratification, temperature, and flow are often observed in long-term case studies. For example, in Muskegon Lake, a Great Lakes estuary, less winter and spring precipitation and warmer winter inflow temperatures were followed by more stable stratification with increased hypoxia in the summer (Dugener et al. 2023). Deep water warming increased DO depletion under ice also in 14 mountain lakes (Perga et al. 2023). In addition, heatwaves (Section 4.2.1) can induce and increase hypoxia. In large but shallow Kasumigaura, the combined effects of weak winds and high solar radiation or high air temperatures induced hypoxia during a heatwave in 2022 (Shinohara et al. 2023).

Table 4.2 Studies indicating or predicting climate change effects on chemical and biological characteristics in lakes, including hypoxia.

	Potential limnological causes for cyanobacteria						
	Cyanobacteria (biovolume, percentage, or dominance)	*Water trans-parency, Secchi*	*Lake water nutrient concentration*		*Hypoxia, redox-metals, DOC*	*Internal P load as cause for cyanobacteria*	
			TN	*TP or SRP*		*Is considered by authors*	*Is likely*
Count of studies:	**41**	**8**	**16**	**24**	**14**	**27**	**11**
	Increased air temperature						
	Chlorophyll						
			+	+	DOC: +	+	
	Chlorophyll						Increased stratification increases the duration of anoxia and internal P load but delays nutrient mixing into surface mixed layer.
	Gloeotrichia						Increased open-water season increases the potential of internal load because of increased temperature and anoxia.
			-	+	Hypo-limnetic anoxia	+	

Study characteristics				
Number of lakes, characteristics	*Location*	*Period/Model*	*Result*	*Source*
343	North of 40° latitude	1964–2019	Earlier, chlorophyll increased in the spring; no change or a later increase in the fall	Adams et al. 2022
Many global lakes			Classic review and analysis	Adrian et al. 2009
230, eutrophic	Mediterranean (54), Continental (128), and Boreal (48) climatic zones of Europe	Warmest two-week period in 2015; space-for-time approach	Light and stratification control chlorophyll concentration.	Donis et al. 2021
7	New Hampshire, Maine, the US	Since early 1800s; paleolimnology	Ice-out date is related to *Gloeotrichia* pigments (or biomass) when watershed-to-lake area ratio is small.	Ewing et al. 2020
Lake Mendota, Wisconsin, hyper-eutrophic; Sunapee, New Hampshire, oligotrophic	The US	Long-term observations, 11 model years	Decreased TP retention and increased TN retention due to increases in temperature, stratification, and hypolimnetic anoxia	Farrell et al. 2020

(Continued)

Table 4.2 (Continued)

Potential limnological causes for cyanobacteria						
Cyanobacteria (biovolume, percentage, or dominance)	*Water trans-parency, Secchi*	*Lake water nutrient concentration*		*Hypoxia, redox-metals, DOC*	*Internal P load as cause for cyanobacteria*	
		TN	*TP or SRP*		*Is considered by authors*	*Is likely*
Cyanobacteria				+	×	
Microcystin +		No influence	No influence			
+ "Peak summer bloom intensity"			Summer tem-perature +			
				DO: −; Hypoxia: +		

Study characteristics				
Number of lakes, characteristics	*Location*	*Period/Model*	*Result*	*Source*
Estuary	Coastal Stockholm Archipelago and Baltic Sea	Since 7000 before present (BP); pale-olimnology	Cyanobacteria are correlated with hypoxia. A period of significant cooling (2°C) coincided with the recovery from hypoxia after 1500 common era (CE); eutrophication and warming, related to modern human activity, led to the return of hypoxia in the 20th century.	Funkey et al. 2014; Helmond et al. 2020
Six polymictic (five eutrophic – hyper-eutrophic, one mesotrophic)	Qu'Appelle River drainage basin, Saskatchewan, Canadian Prairies	2006–2016	Warmer surface water temperatures led to higher microcystin concentrations; high nutrient concentrations had no influence.	Hayes et al. 2020
71 large lakes (<160 km^2), satellite	Global	5–28 years (21 yrs on average) for 1984–2012	Trends of summer air temperature, precipitation, fertilizer-use do not explain bloom intensity, except that decreased blooms are in lakes that warmed less.	Ho et al. 2019
393 lakes with ~45 148 temperature and DO profiles	Global: Northern temperate lakes, New Zealand lakes	1941–2017	DO decline in surface waters is caused by reduced solubility in warmer water temperatures, except for DO increases due to increased productivity. DO decline in deep waters is associated with stronger thermal stratification and loss of water clarity.	Jane et al. 2021

(*Continued*)

Table 4.2 (Continued)

Potential limnological causes for cyanobacteria						
Cyanobacteria (biovolume, percentage, or dominance)	*Water trans-parency, Secchi*	*Lake water nutrient concentration*		*Hypoxia, redox-metals, DOC*	*Internal P load as cause for cyanobacteria*	
		TN	*TP or SRP*		*Is considered by authors*	*Is likely*
				DO: –; Anoxia (0.5 mg/L): + Hypoxia (5.0 mg/L): +		
Florida lakes higher than Danish, dependent on TP in the summer	Summer and winter Florida lakes lower than Denmark	Summer similar; winter Danish higher	Similar, summer Danish higher	Florida more hypoxic	Summer internal load; winter internal load only in Florida	
Biovolume: + (TP, temperature); dominance: + (temperature)	-	+	+		-	+
Chlorophyll in phytoplankton-rich lakes: +62%; in phytoplankton-poor lakes: −38%					×	×
Gloeotrichis echinulata						
Microcystis aeroginosa		+	+			× (P is needed in addition to increased temperature)

Study characteristics				
Number of lakes, characteristics	*Location*	*Period/Model*	*Result*	*Source*
~430 global lakes	Global: Northern temperate lakes, New Zealand lakes	1941–2018	DO depletion attributed to an increased stratification from climate warming. Increases in the proportion of the water column below thresholds ranged 0.9%–1.7% per decade.	Jane et al. 2023
1 656 shallow lakes, subtropical and temperate	1 025 in Florida, 631 in Denmark	1989–2010; water temperature space-for-time approach	Warm lakes have more fish, less zooplankton, more cyanobacteria biomass; both have summer internal load.	Jeppesen et al. 2020
143 (60; 83) shallow	South America; Europe	2000–2001, 2004–2006; space-for-time	Predictions of cyanobacteria biovolume (but not of chlorophyll) by nutrient concentration (TP or TN and TP) are significantly increased by temperature.	Kosten et al. 2012
188, large lakes	Global	2002–2016	Phytoplankton-poor lakes were chlorophyll-poorer in warm years, and phytoplankton-rich lakes were richer in warm years; enhanced stratification decreases chlorophyll in nutrient-poor lakes with internal P load as major (only) P source.	Kraemer et al. 2017
Lake Sunapee, oligotrophic, clear, dimictic	New Hampshire, the US	2009–2016	Water temperature is the most important physical forecaster of blooms.	Lofton et al. 2022
Incubation of natural lake water plankton	Water from urban Dutch lakes	2011; laboratory	Microcystin increases at increased temperature only if nutrients are increased, max at 25°C plus N and P	Lürling et al. 2017

(Continued)

Table 4.2 (Continued)

Potential limnological causes for cyanobacteria						
Cyanobacteria (biovolume, percentage, or dominance)	*Water transparency, Secchi*	*Lake water nutrient concentration*		*Hypoxia, redox-metals, DOC*	*Internal P load as cause for cyanobacteria*	
		TN	*TP or SRP*		*Is considered by authors*	*Is likely*
					+	
Dolichospermum, Planktothrix, Microcystis						
				+ (Hypoxic factor)	SRP: + in correlation with hypoxia	
No change in chlorophyll		Decline, due to declining atmospheric deposition	No change, P limitation			
Microcystin +		Cyanobacteria associated with NO_x in early summer	Blooms associated with SRP in late summer			
Chlorophyll					×	

Study characteristics				
Number of lakes, characteristics	*Location*	*Period/Model*	*Result*	*Source*
Lake Rotorua, polymictic, eutrophic	New Zealand	Combined catchment and lake model	Lake water quality effects caused by direct climate change impacts were much larger than indirect impacts to the catchment, because of sedimentary legacy P causing internal P loading.	Me et al. 2018
Ten pre-alpine lakes	European perialpine region, 1945 to 2752 m a.s.l.	1900s–2015; paleolimnology	Raised temperatures affect the strength of the thermal stratification (normalized annual stability) inducing nutrient fluctuations, which favoured buoyancy-controlling cyanobacterial taxa.	Monchamp et al. 2018
Lake Zurich, oligotrophication due to a decreased external P loading	Switzerland	1972–2010		North et al. 2014
2 913	Northeast US (Lake Multi-Scaled Geospatial and Temporal Database of the Northeast U.S., LAGOS-NE)	1990–2013	Chl responds to nutrient and climate changes (air temperature increase); water quality has degraded in 10% of lakes and 18% of regions across Mid-US since 1990 despite management efforts.	Oliver et al. 2017
Lake Peipsi, microcystin producers	Estonia/Russia		Positive relationship between expressed MC toxicity (MC quota per mcyE gene) and late summer water temperature.	Panksep et al. 2020
Lake of the Woods, six sites	Ontario/ Manitoba, Canada	1900–2000; paleolimnology	Positive correlation with mean annual air temperature and total annual precipitation but not TP.	Paterson et al. 2017

(Continued)

Table 4.2 (Continued)

Potential limnological causes for cyanobacteria						
Cyanobacteria (biovolume, percentage, or dominance)	*Water trans-parency, Secchi*	*Lake water nutrient concentration*		*Hypoxia, redox-metals, DOC*	*Internal P load as cause for cyanobacteria*	
		TN	*TP or SRP*		*Is considered by authors*	*Is likely*
At fall turnover: *Microcystis, Aphanizomenon, and Dolichospermum*			+; + SRP		+	
Chlorophyll and biovolume			+ (TP<100 µg/L)			At least in humic lakes
Increase with temperature, decrease with TP load reduction		Lower or no change	Increase		+	
Chlorophyll					Prolonged and warmer ice-free season +	In 7 of 11 lakes, End-of-Summer TP >> than under ice (2011–2013)
+		+	+			

Study characteristics				
Number of lakes, characteristics	*Location*	*Period/Model*	*Result*	*Source*
Rostherne Mere, deep, monomictic, eutrophic	The UK	2 051–2 060; 2 091–2 100, modelled based on observed data for 2016	Increased stratification increases fall turnover cyanobacteria	Radbourne et al. 2019
495	Central and northern Europe	July, August, September 2000–2009	Importance of alkalinity, colour, mixing state in cyanobacteria biovolume response to TP, air temperature, and residence time (<1 yr) increase.	Richardson et al. 2018
Søbygaard, shallow, hyper-eutrophic	Denmark	1989–2010, observed and modelled	"At high external P loadings, the sediment P pool will quickly rebuild to a degree at which it will display negative P retention during summer and help maintain a large phytoplankton population."	Rolighed et al. 2016
23	Athabasca Oil Sands Region, Alberta, and Saskatchewan, Canada	1900–2014; paleoli-mnology	Positive relationships with annual and seasonal air temperatures, so that warmer air temperatures and likely decreased ice covers enhanced aquatic primary production.	Summers et al. 2016
21	Finland	NAO: 1970–2014; sediment: 1986–2014	Internal P load increase due to increasing temperatures in surface and near-bottom water layers, and increased stability.	Tammeorg et al. 2018
108 sedimentary paleoli-mnological time series and 18 decadal-scale monitoring records	North-temperate to subarctic in North America and Europe and in alpine Pyramid Lake, Himalayas	Since 1800	Cyanobacteria increase was best explained by P and N concentrations, while temperature was of secondary importance.	Taranu et al. 2015

(Continued)

Table 4.2 (Continued)

Potential limnological causes for cyanobacteria						
Cyanobacteria (biovolume, percentage, or dominance)	*Water transparency, Secchi*	*Lake water nutrient concentration*		*Hypoxia, redox-metals, DOC*	*Internal P load as cause for cyanobacteria*	
		TN	*TP or SRP*		*Is considered by authors*	*Is likely*
Planktothrix rub.			-	+	Decreased and stayed in hypolimnion in some springs	
Warming and Precipitation, Drought						
+, Mainly Nostocales; some Oscillatoriales	-	+	N.s. + trend	+ NH_4; − bottom DO		Likely
+ (Especially in dominance)					×	
Microcystis, Aphanizomenon, Dolichospermum						
Summer temperature and duration +			+			×
+	-	Variable	+		+	
Limnothrix	-		TP: no change; PO_4: +		×	

Study characteristics				
Number of lakes, characteristics	*Location*	*Period/Model*	*Result*	*Source*
Lake Zurich, warm monomictic (January to April)	Switzerland	1977–2016	Indirect effect: re-oligotrophication by increased thermal stability leading to diminished P distribution through the water column, despite increased internal P load from anoxic sediments.	Yankova et al. 2017
40 reservoirs 12–100 years old, 2.0–12 m mean depth	Semi-arid northeastern Brazil	2007–2008	Drought-induced cyanobacteria proliferation	Brasil et al. 2016
Esthwaite; review of ten mainly eutrophic lakes	The United Kingdom, Sweden, New Zealand		Water temperature: +, flow: −, TP load: +	Elliott 2010, 2012
Great Salt Lake, hypersaline (11–15%), shallow	Utah, the US	1984–2016	Earlier and increased chlorophyll maxima	Hansen et al. 2020
1 260 observations of lakes and reservoirs	Continental United States	2007, 2012	Summer temperatures drive total phytoplankton abundance, the summer duration drives cyanobacterial abundance; increased temperature is associated with increased TP.	Ho and Michalak 2020
7 lakes (Including Lakes Kinneret and Võrtsjärv). Review of 114 papers	Estonia, Greece, Turkey, Italy, Israel, Brazil	Within 1960–2012	In general, water quality decreased with decreasing lake level from decreased precipitation and/or enhanced water abstraction.	Jeppesen et al. 2015
Võrtsjärv	Estonia	1978–2012	Air temperature: +, precipitation (January to March): +	Jeppesen et al. 2015; Nõges et al. 2020

(Continued)

Table 4.2 (Continued)

Potential limnological causes for cyanobacteria						
Cyanobacteria (biovolume, percentage, or dominance)	*Water trans-parency, Secchi*	*Lake water nutrient concentration*		*Hypoxia, redox-metals, DOC*	*Internal P load as cause for cyanobacteria*	
		TN	*TP or SRP*		*Is considered by authors*	*Is likely*
+	-		+			Likely
	Decrease				×	
Chlorophyll, cyanobacteria: +		+	+	Fe, Mn, DOC: +	Related to redox and resuspension	
Summer air temperature: + summer median chlorophyll		TN external load variability has no effect	Three-year previous external TP load: + annual maximum chlorophyll			
Multiple climate drivers (warming, precipitation, stilling, wind)						
				Fe: 28% increase; 4% decrease		

Study characteristics				
Number of lakes, characteristics	*Location*	*Period/Model*	*Result*	*Source*
Apopka	Florida, the US	1985–2018	Drought-induced cyanobacteria proliferation	Ji and Havens 2019
366	Midwestern US: Michigan, Minnesota, Wisconsin (206); Northeastern US: Maine, New York, Vermont (160)	1981–2010	Secchi declines with warmer summers in the midwestern region and wetter summers in the northeastern region. Summer precipitation increases water column turbulence and deepens the thermocline, which may lead to internal nutrient loading that increases productivity, decreasing water clarity.	McCullough et al. 2019
Lakes, reservoirs (some rivers)	Global	1930–2010	Drought increased internal load related to redox P-release in deep lakes and to resuspension in shallow lakes; pH: +, temperature: +	Mosley et al. 2015
Lake Sunapee, oligotrophic, clear, dimictic	New Hampshire, the US	31-year simulation (1982–2015); based on observed data for calibration (2005–2009) and evaluation (2010–2013)	"The interaction between air temperature and nutrient load imply that even more ambitious nutrient load reductions may be required in a warming climate to achieve water quality goals."	Ward et al. 2020
340	Ten countries in northern Europe and North America	1990–2013	Partially related to hypolimnetic hypoxia and could be followed by internal P load increase.	Björnerås et al. 2017

(Continued)

Table 4.2 (Continued)

Potential limnological causes for cyanobacteria						
Cyanobacteria (biovolume, percentage, or dominance)	*Water trans-parency, Secchi*	*Lake water nutrient concentration*		*Hypoxia, redox-metals, DOC*	*Internal P load as cause for cyanobacteria*	
		TN	*TP or SRP*		*Is considered by authors*	*Is likely*
+ Cyanobacterial akinetes; *Dolichospermum* blooms starting 2014.	-			+, increase of ice-free days	Prolonged stiller and warmer ice-free season; increased hypolimnetic anoxia	
Chlorophyll					×	
+		-	+	+	+	
Chlorophyll + 004 µg/decade or −0.01 µg/ per area and decade						
Gloeotrichia echinulata					Increased nutrients from the bottom water	
Present					Upwelling events and wind mixing	
+			+			

Study characteristics				
Number of lakes, characteristics	*Location*	*Period/Model*	*Result*	*Source*
Dickson Lake, recently oligotrophic	Algonquin Park, Ontario, Canada	Paleolimnology	"Late ice-out and a quick onset to stratification in 2014 may have resulted in incomplete spring mixing, early onset of hypolimnetic anoxia, and increased internal nutrient loading, that . . . may have fueled the algal blooms in this remote lake."	Favot et al. 2019
Four eutrophic lakes	Middle-lower reaches of Yangtze River, China	2000–2010	Increased air temperature and wind speed and decreased precipitation enhanced cyanobacteria.	Huang et al. 2021
Lake Champlain, 15 sites, in 5 deep and shallow basins	Vermont, the US; Quebec, Canada	1992–2012	Increased air temperature, flows, and stilling led to increased internal load, because of increased stratification at the deep sites and prolonged anoxia at the shallow sites.	Isles et al. 2017b
344 large lakes (742 million chlorophyll estimates merged over 6 satellite sensors)	Global	1997–2020	Chlorophyll increased in 65% of the lakes but decreased in 56% of lake area because of a downward trend in larger lakes, possibly due to external load changes, invasive mussels, and stabilization.	Kraemer et al. 2022
Erken, dimictic	Sweden	RCP 8.5 simulation	Wind effects depend on pre-event mixing conditions. High wind speeds decreased phytoplankton concentration compared to medium speeds.	Mesman et al. 2022
~40 oligotrophic lakes (GLEON data)	Global		Climate enhanced the upwelling of hypolimnetic P-enriched waters and increased recruitment	Reinl et al. 2021
2 561	Global		TP explained 42%, climate variables 38%, and geomorphometrics 20% of chlorophyll variation.	Shuvo et al. 2021

(*Continued*)

Table 4.2 (Continued)

Potential limnological causes for cyanobacteria						
Cyanobacteria (biovolume, percentage, or dominance)	*Water trans-parency, Secchi*	*Lake water nutrient concentration*		*Hypoxia, redox-metals, DOC*	*Internal P load as cause for cyanobacteria*	
		TN	*TP or SRP*		*Is considered by authors*	*Is likely*
Planktothrix abundance increased after 1980; blooms after 2013 in Jackfish Lake.		+		+	Prolonged stiller and warmer ice-free season; increased hypolimnetic anoxia	+
Blooms			+		+	
Aphanizomenon, Anabaena, Microcystis						
+					+, also internal N-load during different seasons	

Study characteristics				
Number of lakes, characteristics	*Location*	*Period/Model*	*Result*	*Source*
Seven, including Jackfish Lake	Subarctic Yellowknife, Northwest Territories, Canada	Monitored since 2014; paleolimnology since pre-20th century	Warmer water temperatures, longer open-water growing seasons, and stronger thermal stability in combination with increasing eutrophication from urbanization.	Sivarajah et al. 2021
Meta-analysis of 28 studies on mostly shallow lakes	Global	1976–2012	Increased storm events drive internal P load in shallow lakes, where a larger fetch elicits a larger internal load.	Stockwell et al. 2020
Shallow, hyper-eutrophic Müggelsee	Germany		Highest cyanobacteria biomass during prolonged stratification and hyper-eutrophic TP and TN concentrations.	Wilhelm and Adrian 2008; Wagner and Adrian 2009; Thayne et al. 2022
Lake Chaohu; subtropical, shallow, eutrophic	Middle reaches of Yangtze River, China	2015–2016	Internal P load increases from July to November; internal N load has a different seasonality.	Yindong et al. 2021

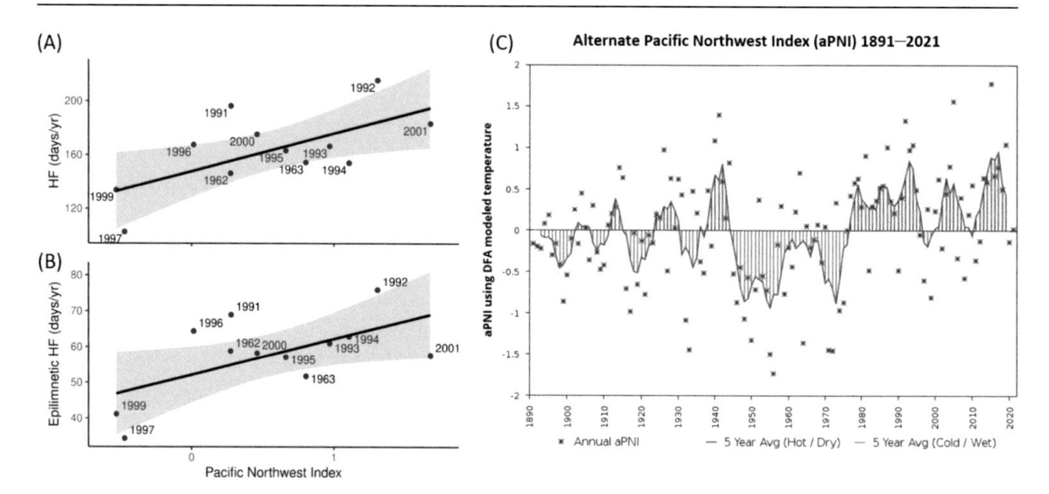

Figure 4.10 (A, B) Hypoxic factors (*HF* refers to the whole lake in A, and epilimnetic *HF* refers to the epilimnetic sediment area for a threshold of 6.5 mg/L DO in the overlying water in B; Section 2.2, Box 2.1) related to the Pacific Northwest Index (PNI) in Brownlee Reservoir ($n = 11$, $p < 0.05$, with regression lines and 95% confidence bands *redrawn from (Nürnberg 2002)*) and (C) records of the (alternate) PNI (1891–2021) showing recent increases in hot and dry conditions (red), *(downloaded on April 4, 2024 from www.cbr.washington.edu/dart/pnii)*.

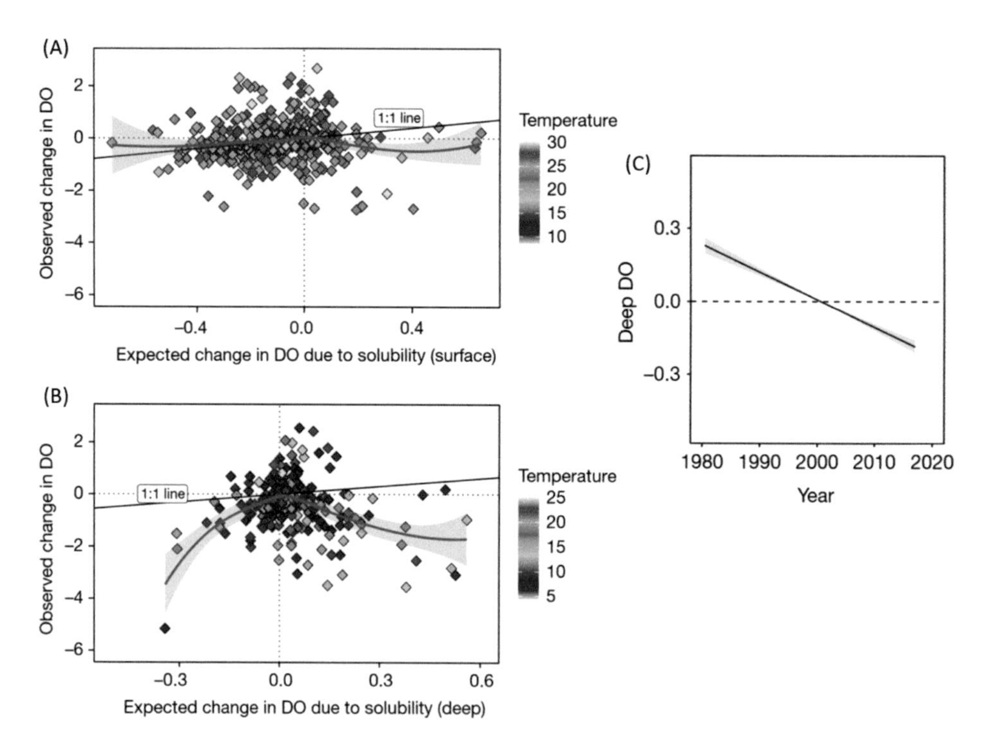

Figure 4.11 Observed changes in dissolved oxygen concentration (DO) versus expected changes due to increased solubility related to temperature increases for surface (A, $n = 415$) and deep (B, $n = 259$) temperate global waters. Changes in deep DO for all lakes between 1980 and 2017 (C). *Figure 2 of Jane et al. (2021), with permission.*

Lake and reservoir locations including latitude and altitude also control water temperature and related oxygen depletion with expected diverse climate change effects. This is evident by the need for re-parametrization of the model to predict anoxic factor in warmer low-latitude lakes (Carneiro et al. 2023). The original model developed on temperature lakes severely under-predicts *AF* in semi-arid, tropical stratified wet season reservoirs (Box 2.1).

Further evidence of changes in the oxygen state of bottom waters and sediments related to climate change includes paleolimnological observations (Table 4.2). In a remote and oligotrophic lake on the Precambrian Shield, for example, decreased DO over the last 20 years was inferred from chironomid remains (Favot et al. 2019). In the coastal waters of the Baltic Sea, climate warming has been identified as important driver of hypoxia in the recent anthropogenic period, while climate cooling prohibited hypoxia increases in 1000 BCE – 1500 CE, despite increased human activity (Helmond et al. 2020).

There are also many indirect indications of increased hypoxia in lakes. They include brownification (including the increase in organic matter) and the increase in chemicals in bottom waters that indicate reduced conditions (including ammonium, phosphorus, and metals), as discussed in Sections 4.3.2 and 2.6.

4.3.2 *Increased internal P loading*

The well-documented increase of lake bottom water hypoxia in the wake of climate change (Section 4.3.1) benefits processes that are enhanced under these conditions. This means more redox-related nutrient and metal release (Section 2.6) from bottom sediments of lakes and wetlands, possibly with associated Fe and DOC (dissolved organic carbon) increase in the water thus leading to brownification, as reviewed by Blanchet et al. (2022). As expressed by Woolway et al. (2021b), stratification, anoxia, and internal P loading are all related: “It is very likely that [the predicted] 33.3 ± 11.7 day prolongation in stratification will accelerate lake deoxygenation with subsequent effects on nutrient mineralization and phosphorus release from lake sediments”.

Many climate change effects support and positively influence redox-related sediment P release as well as microbial P liberation from organic compounds in nutrient-rich systems (Saar et al. 2022). Most of the physical variables related to climate change (discussed in Section 4.2) are theoretically expected to increase sediment P release based on its characteristics presented in Section 2.4.2.

1. Increased sediment and bottom water *temperature* (especially in shallow systems) due to increased air temperature and increased intensity and duration of *heatwaves* (Section 4.2.1) positively affects internal P loading by increasing (a) sediment oxygen demand that lowers the redox potential, which is a prerequisite for anoxic P release, thus increasing spatial and temporal extent of anoxia (anoxic factor) and (b) the areal anoxic P release rates.
2. Enhanced *stratification* due to climate warming and atmospheric stilling (Section 4.2.3) increases the duration and areal extent of quiescent conditions that favour anoxia at the sediment–water interface (increase of anoxic factor). These conditions extend the temporal and spatial conditions favouring redox-related P release. But an earlier onset of stratification can also lead to lower bottom temperature during the open-water season in deep lakes, which would delay oxygen depletion and diminish P release rates.
3. Increased *storm* events (Section 4.2.3) negatively affect thermal stability which can have opposing effects on internal P loading. (a) Decreased stability can decrease the duration of the P release period potentially decreasing internal load. (b) Increased or intermittent mixing

and thermocline erosion distribute sediment-released P throughout the water column increasing its accessibility to phytoplankton.

4. *Decreased precipitation* (increased variability of hydrology, Section 4.4.2) causes enhanced *drought* periods that increase temperature (point 1) and water column stability (point 2). These hydrological changes affect internal load and cyanobacteria.

Further influences like the timing of climate change impacts over the year (seasonality) and specific lake characteristics determine the severity of the climate change effects on specific systems.

- *Importance of seasons and pre-conditions:* most climate effects on internal P loading differ with respect to season: for example, increased temperature extends the stratified period in deep lakes and thus enhances internal load especially if temperature increases occur in late summer and fall. Storm events and increased mixing have the largest effect during stratified conditions.
- *Importance of lake characteristics*: many lake characteristics impact internal P loading in different ways. Characteristics include trophic state; extent of browning and DOC content; depth and morphometry, hydrology, and mixing state. Especially important is the different response to climate variables by shallow compared to deep lakes.

Twenty-seven studies listed in Table 4.2 explicitly report internal P load increases from climate change variables, and an additional 11 report conditions that likely support internal P loading. Decreased P retention in mass balances (Farrell et al. 2020), in combination with external P input large enough to provide a consistent sediment P pool (Rolighed et al. 2016), is explained by climate-change-related increases in water temperature, stratification, and hypoxia that would cause an increased sediment P release. In a study featuring eutrophic Lake Mendota, Wisconsin, and oligotrophic Lake Sunapee, New Hampshire, both in the United States, nitrogen reacted opposite to P so that N export (and N epilimnetic concentration) was decreased (N retention increased), while P export (and P epilimnetic concentration) increased during climate warming (Farrell et al. 2020).

In this context, (Jeppesen et al. 2014) conclude in their review on nutrient-rich, shallow lakes "there seems to be strong evidence that climate change will enhance eutrophication in mesotrophic and eutrophic lakes as a result of physicochemically and biologically induced higher internal loading." Similarly, (Me et al. 2018) states that "polymictic lakes may be particularly vulnerable to eutrophication associated with climate change due to increased internal nutrient loading, which will lead to a biological response of increased algal biomass, while changes in external loads will have lesser relative impact."

Internal load was considered the reason why "anoxia begets anoxia" (Lewis et al. 2024b) in a self propagation of hypoxic conditions in an analysis of 656 global temperate lakes. Here, increased anoxia, determined as anoxic factor, was positively correlated with hypolimnetic TP and, although less, with epilimnetic TP and chlorophyll, reflecting the potential of increasing redox-related internal load on lake productivity (Section 5.1.4).

An obvious indicator that internal P load is increased by climate change and not by external load alone is the seasonality of certain climate effects including lake warming (Section 4.2.1). Correlated increases in air and water temperature mostly in the summer and fall also yielded increases in bottom water temperature and water column stability supporting internal P loading (Section 4.2.1, Winslow et al. 2017). In contrast, increases in external P or in other nutrients (N, Si) occur at different times and hence can be distinguished from internal P sources that cause increased P concentration in the summer (in mixed lakes) and fall (at fall turnover in stratified lakes).

Seasonal patterns of internal P and N loading confirmed the different timings as compared to external loading in eutrophic Lake Chaohu, China, in 2015–2016 (Yindong et al. 2021). Based on modelled separation of organic P and N compounds from inorganic compounds, the increase in TP (including redox-related inorganic) internal P loading lasted from July to November, organic P spiked in spring and lasted until the end of September, while modelled NO_x internal load occurred in the spring, organic N spiked in May, and NH_4 increased after until July. Mainly redox-related TP was released from the sediments and available for algal uptake, while the various N-forms were not. Scenarios of climate change (RCP2.6 – RCP8.5_2050) simulated increased internal TP loading and cyanobacteria concentration from August to November, while impacts in spring and summer depended on N-forms but were less clear.

A negative influence of late summer/fall climate warming on annual P retention was determined in a mass balance study on sediments of 21 Finnish lakes (Tammeorg et al. 2018). In these lakes, the climate index of October-NAO was positively correlated with temperatures in surface and near-bottom water layers and their difference (increased stability) but negatively with P accumulation in the sediment. These trajectories indicate an increasing effect of climate warming on internal P loading.

When determining storm impacts on the phytoplankton community dynamics, Stockwell et al. (2020) explicitly mention the increase of internal P load during wind events (Figure 4.12) and less from increased rain events (Table 4.1, Table 4.2). While considering wind impacts on surface water temperature and light, they concluded that:

> Wind events are expected to have the greatest impact on internal nutrient loading in lakes with greater fetch, stronger antecedent stability, and higher productivity. In particular, strong antecedent stability is expected to facilitate the buildup of nutrients in hypolimnetic waters (deeper lakes) and nutrient release through sediment anoxia (shallower lakes) . . ., although well-oxygenated hypolimnia likely result in little effect.

Long-term monitoring in Lake Champlain (Isles et al. 2017b) explored the differing effects of climate change on deep stratified and shallow polymictic regions. Slightly increased flows and air temperature combined with decreased wind speed at various months over the growing period were related to decreased N and much increased TP concentration. Patterns differed between deep and shallow sites, where TP increases were higher at shallow sites, especially in late summer and fall (October), when air temperature increase was the highest of the growing period, which would enhance sediment P release rates and prolong and extend conditions of hypoxia.

In shallow, historically polymictic lakes increased periods of stratification associated with increased hypoxia (Favot and et al. 2024; Wilhelm and Adrian 2008) enhance internal P load that then is distributed into the mixed layer fuelling phytoplankton growth (e.g., Figure 2.11B). Wind speed and associated shear stress enhance internal P load in shallow lakes especially when the water level is low (Tammeorg et al. 2013). In the polymictic Estonian lake, Võrtsjärv, shear stress was significantly correlated with internal load for 33 years of data (Figure 4.13, (Tammeorg et al. 2022a)). Similarly, wind speed correlated with seasonal average sediment P release, chlorophyll, and trophic state in Brazilian reservoirs during the severe dry periods (Rocha and Lima Neto 2022) (Section 5.7.1).

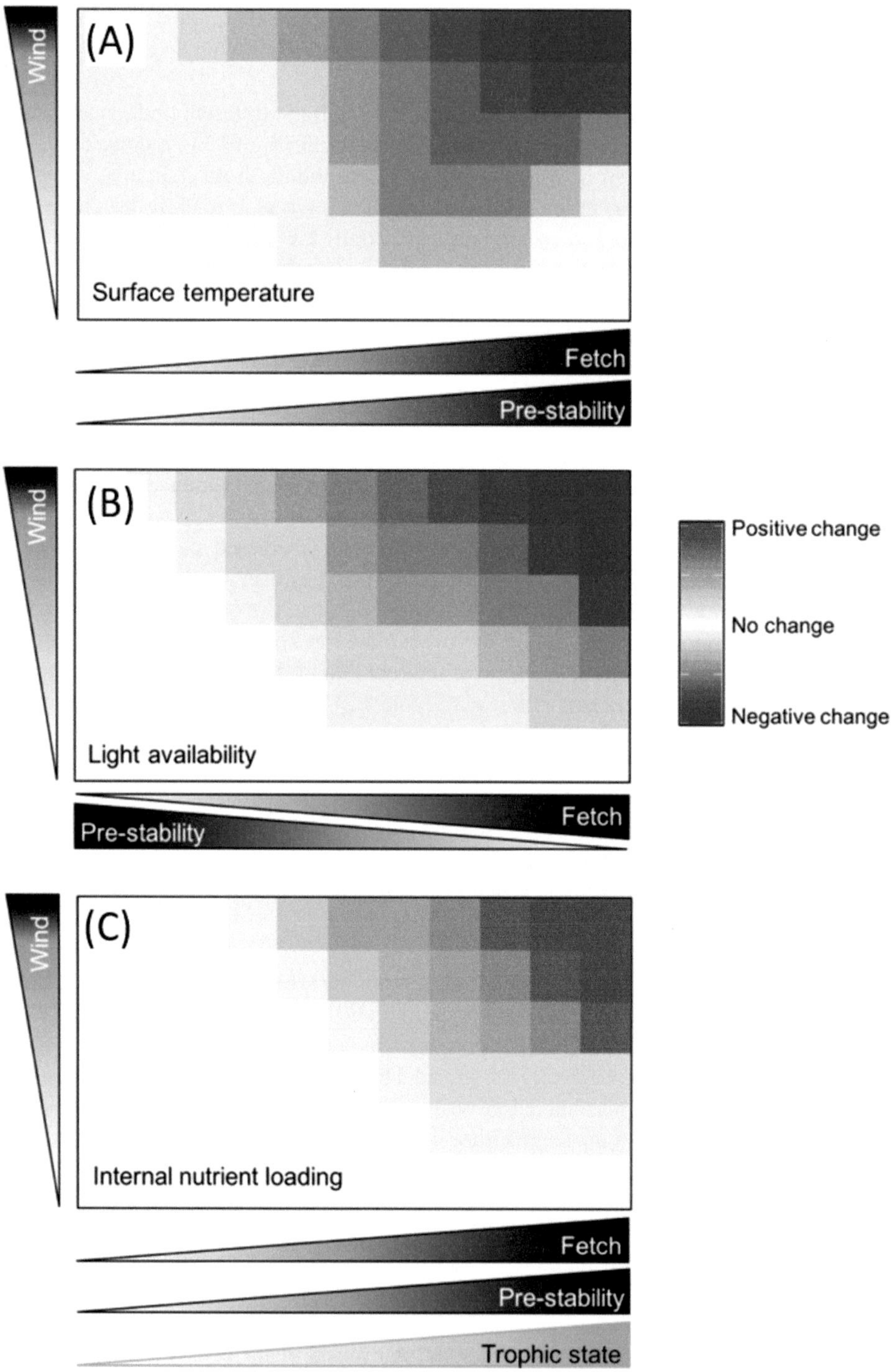

Figure 4.12 Enhancing effects of wind on temperature (A), light (B), and internal load (C). *Figure 3 of Stockwell et al. (2020), open access.*

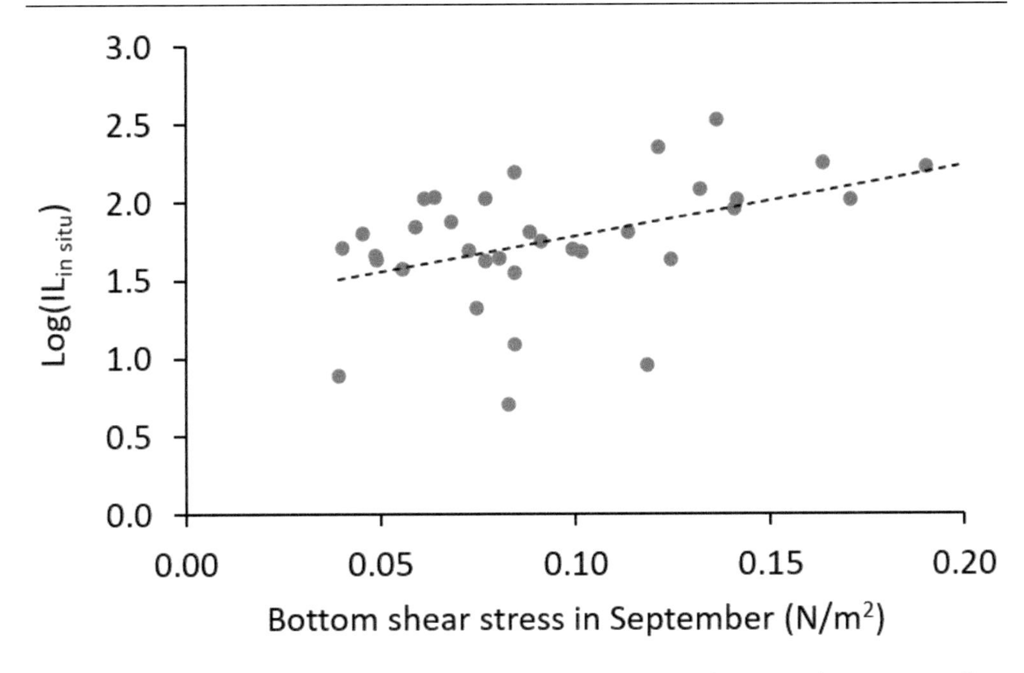

Figure 4.13 Growing period internal P load ($IL_{in\ situ}$) as a function of bottom shear stress in September in Võrtsjärv (R^2 = 0.340, p = 0.002, n = 33. Trend is shown with a dashed line. *Revised from Tammeorg et al. (2022a), open access.*

Climate warming and external P load increase P fractions in the sediments that produce internal P loading (legacy P). A synergistic effect of warming and P input (external load) on sediment P composition and by extension, internal load, was determined in large (1.9 m diameter, 1.5 m deep) mesocosms involving organic rich sediments from Danish lakes overlain by 1 m of gently stirred water (Saar et al. 2022). Nutrient enrichment (presenting the scenario of increased external P input from the catchments due to higher precipitation in north-temperate regions) produced the highest sediment TP, TFe, and mobile P fractions compared to the low-nutrient treatment that exhibited the highest organic content (LOI), Al-P, Humic-P, and NaOH-Al but the lowest mobile P fractions. Warming increased sediment concentrations only in nutrient-enriched treatments and was negatively related with organic matter components so that "higher loading and higher retention increases the risk of problematic internal P loading in the future."

Indications that climate change coupled with former external load causes increased internal P load can be especially obvious after external P load abatement. Previous external P load decrease has led to oligotrophication in warm-monomictic Lake Zürich, Switzerland, but the historic large load had accumulated legacy P in the sediment (North et al. 2014). When hypoxia, determined as anoxic factor, increased due to climate warming, an increasing amount of sediment P was released as internal load. (Evident from a significant correlation of AF with hypolimnetic SRP.)

The relationships among climate change variables, redox-related processes, and brownification are complex and often synergistic. Further, climate effects on the (terrestrial) catchment basin exist that change organic and nutrient export to the receiving waters, besides climate effects that directly change conditions in the lake such as lake water warming and increased variability in flow, lake level, and stability. Many studies that describe an upsurge in internal P load in the wake of climate change connect this increased fertilization with cyanobacteria proliferation and are discussed in Section 5.1.

4.4 Enhancement of cyanobacteria blooms by climate change

Numerous studies demonstrate the direct effects on the biology of lakes from air temperature increases, solar radiation strengthening, precipitation changes, and from other factors related to climate change. These effects may be particularly disturbing because biological variation can produce more extremes faster than meteorological, physical, or chemical variability. This was determined from a quantitative comparison of 595 time series on a large data set of 11 lakes in Northern Wisconsin, USA (Batt et al. 2017). Biological variables produced larger and more frequent "surprises" than the other categories of variables. The detected biological patterns were explained by "directional environmental change in climate, nutrient inputs, habitat loss, or other factors" and thus may become increasingly frequent as they were partially caused by climate change.

Important climate change effects on freshwater biology involve the proliferation of cyanobacteria as shown in this section (cyanobacteria increase: 41; Secchi water transparency decrease: 8; Table 4.2). Additional severe changes in freshwater biology are described in Section 4.5.

A review of emerging threats in the Anthropocene lists climate change and harmful algal blooms as 2 of 12 threats for freshwater biodiversity (Reid et al. 2019). Climate-change-related effects potentially causing the increase in cyanobacteria include: (1) warming and eutrophication, (2) hydrological intensification (increased extremes of drought and storm events), (3) increased lake stability and stratification, and (4) brownification. Related studies are listed in Table 4.2, some of which are further discussed below.

As climate-change-related warming affects growth rate, abundance, and composition of lake plankton, shifts in phytoplankton species composition to cyanobacteria have been repeatedly observed. An early review of the climate effects on cyanobacteria predicted that "these [climate change related] processes are likely to enhance the magnitude and frequency of these [bloom] events" (O'Neil et al. 2012). Another review came to the same conclusion (Carey et al. 2012), namely that

> cyanobacterial eco-physiological traits, specifically, the ability to grow in warmer temperatures; buoyancy; high affinity for, and ability to store phosphorus; nitrogen-fixation; akinete production; and efficient light harvesting, vary amongst cyanobacteria genera and may enable them to dominate in future climate scenarios.

In reversed logic, such shifts were declared as an indicator of climate change (Adrian et al. 2009), and harmful algal blooms were considered a climate change co-stressor in marine and freshwater ecosystems (Griffith and Gobler 2020). Higher temperature, more frequent storms, longer periods of drought, and stable stratification were registered in a large overview to positively affect cyanobacteria indicating connections with climate change (Chorus and Welker 2021).

The evaluation of satellite observations for 71 large worldwide lakes with 5–28 years (1984–2012) of observations each indicates that the apparent increase of near-surface phytoplankton blooms (including harmful algal blooms by cyanobacteria) is real (Ho et al. 2019) (Figure 4.14) and not (just) caused by increased monitoring. Lakes with a long-term decrease in bloom intensity were rare. While there were no temporal trends consistent with temperature, precipitation, or fertilizer-use, a decrease in bloom intensity was only observed in lakes that had "warmed less compared to other lakes".

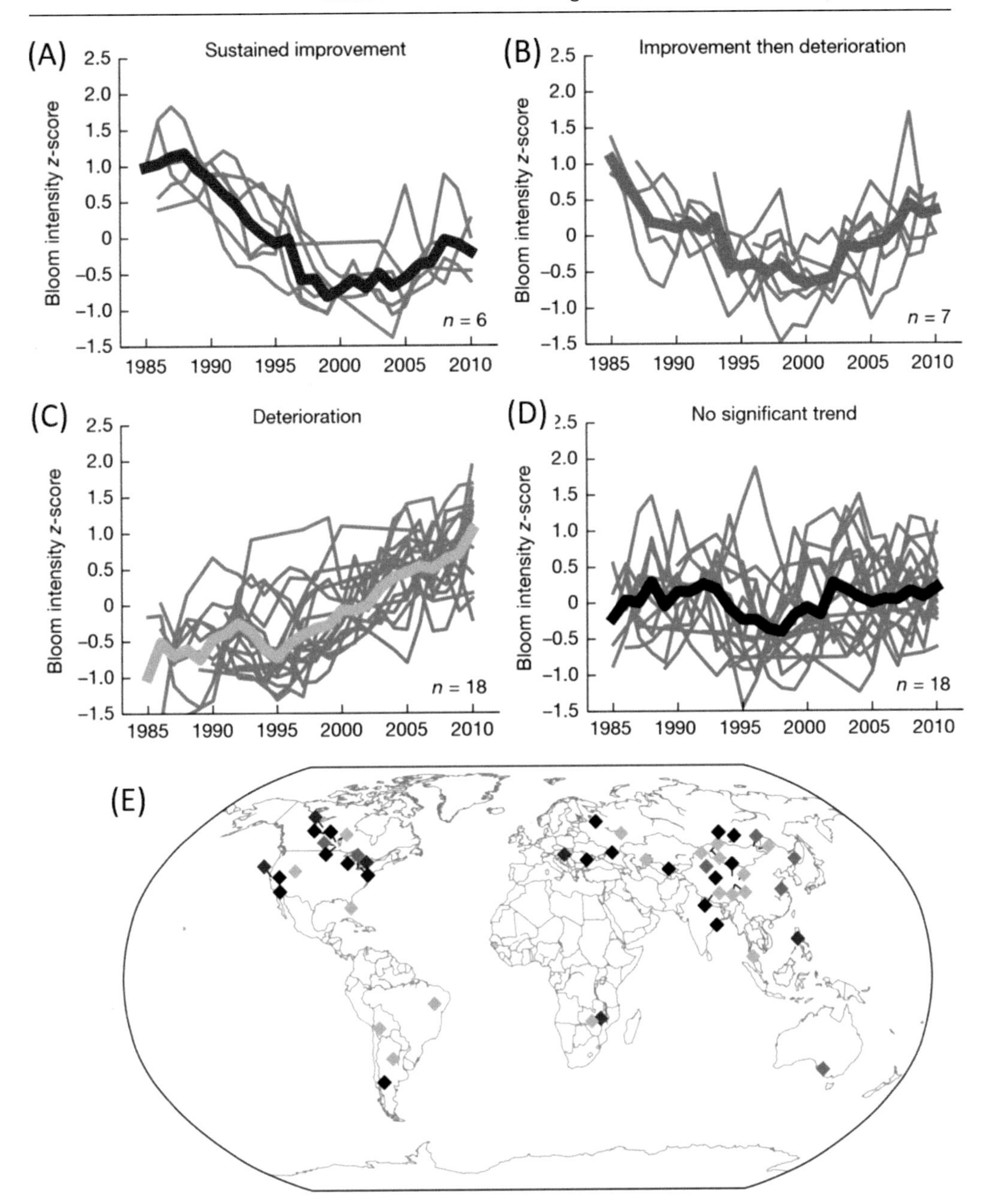

Figure 4.14 Lake bloom histories follow one of four pathways. (A) Sustained improvement, (B) improvement then deterioration, (C) deterioration, and (D) no significant trend. Shown are time series for lakes with at least 14 years of data (n = 49) categorized by historical pathway. Grey lines show five-year moving averages of normalized bloom intensity, with coloured lines showing pathway averages across lakes. Bloom intensity z-scores for each lake were calculated using its own historical mean and standard deviation. (E) Global distribution of the four lake pathways. *Figure 2 of Ho et al. (2019) with permission.*

However, a singular study failed to report an increase in cyanobacteria. This study on 323 lakes located in the contiguous United States is titled "no evidence of widespread algal bloom intensification in hundreds of lakes" (Wilkinson et al. 2022) since only 10% experienced bloom intensification. Most of these lakes (64% of those with declining chlorophyll concentration) had undergone external load abatement since the mid-seventies which would have decreased nutrients available for phytoplankton, and many had been very productive so that perhaps added climate stressors could not further increase blooms.

The trend of chlorophyll changes over time may be different for large versus smaller lakes. The analysis of 742 million chlorophyll-a estimates from six satellite sensors in 344 globally distributed lakes detected variable changes between 1997 and 2020 (Kraemer et al. 2022). Chlorophyll estimates increased in 65% of the lakes but decreased in 35%, mainly large, lakes. Therefore, normalized for (pixel) area, global chlorophyll concentration per area increased in only 44% but decreased in 56%. Some of the decrease in large lake was attributed to efforts of external load abatement, invasive dreissenid mussels, and climate change-induced stabilization that prevents deep-water nutrient entrainment.

Much support for the reality of increased cyanobacteria is provided by paleolimnological observations. Based on 108 sedimentary time series and 18 decadal-scale-monitoring records in north temperate-subarctic lakes (North America and Europe) it was found that "(1) cyanobacteria have increased significantly since ~1800 CE, (2) they have increased disproportionately relative to other phytoplankton, and (3) cyanobacteria increased more rapidly post ~ 1945 CE" (Taranu et al. 2015).

In a database for 2 561 lakes worldwide, 60% of the variation in chlorophyll *a* concentrations was explained by various variables, where TP explained 42%, a combination of climate variables accounted for 38%, and geomorphometrics accounted for 20% (Shuvo et al. 2021). In particular, there were strong effects of lake characteristics that likely affect internal loading, including summer and spring temperature, water residence time, lake volume, and mean depth.

Thus, many specific climate-change effects support and increase cyanobacteria abundance and bloom frequency. The following variables, mainly related to the physical effects of climate change (discussed in Section 4.2), have been shown to or are theoretically expected to increase cyanobacteria because of their characteristics as presented in Section 3.5.

1. Increased water *temperature* caused by increased air temperature and wind stilling; increased intensity and duration of *heatwaves* (Section 4.2.1).
2. Enhanced *stratification* due to climate warming and atmospheric stilling (Section 4.2.3) increases the duration and areal spread of quiescent conditions that favour cyanobacteria.
3. *Light* provides a competitive advantage by cyanobacteria over other phytoplankton: there is enhancement of surface bloomers by light tolerance and enhancement of deep-water bloomers by shade tolerance.
4. Increased *storm* events (Section 4.2.3) have variable effects dependant on the lake depth and mixing status. Decreased thermal stability in stratified lakes can have a negative effect on surface-blooming cyanobacteria. Increased sediment disturbance and mixing can have a positive effect on cyanobacteria in polymictic lakes because of increased contact with nutrients.
5. Decrease in *precipitation* (an increased variability of hydrology, Section 4.4.2) causes enhanced drought periods that increase water temperature and stratification.
6. Increase in *lake browning* (Section 4.3.2) alters light regime, water temperature, stability, and stratification.
7. Increase in *internal P load* provides bioavailable P at an opportune time for cyanobacteria (further explored in Section 5.1).

Climate change affects cyanobacteria proliferation simultaneously, and the potential drivers ought to be considered simultaneously. For example, higher than long-term average air temperature together with lower precipitation during summer correlated with longer anoxia and hypolimnetic P increases in a mesotrophic drinking water reservoir (Nürnberg, unpublished data 2017–2022). Cyanobacteria coincided with internal P load and only proliferated in years with extreme climate conditions.

The following subsections describe the effects of some of the climate drivers on cyanobacteria separately.

4.4.1 Temperature and nutrients

Summer air temperature and water temperature are the most often studied variables, and most studies found a supporting influence of them on general phytoplankton, or specifically on chlorophyll *a*, which mainly represents cyanobacteria in summer and fall.

While it is often accepted that the supply of nutrients is the principal influence on the proliferation of cyanobacteria (Section 3.3), the nutrient effect is modified by temperature changes due to climate change (Section 3.5.1). In the paleolimnological study mentioned above, "variation among lakes in the rates of [cyanobacteria] increase was explained best by nutrient concentration (phosphorus and nitrogen), and temperature was of secondary importance" (Taranu et al. 2015).

A data set of 1 260 monitoring summer observations of US lakes and reservoirs on phytoplankton abundance, species dominance, and toxicity determined the effect of potential climate-related drivers specifically on cyanobacteria (Ho and Michalak 2020). While summer temperatures controlled total phytoplankton abundance, the length of the summer (stratification) drove cyanobacterial abundance. The authors specify internal P loading that increases with temperature as the reason for increased lake TP concentrations driving cyanobacteria proliferation (Section 5.1.3).

Conversely, a review of ten lake studies determined elevated temperature as the most important driver that increased biomass and dominance of cyanobacteria in all but two lakes (Elliott 2012). In three of these lakes, nutrients had also increased. A subsequently developed model showed an increase in relative cyanobacteria abundance with increasing water temperature, decreasing flushing rate, and increasing nutrient loads. The authors further addressed the importance of internal load.

> [U]nder decreased flushing, nutrient load (i.e., of phosphorus, nitrogen and silica) via the inflowing rivers was reduced leading to increasing reliance of internally released phosphorus to support the summer and autumn growth, which, . . . gave the nitrogen-fixing cyanobacteria an advantage.

Based on 2007 monitoring results of 1 076 natural freshwater US lakes, ponds, and reservoirs (at least 1 m deep and 1 ha large) covering a wide range of trophic states, the relative effects of nutrients and temperature on cyanobacteria biovolume differed with trophic state and for different cyanobacteria taxa (Figure 4.15 (Rigosi et al. 2014)). Eutrophic and hyper-eutrophic lakes exhibited a significant interaction between nutrients and temperature that the authors attributed to light limitation. But it is also possible that this interaction is partially caused by internal P loading, which is increased at elevated trophic state and high temperature (Section 2.4.2).

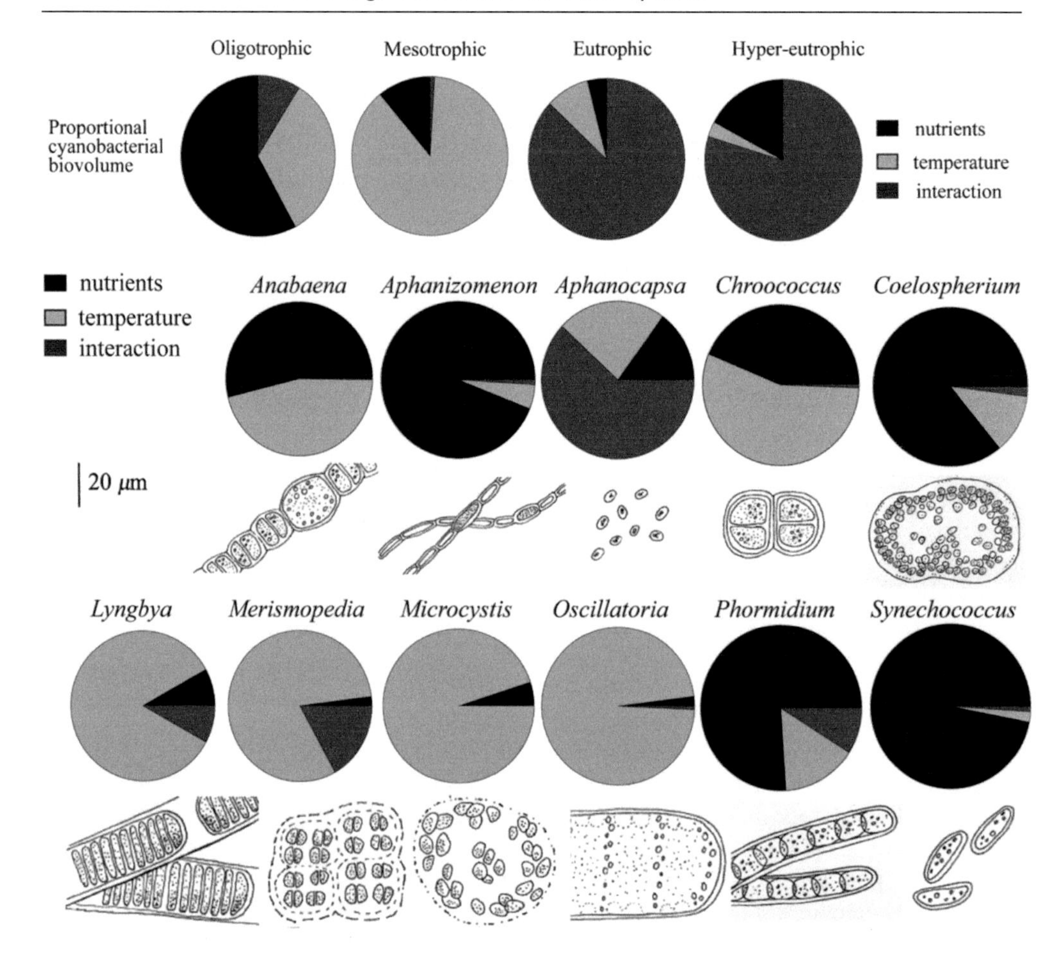

Figure 4.15 The relative effect of nutrients (TN and TP), temperature, and their interaction on the proportional cyanobacterial biovolume in lakes of different trophic state (top row) and on 11 specific cyanobacterial taxa (subsequent rows with morphological drawings) in a data set of 1 076 US lakes (*Anabaena* = *Dolichospermum*). *Figures 2 and 3 of Rigosi et al. (2014), open access.*

The impact of temperature increases is most obvious in hyper-eutrophic systems, where nutrient limitation is absent. In the lake Nieuwe Meer (1.3 km^2, 30 m maximum depth), a recreational lake in Amsterdam, the Netherlands, the impact of temperature, wind stilling, and cloudiness on the proliferation of a cyanobacteria bloom (*Microcystis*) was evaluated in an exceptionally hot summer (August 2003) (Jöhnk et al. 2008). Using models describing phytoplankton growth rates, and long-term average meteorological effects, conditions of the extreme hot, still, and sunny year explained the observed large cyanobacteria bloom. (In this lake, artificial mixing was used to prevent stratification, except in the study year when mixing was interrupted throughout the summer. Section 4.4.3) *Microcystis* blooms in the hot summer 2003 were explained by: (1) the cyanobacteria's high physiological temperature optimum compared to diatoms and green algae, combined with a much higher response to temperature increase (large Q_{10} value) that is supported by increased light availability, (2) a strong thermal stratification during still and hot conditions that favoured the buoyant *Microcystis* so that dense surface blooms were shading other phytoplankton species, and (3) reduced viscosity at very high temperatures, which supports the flotation of buoyant species and sinking velocity of sinking species (Section 3.5). The observed trend of increased summer temperature (1900–2004) in that study (Jöhnk et al. 2008) and since then (Section 4.2) forecasts further such beneficial conditions for cyanobacteria proliferation.

Cyanotoxin prevalence was related to temperature in six hyper-eutrophic lakes of the Qu'Appelle River, Saskatchewan (Hayes et al. 2020), known for consistent internal P loading. Modelling of 11 years determined that duration and maxima of microcystin toxicity were related to water temperature but not nutrient levels during the growing period (May through August).

4.4.2 Precipitation, hydrology, and drought

A detailed early review investigated the potential effects of changes in rainfall patterns in a changing climate on cyanobacteria blooms (Reichwaldt and Ghadouani 2012). The authors connect the proposed "possible" effects on cyanobacteria biomass to mechanisms and consequences in a comprehensive list. Most changes in the rainfall patterns will lead to favourable conditions for cyanobacterial growth:

- Increased rainfall intensity and flushing destabilize water column by water inflow and wind: decrease of cyanobacteria.
- Increased nutrient input through surface and subsurface flows: increase of cyanobacteria.
- Increased sediment resuspension creating light limitation in the water column: increase of surface-blooming and low-light-adapted cyanobacteria taxa.
- Longer water residence time and dry periods, drought, lower lake level, and increased stratification: increase of cyanobacteria.
- Increased anoxia: increase of cyanobacteria.

Precipitation and related factors, including water residence time and lake level as predictive variables, have gained much attention in the cyanobacteria literature (Table 4.2). For example, a meta-analysis of 28 studies on global, mostly shallow, lakes (that determined several physical and chemical storm effects as discussed in Sections 4.2.3 and 4.3.2) found subsequent biological (cyanobacteria) effects of three possible storm manifestations: (a) dry storm (wind), (b) rain only (rain), and (c) wet storm (wind and rain) (Stockwell et al. 2020). Rain mostly affected nutrients, hydrology, and temperature, which decreased biomass and altered community composition, while rain associated with wind increased chlorophyll and productivity and affected other cyanobacteria-related variables (Figure 4.16). Inconsistent effects of precipitation can be explained by their dependence on preceding conditions in the lake and catchment basin, including previous rain events, seasons, temperature, watershed geochemistry, usage, and areal ratios (watershed to lake surface), and morphometric and hydrological conditions in the lake (Stockwell et al. 2020).

Evidence for the impact of precipitation was also mixed in a data set of 1 260 summer observations of US lakes and reservoirs mentioned earlier (Ho and Michalak 2020). "While increased nutrient runoff from precipitation could support blooms, lake nutrient concentrations could also be reduced through greater flushing due to precipitation." Further, precipitation affected eutrophication differently for seasonal and annual averages, so that its impact varies with the period of observation.

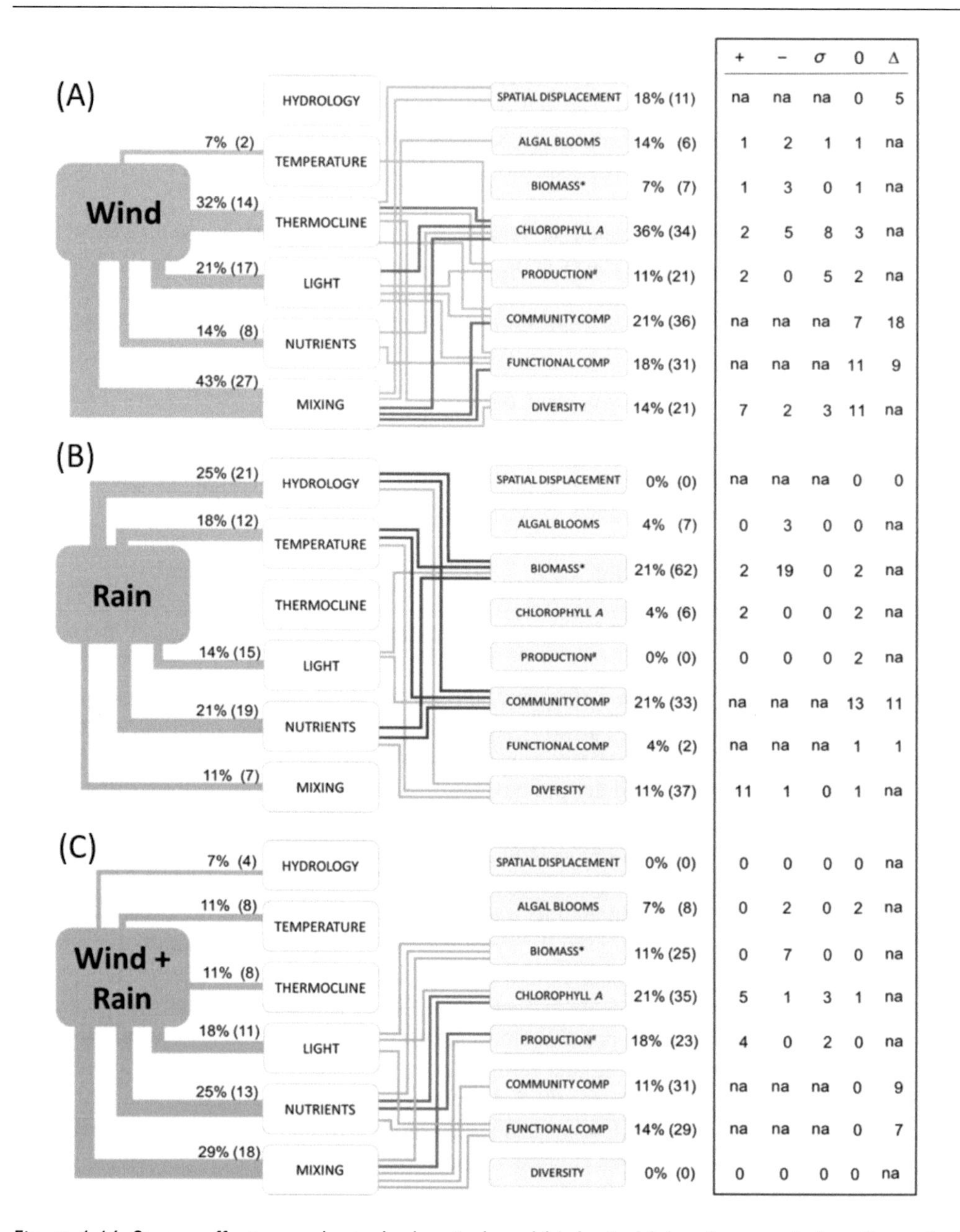

Figure 4.16 Storm effects on physical, chemical, and biological lake characteristics. *Figure 1 of Stockwell et al. (2020), open access.*

Note to Figure 4.16 *(Revised from Stockwell):* summary of the systematic review linking three types of storm events, wind (a), rain (b), and wind plus rain (c), to six variables related to lake chemical and physical condition (in centre). The connectors between different variables represent the links described by the authors in the studies or supported by data presented in the publications. The width of the connectors between weather events and lake conditions is proportional to the percent occurrence of each link in the studies which met our criteria. The percent occurrence and the total number of reported links (in parentheses) are located above the connectors. To the right of the physical/chemical conditions are phytoplankton-related variables. The numbers to the right of these variables represent the percent occurrence and the total number of links (in parentheses) in which each phytoplankton-related variable was found, with at least 16% (darker) or 9% of the studies included.

Chlorophyll dependencies on climate-related variables confirmed that increased air temperature and wind speed and decreased precipitation and lake level fluctuation (where high fluctuation indicated riverine influences) enhanced cyanobacteria in highly eutrophic reservoirs (Huang et al. 2021).

Further, geographic differences that aligned with land use were responsible for different sensitivities to precipitation (and temperature) in 160 northeastern (85% forest and 8%, agricultural cover, median TP 9 μg/L) versus 205 midwestern US lakes (43% forest, 23% agriculture, median TP 20 μg/L) (McCullough et al. 2019). In the northeastern lakes, precipitation (height of summer rain) was the most important predictor, while in the midwestern lakes, summer mean and maximum and fall minimum air temperatures were the best predictors of lake clarity. At increased total summer precipitation, water clarity (representative of cyanobacteria) decreased in 40% of the northeastern lakes with lower TP concentration, indicating increased nutrient and organic matter input. In the more-enriched and productive Midwestern lakes, elevated temperature was more important in predicting cyanobacteria. Besides direct physiological effects, temperature effects were potentially caused by increasing the already occurring internal load by increasing sediment P release rates and the period and area of anoxia.

Other studies found an increase in water clarity after precipitation events in more eutrophic lakes, indicating a dilution or growth disruption effect. For example, in a shallow eutrophic (nutrient-replete) urban reservoir in subtropical China, internal loading from sediment would occur during quiescent and dry periods when dissolved oxygen concentration was decreased in the summer. But after short-term summer rainfall events (within three days), dilution by precipitation and changes in wind speed and sunlight reduced water temperature and decreased cyanobacteria (Luo et al. 2022). In contrast, enhanced light penetration increased cyanobacteria after rain events in the colder winter months.

The varying responses in different lakes to precipitation and storm events can be further explained by internal P loading processes in lakes of different trophic states (Section 5.1). In deep oligotrophic northeastern lakes, increased water column turbulence and deepening of the thermocline occurred as a consequence of increased precipitation, "which may lead to internal nutrient loading that increases productivity, decreasing water clarity" (McCullough et al. 2019).

Entrainment of cyanobacteria (mainly *Dolichospermum circinale*, which formed deep cyanobacteria maxima) was observed in an oligotrophic deep Northern German lake after an extreme storm event that triggered thermocline erosion (Kasprzak et al. 2017). While the authors did not directly use sediment-derived P as explanation, the large increase in cyanobacteria biomass was attributed to high P storage within the cells, which was likely accumulated from the higher P concentration in the lower water layers exposed to the sediments.

Droughts from severely decreased precipitation are predicted for many regions by climate change. Droughts are often associated with increased cyanobacteria. For example, in polymictic Estonian Võrtsjärv, a 54-year monthly phytoplankton time series determined that decadal changes in water level and wind contributed 8% and 20% variance in cyanobacteria, while seasonal variability contributed a large 45% of the variance (Janatian et al. 2021). In particular, water level and light (photosynthetically active radiation, PAR) were negatively correlated with cyanobacteria, while temperature did not have any effect, but wind speed, after detrending with respect to year, and TP had positive effects.

The impact of droughts is especially obvious in reservoir systems where large intra- and interannual variations in water level are associated with varying blooms (Sections 3.5.3, 5.1.9, 5.7.1). For example, in 40 impounded Brazilian lakes, drought-induced water-level and discharge reduction favoured cyanobacteria blooms (Brasil et al. 2016).

A review of the effects of water-level fluctuations on cyanobacterial blooms in 13 north-temperate and subtropical lakes noticed an increase in cyanobacteria after drawdown to a decreased water level in seven of eight lakes but in only three out of five lakes after water-level rise (Bakker and Hilt 2016). The cyanobacteria increase after drawdown was attributed to increased P concentrations, partially due to internal load enhanced by increased sediment temperature and stagnant water. The potential benefit of raised lake levels in cyanobacteria management is explored in Section 6.3.

4.4.3 Thermal stratification

Increased stability in deeper lakes with changing climate could prevent sediment-derived P from reaching the mixed layer, but this possibility is affected by lake depth and may be counteracted by increased storm events. In a space-for-time substitution approach (Section 4), 230 nutrient-rich European lakes were sampled in the regionally expected warmest two-week period of the same summer (2015) (Donis et al. 2021). Overall, stratification metrics and light were better predictors for phytoplankton biomass than nutrient concentration and surface water temperature. In a shallow lake subset, N concentration was also correlated with chlorophyll, which could be explained by seasonal N limitation caused by elevated internal P load from shallow lake sediments (Box 2.2). The authors stressed the importance of including terms of stability in multi-lake surveys because they determine the prevalence of specific phytoplankton (i.e., cyanobacteria) exhibiting buoyancy control and low-light adaptation (Table 3.1).

Cyanobacteria are best equipped to reach nutrient-rich bottom waters despite potentially increased stability (Section 3.5.3). When temperature (heat) effects were separated from stability (energy) effects, the quiescent effect seemed more important in a meta-analysis (Ho and Michalak 2020). It was determined that the length of summer drives cyanobacteria abundance and not temperature in the study lakes. Climate change was probably the cause of two hot Central European summer heatwaves (Huber et al. 2012). Cyanobacteria only proliferated in the summer with prolonged quiescent and thermally stable conditions (2006) in shallow, eutrophic Müggelsee, Germany (Section 3.5.3). Also in shallow Nieuwe Meer, changes in *Microcystis* were explained mainly by temperature and stability effects related to climate (Section 3.5.3 and 4.4.1) (Jöhnk et al. 2008).

Similarly, increased quiescent conditions supported cyanobacteria which can outcompete green algae because of their ability to regulate buoyancy in ten deep European perialpine lakes despite a general decrease in TP concentration but increase in nitrate concentration, during oligotrophication (Monchamp et al. 2018). The authors attribute the increase in cyanobacteria to increased thermal stratification due to raised temperatures and nutrient fluctuations. Involved were the cyanobacteria groups, Chroococcales, Nostocales, and Oscillatoriales, which can float and sink periodically to access light (at the surface) and nutrients (below the photic zone, e.g., from internal loading) during strong and extended water stratification (Section 3.5.3).

The timing of any stability change is important and controls the distribution of the sediment-derived P and its ultimate fertilizing effect. Instead of increased productivity, climate-change-induced winter stratification under ice caused re-oligotrophication in deep Central European perialpine lakes (Yankova et al. 2017). In one case, Lake Zürich, Switzerland, warmer air temperature has been strengthening thermal stratification preventing holomixis (complete turnover) that used to occur in the fall and spring since 2013. Decreased surface and bottom water exchange meant that bottom water was not replenished with oxygen, that P-enriched hypolimnetic water (even with increased internal P load) was not distributed to the surface to support (previous) spring diatom blooms, and that the low-light adapted *Planktothrix rubescens* was favoured and prevailed at the surface in winter and spring, because of its vertical migration potential (Knapp et al. 2021).

4.4.4 *Lake browning*

Coloured lakes have specific characteristics compared to clear lakes (Section 2.6, Table 2.6) impacting nutrients, productivity, phytoplankton biomass, transparency, bacteria biomass, as well as the abundance and bloom potential of cyanobacteria (Nürnberg and Shaw 1998). These interactions lead to enhanced sensitivity regarding climate change.

Phytoplankton increase with future browning is predicted in subarctic lakes (up to 76% for a 1.7 mg/L increase in DOC) where current DOC values are low (Isles et al. 2021). In more coloured boreal lakes, increase was predicted to be less, based on a data analysis on almost 5 000 lakes between 1996 and 2018.

In particular, reactions to climate change and nutrients by cyanobacteria and chlorophyll depended on lake characteristics, such as colour, alkalinity, and mixing state in 495 central and northern European lakes (Richardson et al. 2018) (Table 4.2). Overall, temperature effects on cyanobacteria were highest in lakes at higher latitudes. In polymictic lakes, the degree of colour (natural humic acid content) determined whether only water residence time affected cyanobacteria (clear lakes), or whether temperature and TP concentration were additional determinants (humic lakes). In stratified lakes, TP concentration was generally the most important predictor of cyanobacteria, with water retention also important. Dependencies were different for chlorophyll, where temperature was less important compared to TP. This difference can be explained by the fact that many lakes had a relatively low trophic state so that chlorophyll did not solely represent cyanobacteria but included eukaryotic phytoplankton as well.

Temperature impact on colour was found to be more important than precipitation changes in a high-emission climate RCP 8.5 scenario. Temperature increase was modelled as the main driver of 25% increased colour in 2050–2079 in two Scottish lakes unaffected by acid rain (Haaland et al. 2023). Increasing air temperature combined with increasing solar radiation from decreasing cloudiness in spring and fall increased the period of elevated water temperature (>20°C) specifically in a lake with tree dieback in its Bohemian Forest catchment (Czech Republic) (Kopáček et al. 2024). Tree dieback increased DOC and TP leaching and thus was associated with brownification, algal and cyanobacterial production, and decreasing water transparency.

The TP threshold for cyanobacteria to occur was different for different levels of humic acids. Clear lakes had a far lower threshold (10 μg/L TP, involving *Planktothrix*, trait R) than coloured Finnish lakes (50 μg/L TP, involving *Microcystis*) in a study on 7 219 phytoplankton samples from 2 186 lakes (Vuorio et al. 2020), stressing the importance of close consideration of lake water characteristics and cyanobacteria taxa. Similarly, a lower TP threshold in clear (10–20 μg/L) than in humic lakes (20–30 μg/L) was determined in 588 northern European lakes (Lyche Solheim et al. 2024). Light limitation due to humic acids was identified as the likely mechanism underlying this response, although a model study determined only a moderate decrease in light adsorption by phytoplankton in response to a "typical degree of browning" (Ahonen et al. 2024).

Increased browning may also affect cyanobacteria species and their toxicity. For example, a change in species was determined in response to humic acids in a mesocosm study (Ekvall et al. 2013). Because of the high specific cyanotoxicity of a new species (*Microcystis botrys*), toxicity increased despite constant cyanobacterial biomass.

4.5 Selected biological changes in freshwater from climate change

Besides the phytoplankton community and cyanobacteria, many biological entities are affected by climate change. Particularly interesting are effects related to temperature and hypoxia increases that also drive increases in internal loading and thus amplify water quality deterioration.

4.5.1 *Ecological networks, freshwater community structure, food web, and extinction risk*

Disruptions of ecological networks as a result of climate change are projected to be abrupt and widespread in marine and terrestrial species (Trisos et al. 2020). Loss of global diversity is forecast to start in the tropics first and then to move to higher latitudes with increasing effects for increasing temperature projections. For example, for global warming "kept below 2°C, less than 2% of assemblages are projected to undergo abrupt exposure events of more than 20% of their constituent species", but scenarios for 2–4°C threaten 15% of assemblages. Similar predictions apply to freshwater ecological networks.

A review of paleolimnological records and monitoring observations revealed recent climate change as the main driver for consistent changes in diatom species, leading to smaller species (Rühland et al. 2015). Warmer climate and strengthening of thermal stratification consistently favour small-celled, more buoyant taxa that become dominant over faster settling, larger-celled diatoms. The evidence of the climatic trends has been strong at high-latitude (arctic) and alpine regions, but evidence has emerged also in temperate lakes where less extreme temperature trends and other confounding factors (exposure to more anthropological influences) may obscure the signs.

Studies on individual lakes explored climate dependencies of food webs. For example, warming lake temperatures by up to 2°C, rather than changes in trophic state or fishing effort, were identified as the cause for restructuring of the pelagic food web in a large European lake (Lake Maggiore, Italy) over 28 years (Tanentzap et al. 2020). As the lake warmed, the food web shifted toward the zooplankton, spiny water flea (*Bythotrephes longimanus*), thus limiting the abundance of their prey, the important phytoplankton grazer *Daphnia*. Warmer temperatures caused these food web changes by significantly extending the predator's periods for population growth.

Changes in zooplankton feeding pressure on phytoplankton biomass with increasing temperature have also been observed elsewhere. The timing of the clear water period, when phytoplankton is grazed by zooplankton producing a clear water phase in the spring, started earlier when water temperatures, following air temperatures, warmed earlier in the spring in shallow eutrophic Müggelsee (1979–1998, Gerten and Adrian 2000).

Extinction threatens biodiversity, as well as community structure and coherence. A synthesis of 131 studies suggests that extinction risk, which was highest in South America, Australia, and New Zealand, will accelerate with future global temperatures under current policies, threatening up to one in six species (Urban 2015). Extinction risks are increasing at an accelerating rate impacting freshwater and terrestrial habitats.

4.5.2 *Fish*

Climate change effects are especially severe where there is little resource limitation. In these cases, increasing temperatures accelerate growth and development of individual organisms ultimately impacting the abundance and composition of species (Adrian et al. 2009). On the other hand, water temperature increases and thermal stratification changes can reduce available habitat, for example, for cold-water species such as northern fish species. Globally, 73% of all documented climate effects on freshwater fish implicated temperature, with changes in flow and precipitation each contributing another 13% and 6% in an analysis of 624 peer-reviewed publications, respectively (Myers et al. 2017).

Climate-change-induced hypoxia (Section 4.3.1) decreases fish species richness. Fish species richness was related to lake area and *AF* (anoxic factor) in summer and mean depth and *AF* in winter in 52 acid and circum-neutral Ontario lakes on the Precambrian Shield (Nürnberg 1995b).

In particular, cold-water adapted species (including coregonids) were negatively affected by *AF* and disappeared from lakes with *AF* more than 32 d/summer or more than 4.4 d/winter.

Changes in water level (due to water abstraction and recent increase of evaporation) were associated with increased hypoxia and the replacement of coregonids with warm-water-adapted cyprinids (including bottom-feeding carp) that increase sediment disturbance with associated internal loading from the sediment in deep natural Lake Vegoritis in a semi-arid region of Greece (Jeppesen et al. 2015).

Climate effects on fish are complicated as they affect individual fish differently depending on the position in their life cycle. A study on 694 marine and freshwater fish determined that spawning adults and embryos consistently had narrower temperature tolerance ranges than larvae and nonreproductive adults and were most vulnerable to climate warming (Dahlke et al. 2020).

4.5.3 Invasive species

Climate change has been shown to facilitate the invasion of non-native species that expand their geographical range as water temperatures get warmer (Rahel and Olden 2008). Introductions based on the expansion of warmer habitat and changes in related limnological characteristics, as well as temperature-related increase of aquaculture, can increase invasion. Northward migration to warming areas on the Northern Hemisphere has been documented for terrestrial and aquatic plant species, as well as for insects, amphibians, fish, reptiles, crustaceans, and mammals (Rahel and Olden 2008). Climate-associated changes in the water besides warming, such as increased stratification and hypoxia, can facilitate invasions by decreasing key native predator species. On the other hand, some climate effects during species invasion can be beneficial. For example, climate warming is modelled to increase the competitive outcome of native over invasive mussel species at South African seashores (Monaco and McQuaid 2019).

A long-time comparison between water temperature and the invasive cyanobacteria *Cylindrospermopsis raciborskii* in large, shallow Lake Balaton, Hungary, explicitly demonstrates the potential effect of elevated temperature (Padisák 1998). Records indicated three years of elevated biomass (only) in years of extreme warmth with long periods of hot and calm weather, while records for 1933–1994 indicated a smooth increase in *Cylindrospermopsis* since the sixties. *Cylindrospermopsis* bloomed after about two weeks of consistent heat of 2°C or more above the long-term average. Padisák concluded that *Cylindrospermopsis*' high germination temperature optimum (21–22°C) was the reason for the blooms in the hot summers, because it is not nutrient dependent in the summer as it can fix N and can assimilate P from the sediments in the overwintering akinetes. Later, starting in the 2000s, *Cylindrospermopsis* was replaced by *Aphanizomenon* that was explained by an increased internal P loading in increasingly stratified, hot, and dry summers (Section 5.1.3) (Istvánovics et al. 2022).

Unexpected influences from introduced organisms complicate the simple increasing response of cyanobacteria to warming. Climate-change-related temperature increases that concurrently increase dreissenid mussel growth, anoxia, and sediment P release rates (Sections 3.6 and 5.1.11) can have an effect up to the acute lethal temperature threshold of 30°C, above which mussel die-off events have been reported (Bahlai et al. 2021). For example, *Microcystis*, which thrives at temperatures above 30°C, was reduced in the extreme hot year 2012 in Gull Lake, potentially a consequence of the lack of additional nutrients provided by sediment anoxia after the disappearance of the mussels.

4.6 Conclusions with respect to climate and climate change

Climate and climate change are known determinants of freshwater functioning in general and cyanobacteria proliferation in particular. Physical, chemical, and biological effects are numerous and diverse as documented in many studies (Table 4.1 and Table 4.2). As described here, much progress has been made by using climate models in combination with lake observations to determine and forecast effects on lake water quality related to climate and climate change.

Besides the direct consequence of a warming and sometimes drying world, important effects of climate change on freshwater are changes in thermal stratification and mixing status of the water bodies. In general, there is a tendency of temporal and spatial expansion of stability from enhanced temperature differences between surface and bottom water layers and lake stilling, and of an associated increase in the extent and duration of hypoxia. These physical and chemical changes directly affect the timing, extent, and biological availability of internal P loading. While much of the information is still varied, not always directly comparable (e.g., sampling different seasons, different parts of the water column), and difficult to interpret, there are indications that climate change positively influences cyanobacteria via increased internal P loading as is examined in the next chapter (Chapter 5).

Significant responses to climate change (involving climatic indices, air temperature, wind, precipitation, and storm events) are reported in worldwide studies on individual lakes and in meta-analyses of large data sets:

- Water temperature changes (usually increase but can decrease in stratified lake bottom water)
- Water flow changes in volume, continuity, and seasonality
- Changes in mixing status and increased lake stratification
- Changes of ice cover and snowpack duration and thickness
- Increased hypoxia and anoxia in response to increased lake stratification
- Redox-related chemical changes: increased Fe and Mn metals, phosphate (internal loading), gases (methane, hydrogen sulfide)
- Increase in organic acid concentration and colour (browning)
- Increase in salinization (Section 5.6.4)
- Changes in community structure, food web, number of invasive species
- Increase of short-term extreme climate events with unpredictable consequences on the freshwater environment

Much of climate variability is due to short-term, seasonal, and local changes besides expected long-term and regional changes that can be captured by climate indices, confusing the pattern in individual assessments.

Freshwater responses largely depend on lake characteristics including:

- Morphometrical and hydrological characteristics
- Thermal mixing status
- Trophic state (nutrient availability)
- Light availability (latitude, organic acids, browning, enrichment)
- Location (altitude, latitude)
- Catchment characteristics (geochemistry, land use, anthropological development)

Many of climate-change-related effects promote internal P loading (Section 4.3.2) and significantly influence cyanobacteria distribution and abundance (Section 4.4).

Chapter 5

Internal phosphorus load, cyanobacteria, and climate change

While the previous chapters describe the characteristics of nutrients, cyanobacteria, and climate change and their effects on freshwater systems separately, this section presents a combined assessment showing the synergy between these water quality determinants. Therefore, this part can be read separately, with occasional reference to the previous chapters.

Explorations of specific limnological conditions in the context of climate change clarify the dependence of cyanobacteria on internal P load throughout Section 5.1. Section 5.2 explores decreases in cyanobacteria biomass in response to internal P load abatement as further evidence for the impact of internal P loading on cyanobacteria proliferation. Modelling exercises that include the response by cyanobacteria to internal P load are reviewed in Section 5.3. Section 5.4 evaluates a reverse hypothesis which has occasionally been argued that internal P loading is enhanced by cyanobacteria. Section 5.5 lists potential challenges in finding links between internal P loading and cyanobacteria proliferation and offers suggestions to avoid such problems, for example, by using novel approaches.

Other explanations, hypotheses, and data interpretations to explain the cyanobacteria proliferation are considered in Section 5.6. Cyanobacteria and internal P load relationships in systems other than lakes, including reservoirs and large rivers, wetlands, and estuaries, are investigated in Section 5.7. Section 5.8 concludes Chapter 5 by summarizing the role of internal P loading in global cyanobacteria proliferation as influenced by climate change.

5.1 Internal P load and cyanobacteria

Many studies seek to determine why the extent and frequency of cyanobacterial proliferation are increasing in a changing climate, even in nutrient-poor lakes or lakes where external phosphorus (P) sources have been steady and nitrogen (N) is not influential (i.e., not limiting).

This section examines the impact of increased internal load on the recent proliferation of cyanobacteria while separating internal P loading effects from more direct climate change effects. The likely influence of increasing internal P load on the increase of eutrophication in general and cyanobacteria in particular has been suspected previously (Gao et al. 2014; Istvánovics et al. 2004; Jeppesen et al. 2014; LeBlanc et al. 2008; Me et al. 2018). Both climate change and increased anthropogenic activities have enhancing, and sometimes additive, effects on internal loading (Section 4.3.2), with increasing external P load leading to increases in the legacy P sediment pool that in turn eventually contributes to internal load (Section 2.4). Therefore, an increase in cyanobacteria and their blooms is not surprising, especially considering the direct positive effects of climate change on cyanobacteria abundance (Section 4.4).

DOI: 10.1201/9781003301592-5

Mechanisms and characteristics of internally derived P are much more capable of supporting cyanobacteria than (the equivalent concurrent amount of) P from external sources, thus outweighing those of external load (Figure 5.1). Chemical characteristics, timing, and source location contribute to a high availability of sediment-derived P. Increase of stability, hypoxia, and drought with climate change preferentially enhances internal P loading. Increasing internal P load is therefore more likely to lead to increasing cyanobacteria than increasing external P load.

These theoretical deliberations are supported by various case studies and model-based research. In hyper-eutrophic Clear Lake, California (area 151 km^2, mean depth 8 m), for example, TP concentration (up to 95% from internal load) was significantly and positively correlated with a satellite-derived cyanobacteria bloom metric between 2004 and 2021 ($R^2 = 0.47, p < 0.01, n = 14$) (Swann et al. 2024). Also, modelling studies determined that the reduction of internal loading is more effective than a similar reduction of catchment fluxes in reducing cyanobacteria

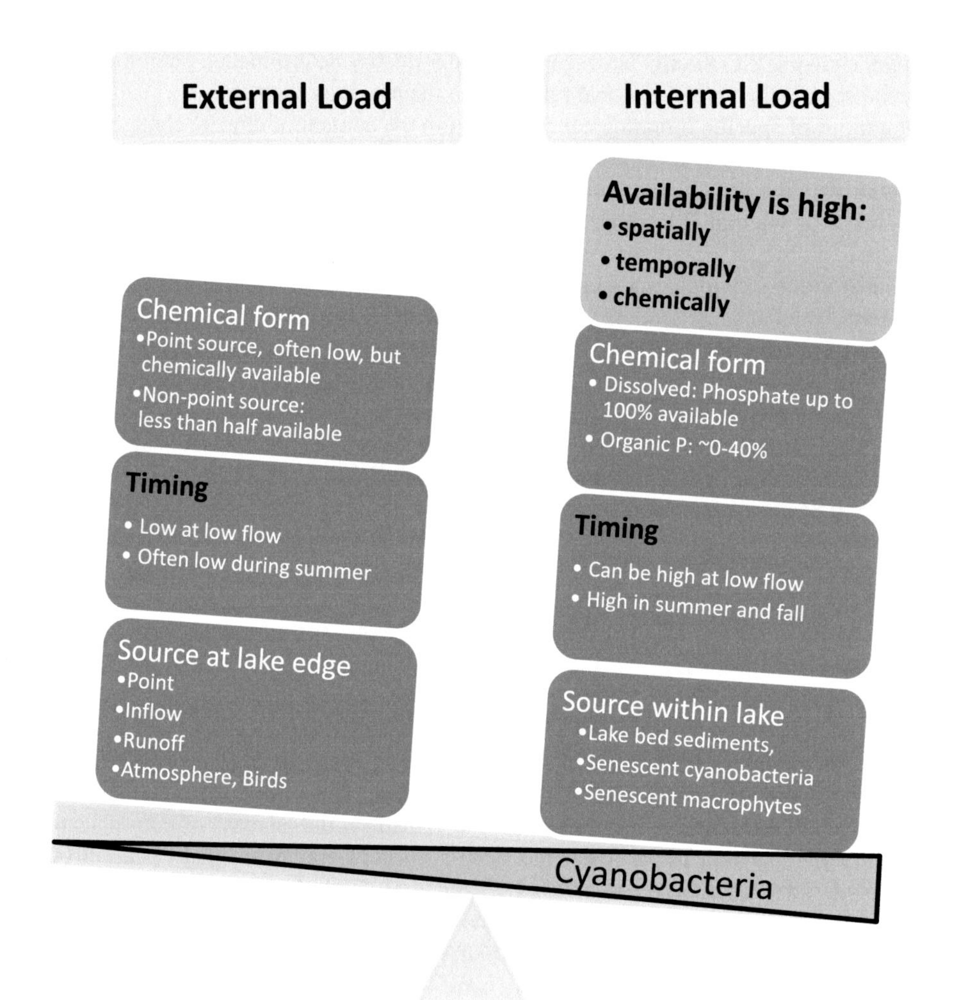

Figure 5.1 Comparison of characteristics of external and internal P sources relevant to cyanobacteria and conditions influenced by climate change (e.g., seasonality and timing).

biomass in large, eutrophic, polymictic Lake Rotorua, New Zealand (area 79 km^2, mean depth 10.8 m) (Burger et al. 2008). Another model identified internal load as the driver of cyanobacteria in a eutrophic tropical Malaysian lake (Tay et al. 2022). The model suggested internal P load abatement as the most effective restoration measure to avoid a predicted "catastrophic regime shift" following disturbed equilibrium conditions.

The effects of internal versus external P sources on phytoplankton taxa also differ seasonally (Steinman et al. 2009). For example, in a hyper-eutrophic Polish lake, external load was most important in spring, whereas internal load (which contributed over 80% to the total P load) supported the cyanobacterial blooms in summer (Kowalczewska-Madura et al. 2022). Similarly, a model exercise for a eutrophic, slow-moving German river determined that reducing external P load reduced spring blooms, besides lowering accumulated sediment P, but that internal load reduction was needed to reduce blooms in late summer and fall (Lindim et al. 2015).

Managing external P load is most beneficial when the external load is high and includes bioavailable sources like wastewater effluents and fertilizer runoff. For example, decreases in external P load from improved soil and water management (53% decrease after 1998), in addition to dreissenid mussel invasion (since 1995), seemed responsible for a concurrent decrease in cyanobacteria, internal P load (Figure 2.13C), and hypoxia in two basins of Lake Simcoe, Ontario, the deep stratified Kempenfelt Bay (Figure 5.2), and the mixed shallower Main Basin

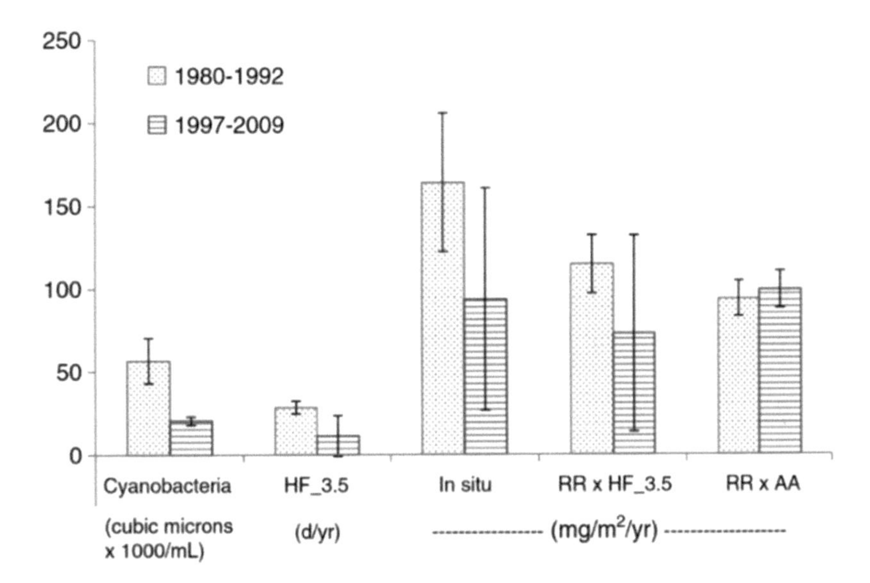

Figure 5.2 Synchronized trends in two periods, with high and low external P load, showing high and low August–October cyanobacteria biomass, hypoxia (hypoxic factor, HF_3.5, for DO threshold of 3.5 mg/L), and internal P load estimates in deep Kempenfelt Bay of mesotrophic Lake Simcoe. The three internal load estimates were (Box 2.3): in situ, from summer–fall euphotic TP increases (L_{int_1}); *RR* × *HF*_3.5, as the product of *HF*_3.5 and P release rate; and *RR* × *AA*, as the product of *AA* and P release rate (L_{int_3}). Standard errors are indicated by vertical lines for seven years (1980–1992, first bar with high external load) and five years (1997–2009, second bar with low external load and dreissenid mussel invasion). *Figure 11 from Nürnberg et al. (2013b), with permission.*

(in-situ internal load decrease of 61% coincided with similar decrease in cyanobacteria) (Nürnberg 2020c; Nürnberg et al. 2013b, 2013a). Here, internal P loading has been on a decreasing trajectory with declining hypoxia and cyanobacteria biomass, although climate change, bottom water stilling, and the fertilizing effect of dreissenid mussels could override the benefits of reduced external P input in the future.

Recent eutrophication literature has debated the relative importance of N versus P for cyanobacteria nutrient limitation and bloom formation (Section 3.3). In many situations, internal loading of P is more directly related to bloom formation, even though N limitation may persist in late summer. In hyper-eutrophic shallow Lake Dianchi, China, internal P loading (77% of total load; at least 7% of lake volume was hypoxic) maintained phytoplankton biomass (mainly cyanobacteria) more than N, despite high organic N fluxes from mineralization processes in the sediment (internal N loading, 72% of total load) (Wu et al. 2017). The authors' dynamic mass balance model for 2002–2009 determined that the limiting nutrient changed from P to N when P was released from the sediment in the summer but then changed back to P, as P settled to the sediment after the growing period (see also Box 3.2). Because the internal organic N load resulted from biomass increase caused by sediment-released P, external load reduction of P was deemed to be more effective than of N.

Generally, internal P load control of lake productivity is less pronounced when P is not limiting (under P-enriched conditions), and N inputs also control productivity and cyanobacteria. Nonetheless, any nutrient addition (often exacerbated by climate change) supports biomass. This means that increasing N, by mineralization, and increasing P, by internal load from redox-related release and mineralization, positively affect biomass, even if the specific nutrient is not limiting at that time.

Climate change may amplify internal P loading in both nutrient-rich and nutrient-poor lakes. An experimental, mesocosm study with sediments from an algae-dominated, hyper-eutrophic Chinese lake (~200 µg/L TP) and a macrophyte-dominated mesotrophic lake (23 µg/L TP) determined that "climate warming does not override eutrophication, but facilitates nutrient release from sediment and motivates eutrophic process" (Wang et al. 2023a). The effect of warming (+4.5°C) on the release of phosphate and ammonium was especially strong in the relatively nutrient-poor sediment from the mesotrophic lake, further indicating the vulnerability of low-nutrient lakes to climate warming via internal loading of P and N.

When evaluating internal P loading effects, cyanobacterial characteristics and their classification, for example, into functional groups or traits (Table 3.1) must be considered. Cyanobacteria taxa respond differently to increased internal P based on variable adaptations to access such P. For example, surface-blooming cyanobacteria respond negatively to increased mixing and low light during destratification as compared to metalimnetic bloomers and vertically migrating cells (Section 3.5.2). As explored in the following subsections, opportunities for cyanobacteria to contact sediment-released phosphate also vary with stratification status, flows, water level, and other variables that directly or indirectly affect the redox state at the sediment–water interface and the transfer of sediment P to cyanobacteria. Direct observations that internal P load enhances cyanobacteria have been reported with increasing frequency; however, the relevance of internal load impacts on cyanobacteria proliferation depends on many limnological and weather-related conditions.

Isles and Pomati (2021) defined six types of temporal phytoplankton bloom patterns (including pro- and eukaryotic blooms), most of which were at least indirectly supported by internal P load. They concluded that internal outweighed external nutrient influences in *seasonally determined*, *single-type composite*, and *multiple-type composite blooms*. Another bloom type, the *persistent single blooms*, was directly linked to hypoxia-driven sediment-released P (and ammonium from mineralization) by access via vertical migration.

5.1.1 Internal load and cyanobacteria relationships

Comparisons of cyanobacteria biomass indicators with estimates of internal P load over the growing period or over several years can serve as simple empirical tests to determine whether internal load positively affects cyanobacteria proliferation. Such correlations are most possible during conditions of P limitation which may not always occur throughout the growing period. Seasonal increases in biomass or pigment concentration may be correlated with P until N or light becomes limiting (Box 3.2). In contrast, long-term mean summer trends may indicate enhanced dependency on other variables, for example, when N external inputs increase over time, impacting cyanobacteria biomass (e.g., in Lough Neagh, Northern Ireland, the UK (McElarney et al. 2021), Section 3.3.3).

Regrettably, internal loads are not often quantified over separate years in lakes where cyanobacteria are studied simultaneously. Nonetheless, several case studies are available and confirm correlations between cyanobacteria and internal P load estimates among years. For example, in Lake Erie, 28 years of Central Basin internal load values were significantly positively correlated with average epilimnetic chlorophyll concentrations (Figure 5.3A) and cyanobacteria biomass measures (Figure 5.3B) at a site 1.2 km off the northern shore (Nürnberg et al. 2019). The extremely hot and dry year 2012 (Zhou et al. 2015) with the highest internal P load displayed the highest cyanobacteria at the northern site. In the same year, maximum levels of the cyanobacterial pigment, phycocyanin, were more than twice that of the two preceding summers at a site in the Western Basin (Gobler et al. 2016) pointing to P release also in the shallower Western basin or exchange between the basins, as discussed next.

Potential causes of increased blooms in the shallow mixed Western Basin of Lake Erie are numerous. They include spring SRP inputs from Maumee River (Kane et al. 2014), total spring TP or water load (Stumpf et al. 2012), as well as internal P load. Despite mixed conditions, potential anoxia near the bottom sediment during quiescent episodes caused sediment P release in as early as 1959–1970 (Mortimer 1987). Hypoxia-related sediment P release has also been observed more recently (Loewen et al. 2007), just as it has been assumed or observed in other parts of the Great Lakes (Tellier et al. 2021) and in many shallow mixed lakes (e.g., Section 5.1.3). A review on Western Basin internal P load estimated its contribution as a small but significant fraction to the target total TP load (6–25%) (Kaltenberg and Matisoff 2020). Similarly, a mass balance analysis estimated internal load contributions to the Western Basin as 13% of total input (Bocaniov et al. 2023).

A recent hydrological–physical study showed that the internal load of the stratified Central Basin of Lake Erie may partially fertilize cyanobacteria blooms in Western Basin in addition to its own bottom sediments (Jabbari et al. 2021). The authors found that increased wave action and eastern flow of P-rich water from the Central to the Western Basin resulted in a greater exchange between the two basins. A connection between these basins explains the significant positive correlation between the Western Basin cyanobacteria index (CI, Stumpf et al. 2012) (Figure 5.3C) or *Microcystis* biovolume (Bridgeman et al. 2013) ($n = 9$, $R^2 = 0.73$, $p < 0.05$) with the Central Basin anoxic factor (*AF*, Box 2.1) and thus with internal load. Because both, internal P loading and cyanobacteria proliferation, coincide in late summer and fall, it is more likely that cyanobacteria take advantage of increased bioavailable P than that they themselves increase internal P loading in the Central Basin, because their settling and subsequent sediment P diagenesis would create a time lag.

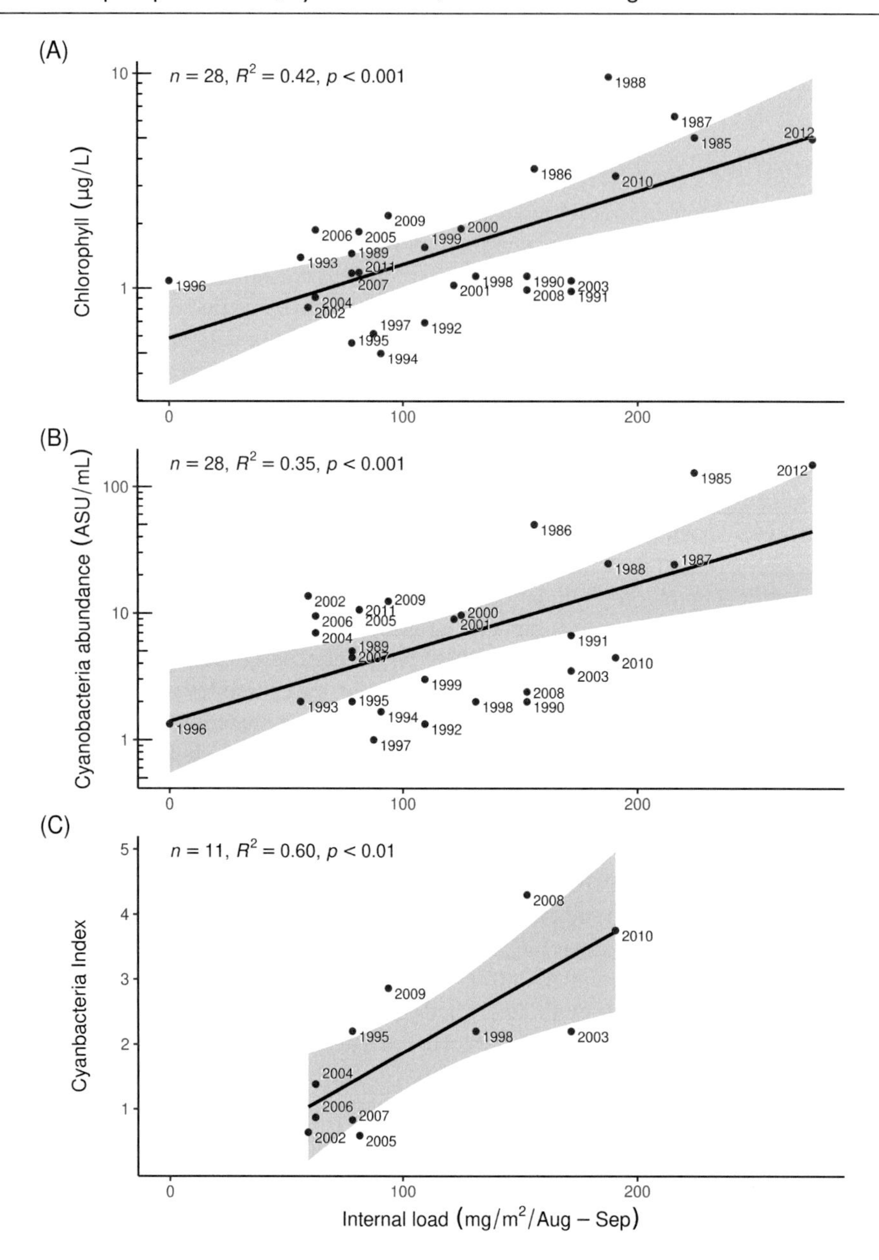

Figure 5.3 Lake Erie, Central Basin, August–September internal P load compared to mean August–October phytoplankton biomass (A, chlorophyll; B, cyanobacteria) at a northern shoreline site for 1985–2012, and (C) compared to the Western Basin Cyanobacteria Index (Stumpf et al. 2012) for 1995–2010. Internal load was computed as $RR \times AF$ (Box 2.3, with RR = 8.0 mg/m^2/d) and was similar to in-situ internal load estimates (Figure 2.13A). Cyanobacteria in B are expressed as areal standard units per millilitre (ASU/mL) with 1 ASU equivalent to an algal cell area of 400 µm^2. Lines of significant regressions and 95% confidence bands are shown. *Redrawn from Nürnberg et al. (2019) with permission and with data from Nürnberg and LaZerte (2015).*

Another large multi-basin lake with signs of occasional hypoxia and sediment P release is the large, polymictic Lake Winnipeg, Manitoba, Canada. Lake Winnipeg has been experiencing increasingly extended cyanobacteria blooms at least partially caused by internal loading (Figure 2.13A) (Nürnberg and LaZerte 2016b). In the various basins of Lake Winnipeg, July–October chlorophyll correlated with internal load estimated from TP mass balances (L_{int_2}, Equation 2.8 and Equation 2.9), with TP concentration under previous ice (Section 5.1.7) and with July–August bottom temperature. Further, the share of cyanobacteria on total phytoplankton biomass was positively correlated with in-situ internal load estimated from TP increases over the summer (L_{int-1}, Equation 2.5, Figure 5.4).

In the large shallow Lake Taihu, internal P load as sediment P release contributed 23–90% of the cyanobacteria's P demand, while the lake was N limited in the summers (2005 to 2018, Xu et al. 2021). Further, cyanobacterial density was significantly correlated with TP concentration (2005 to at least 2012, Xie et al. 2020). Cyanobacteria biomass and P release rates were especially large during the unusually warm early spring of 2017.

Sediment-released P as internal load has been repeatedly implied in the elevated eutrophication of shallow Missisquoi Bay of Lake Champlain (Giles et al. 2016; Isles et al. 2015). Here, the cyanobacteria growth phase was associated with sediment phosphorus release and the depletion of available nitrogen (see also Box 3.2). Seven years of observations (2006–2012) show chlorophyll (mainly by *Microcystis spec.*, *Dolichospermum spec.*, and *Aphanizomenon flos-aquae*) varying with TP but not TN concentrations in the water column (Isles et al. 2015). This relationship cannot be explained by elevated external TP load, thus indicating an internal source. Missisquoi Bay shows much variability among and within years, and varying seasonal

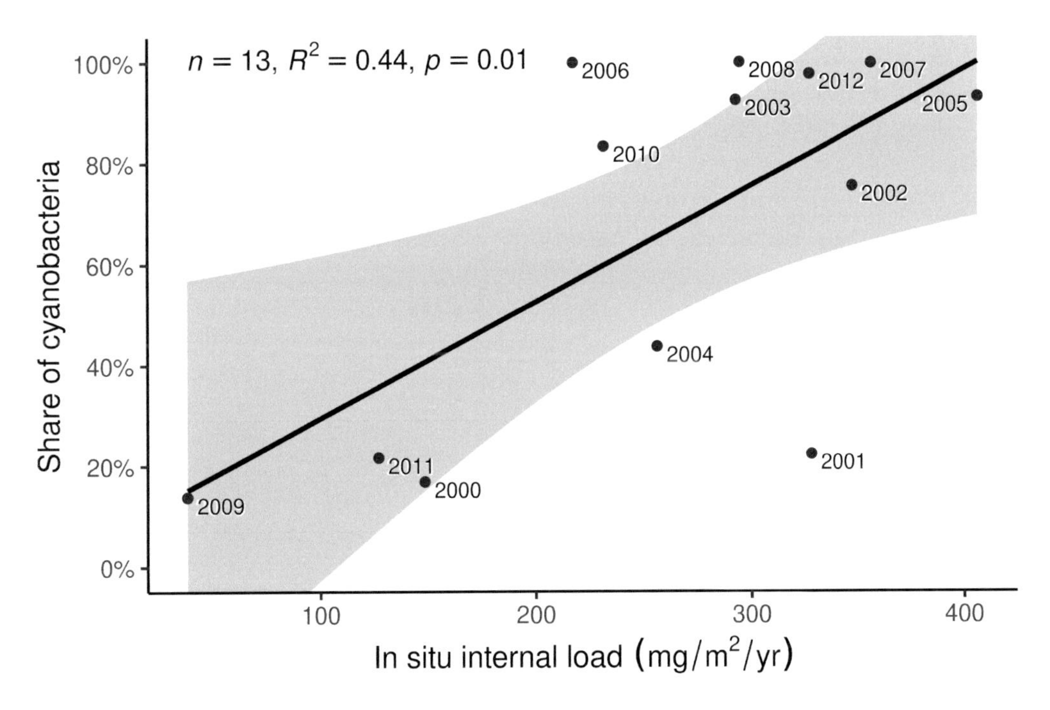

Figure 5.4 Annual average South Basin cyanobacteria (expressed as percentage of total phytoplankton biomass) compared to the whole lake in-situ L_{int_1} for 2000–2012 in Lake Winnipeg, Manitoba, Canada *(based on data from Nürnberg and LaZerte 2016b)*.

increases of TP and chlorophyll from early summer to fall indicate the varying importance of internal loading (Figure 5.5). A large internal P load (shown by increased TP concentration without concomitant external load increases) in the extremely dry year of 2012 was particularly insightful because it coincided with the highest chlorophyll concentration measured during the study years (Figure 5.5, red symbols).

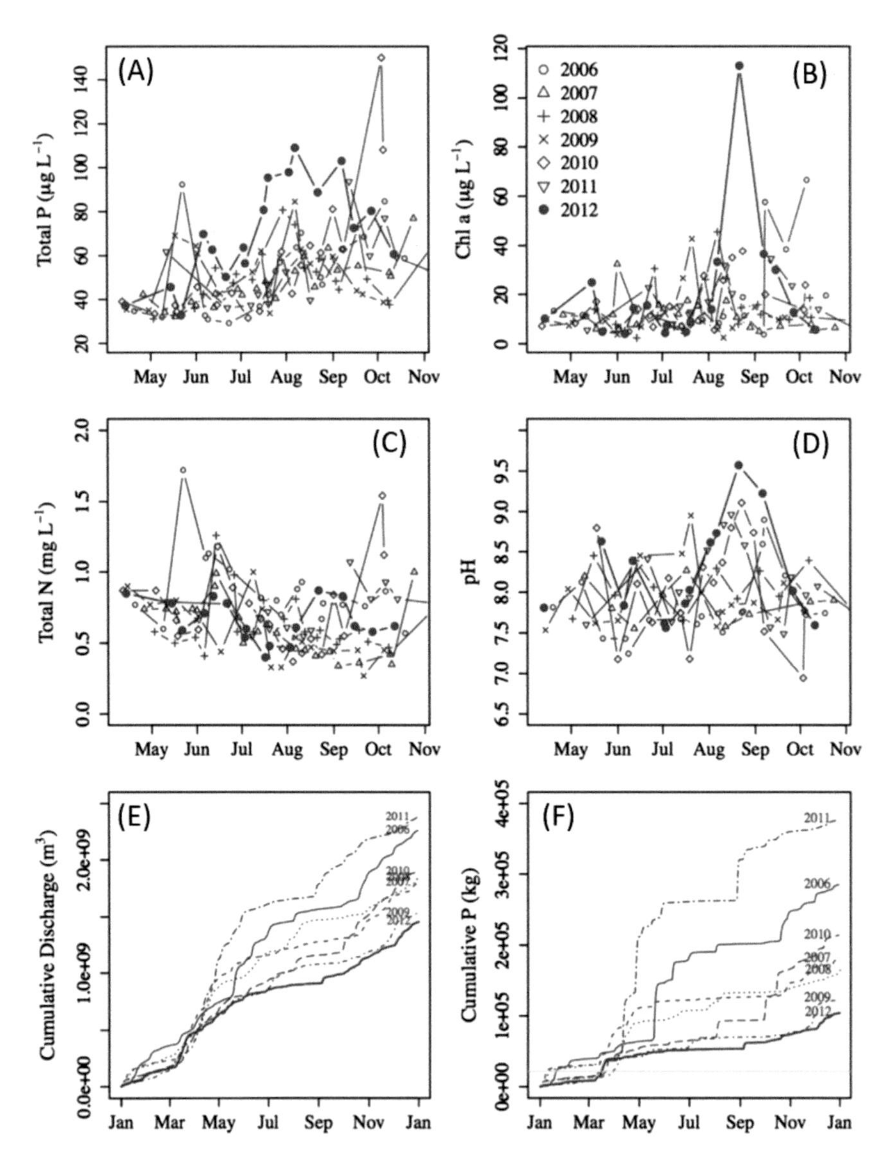

Figure 5.5 Season average TP (A), chlorophyll (B), TN (C), inflow (Cumulative Discharge, E), and inflow TP load (Cumulative P, F) in Missisquoi Bay, Lake Champlain over seven years. The drought year of 2012 is shown in red. An estimate of needed total water column TP to sustain peak bloom conditions is represented by a blue horizontal line on panel F. *Figure 2 from Isles et al. (2015), with permission.*

Based on the seasonal processes described in Section 3.3.3 and Box 3.2, it is no surprise that many studies find within-year correlations of cyanobacteria with sediment-derived P from summer to fall. In Missisquoi Bay, detailed measurements of ambient conditions in 2013 (water column stability, DO, wind speed, surface temperature, and light) help to separate and explain the bloom periods (pre-bloom, onset, main, collapse, and a late season bloom) in relation to the timing of sediment-derived fluxes of P, Fe, and Mn in the year 2013 (Figure 5.6) (Giles et al. 2016). Cyanobacterial blooms cause light limitation at high densities, triggering decreased photosynthetic oxygen production followed by hypoxia in deeper waters. The decrease of phosphate (SRP) concentration compared to Mn and Fe, late in the stratified bloom period was attributed to uptake by bloom-forming cyanobacteria. A subsequent bloom collapse during mixing initiated an SRP increase after P liberation from the senescing cells and decreased uptake. Giles et al. (2016) further suggested that increased reducing conditions from bloom productivity could lead to sediment P release, thus creating a positive feedback loop.

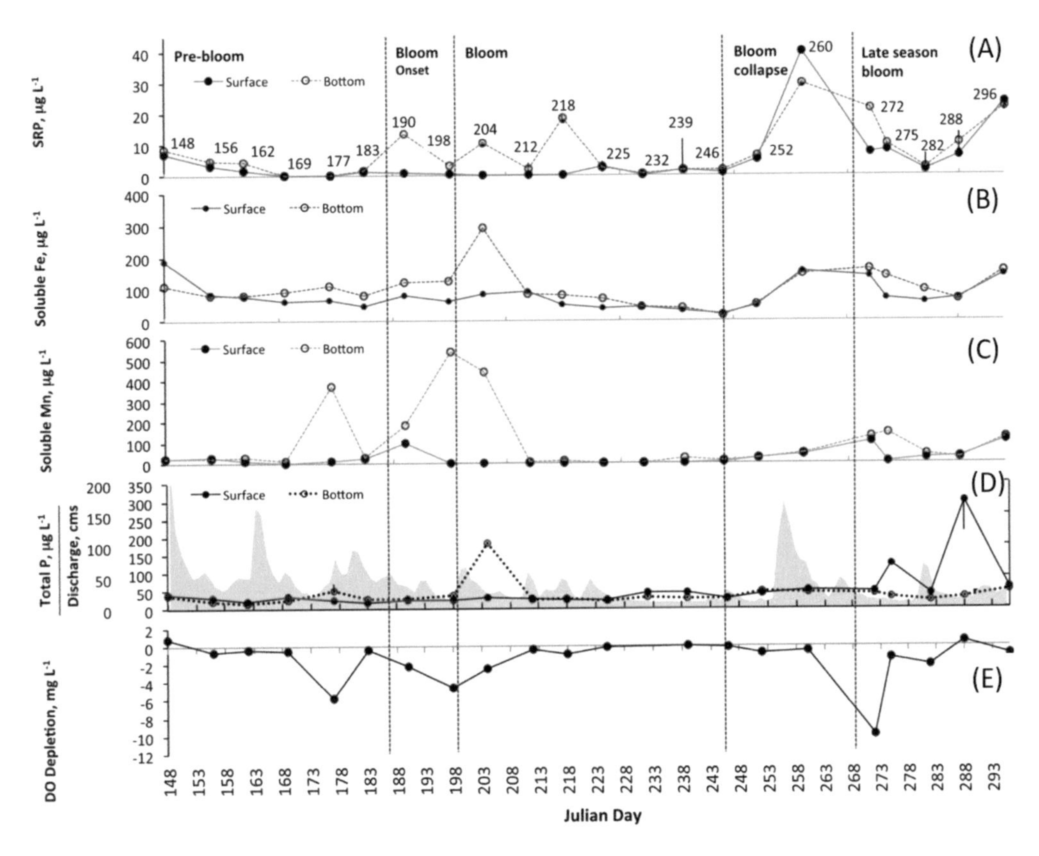

Figure 5.6 Soluble reactive phosphorus (SRP, A), soluble iron (Fe, B), soluble manganese (Mn, C), total P (D), and dissolved oxygen depletion (DO, E) in surface and bottom waters in Missisquoi Bay. The solid blue series in (D) represents Missisquoi River discharge. Solid and dashed lines (A–D) represent surface and bottom depths, respectively. DO depletion is defined as the difference between bottom and surface water concentrations. Vertical dashed lines delineate bloom periods. *Figure 2 from Giles et al. (2016), with permission.*

Internal load was significantly correlated with mean summer chlorophyll concentration in two Finnish lakes before and after an aluminum restoration treatment (Section 6.4.1) (Sarvala and Helminen 2023) (Figure 5.7). Chlorophyll was drastically reduced in the five and ten post-treatment years of the two lakes, especially of cyanobacteria.

The introduction of non-native cyanobacteria species presents additional evidence that internal P load increases cyanobacteria. For example, highly toxic *Cylindrospermopsis raciborskii* used to be classified as tropical and subtropical but has been moving north into temperate North American lakes (Hamilton et al. 2005) to attain a cosmopolitan status (Antunes et al. 2015). This range expansion likely results from climate warming (Sections 4.4.1 and 4.5.3) and increasing mobile sediment P. *C. raciborskii* sedimentary resting cells profit from increased temperature and porewater P concentration under warming conditions, and planktonic forms can quickly adapt its physiology to phosphate availability resulting in accelerated growth rates (Antunes et al. 2015).

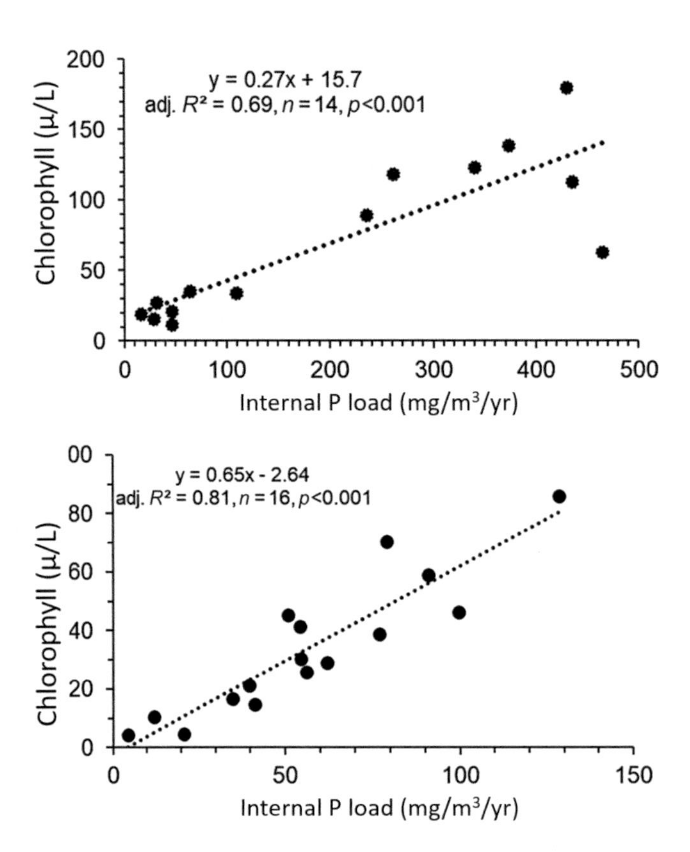

Figure 5.7 Summer in-situ internal P load compared to summer (June–September mean) chlorophyll concentration in two Finnish lakes over 14 and 16 years, including 10 and 5 years, respectively (for top and bottom display) after an aluminum restoration treatment (Section 6.4.1). Internal load was determined from TP increases over the summer (in-situ L_{int}, Box 2.3) and is presented after division by mean lake depth. *Figure 5 from Sarvala and Helminen (2023), open access.*

Positive relationships between cyanobacteria biomass and internal P load are also widely reported in reservoirs. For example, P concentration changes in the water and release rate experiments indicated that "the sediment phosphorus release was high in a lake that has annual harmful algal blooms, and is an important piece of the watershed management puzzle" in an Arkansas reservoir (Haggard et al. 2023). Seven years of hypolimnetic oxygen manipulations in another eutrophic reservoir revealed that anoxia significantly increased internal load and hypolimnetic concentrations of carbon, nitrogen, and phosphorus concentrations (Carey et al. 2022). Consequently, nutrient and DOC export increased during anoxic conditions so that the reservoir became a net source of P and organic carbon. Generally, reservoirs are especially vulnerable to climate variation, when droughts, water level, and temperature changes can affect their trophic state (Section 5.7.1).

Throughout this section, case studies show how internal loading coincides or correlates with cyanobacteria biomass. While these examples do not necessarily imply causal relationships, they suggest that internal nutrient sources are particularly important once external sources have been reduced. These case studies emphasize the relationship between increasing internal loading and increasing cyanobacteria proliferation in the context of climate change.

5.1.2 *Sediment P fractions with related processes and cyanobacteria relationships*

Only certain P fractions in the sediment are involved in internal loading. Besides the (usually small) porewater P concentration analyzed as labile, the reductant-soluble fraction (often called Fe-P or BD-P) of sediment P and the organic fraction (org-P) in enriched sediments represent the main fractions of sediment-derived P (often called mobile P, Section 2.4.2; Table 2.3). Which fractions are most important depends on lake characteristics such as productivity or trophic state and geochemistry. When sediment P fractions are correlated to cyanobacteria in multiple lake studies or in individual lakes during blooms, it suggests that internal P loading supports the cyanobacteria.

The mobile P fraction in 0–6 cm profundal sediments of seven hyper-eutrophic shallow (maximum depth 1.6–8 m) lakes in Iowa, USA, was positively correlated with the long-term mean of summer chlorophyll concentration (Figure 5.8) (Albright et al. 2022). The proportion of the three P fractions, which the authors combined as mobile P, varied between the study lakes: loosely bound P (3–22%), redox-sensitive (10–36%), and total organic P (40–75%), while total sediment P ranged from 738 to 1165 mg/g dry weight. Further analysis of the publicly available data from this study showed chlorophyll concentrations also increased with total sediment P, total organic P, and redox-sensitive P. These relationships indicate that besides the redox-related P release, much P cycling may be mediated by organic pools in these hyper-eutrophic systems. In turn, the mobile pool is replenished by the enhanced settling of phytoplankton cells, which returns P to the sediment pool.

Redox processes were involved in P release and cyanobacteria blooms of wind-mixed Missisquoi Bay of Lake Champlain (Figure 5.5, Figure 5.6), where sediment Fe-P was correlated with cyanobacteria cell counts (Smith et al. 2011). The reductant-soluble P fraction was higher, and the redox front moved further up toward the water interface in the cyanobacteria-rich year (2008) compared to the cyanobacteria-poor year (2007). Reducing conditions also increased overnight in the presence of blooms, further accelerating internal P loading.

The impact of sediment P fractions on water quality was examined in more than 100 New Zealand lakes by comparing sediment geochemical, catchment, and physiographic variables

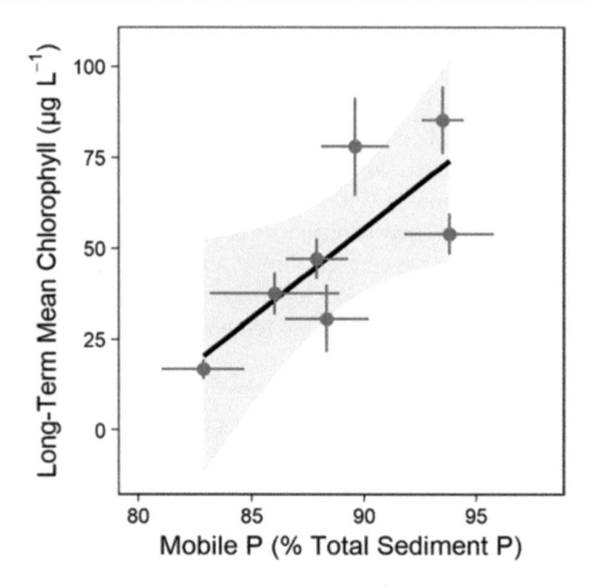

Figure 5.8 Fraction of the total sediment P pool consisting of mobile species (averaged over 0–6 cm sediment depth at the profundal site) and the long-term (2000–2018 or 2006–2018) mean chlorophyll-a concentrations in seven Iowa lakes ($p < 0.05$; $R^2 = 0.52$). The grey shaded area indicates the 95% confidence interval. *Figure 2 from Albright et al. (2022), open access.*

with trophic state indicators (Waters et al. 2023). Reducible P was the most important sediment characteristic tested in the correlation with water quality after sampling depth and lake elevation, being more important than solar radiation and calcium content of the catchment. In addition, the ratio of sediment reducible iron to total organic carbon, which partially reflects the P adsorption capacity due to iron, was negatively correlated with water quality (higher ratios indicating better water quality), while sediment TP was positively correlated. Similarly, sediment P fractions involved in redox-related P release were correlated with the frequency of cyanobacteria blooms in six shallow New Zealand lakes (Waters et al. 2021) (Figure 5.9).

Sediment cores collected from Lake Taihu demonstrated peaks in porewater concentrations of SRP and reduced Fe during a bloom (up to 150 μg/L chlorophyll of mainly *Microcystis spp.*) and over the bloom's collapse during the dark hours (Chen et al. 2018). (Photosynthesis of the phytoplankton produced oxygen that prevented the sediment surfaces from getting reduced during daylight hours.) Significant positive correlations between increasing reduced Fe and SRP up to the time of the collapse demonstrated the classic sediment P release from iron oxyhydroxides, feeding the bloom. However, during the collapse of the blooms, these correlations were not significant, with SRP still increasing, indicating that mineralization from the decomposing cyanobacteria (organic compounds) likely contributed to the P concentration in the water after settling of the cyanobacteria (discussed in Section 5.4.2), besides redox-related release from Fe compounds. Meanwhile, Fe release was less noticeable due to the formation of iron sulfide during periods of algal decomposition. Also in Lake Taihu, water chlorophyll and phycocyanin concentrations were positively correlated

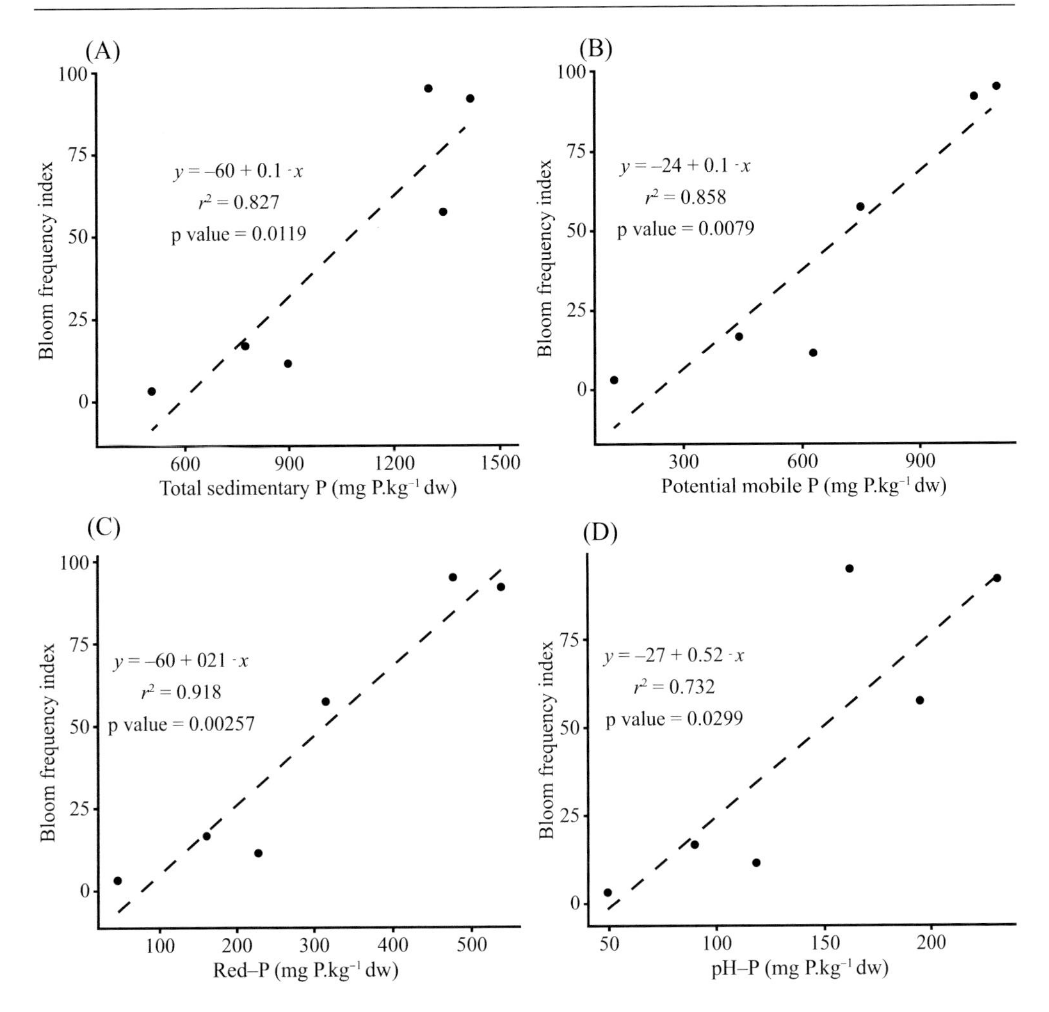

Figure 5.9 Relationships between the frequency of summer bloom events and sediment P fractions in six New Zealand lakes. Fractions according to Table 2.3 with red-P, reducible P and pH-P, Al-bound P. *Figure 6 from Waters et al. (2021), with permission.*

with sediment porewater P and P release in incubations and increased with temperature (Yin et al. 2023).

Aphanizomenon flos-aquae and *Dolichospermum flos-aquae* flourished at nutrient-rich sediment and hyper-eutrophic water sites in the east side of shallow Utah Lake, USA, in 2016 and 2017 (Randall et al. 2019). Internal P loading from reductant-soluble P compounds (50% of sediment TP) and organic sediments (~6% of TP) was proposed to support the observed cyanobacteria blooms in the enriched part of Utah Lake as legacy P.

Relationships between mobile sediment P fractions and cyanobacteria biomass or their indicators like chlorophyll and phycocyanin concentrations support the involvement of internal P loading mainly from redox-related diffusion, aided by the mineralization of organic matter under highly productive conditions.

5.1.3 Effect of mixing state on the availability of internal P load to cyanobacteria

Historically, it was argued that internal loading from sediment could not be a major nutrient source for phytoplankton in stratified lakes because the thermocline was understood as an impenetrable barrier to the diffusion of nutrients from the hypolimnion. Possible migration of epilimnetic phytoplankton was hardly considered, and sediment-released P was supposed to be taken out of the lake cycle by precipitation with Fe as iron phosphate (early textbooks, including the first edition of Wetzel (1975)). It was also argued that internal loading from sediment in shallow lakes could not occur because shallow water layers were thought to be consistently mixed by wind and other forces so that the sediment–water interface would be continuously aerated, not providing conditions necessary for redox-controlled P release. Consequently, earlier influential work did not always explicitly include internal P-loading effects on cyanobacteria (e.g., Paerl and Huisman 2009) compared to more recent work (Paerl 2018, 2016).

Contrary to these arguments, later studies have shown that thermocline erosion and internal wave action can provide hypolimnetic water to the photogenic epilimnion even during stratification in some deep lakes (Section 2.2, Figure 2.2) and that dilution prevents the formation of iron phosphate and limits the adsorption to iron oxyhydroxides (Section 2.4.2). Increasing storms with climate change are expected to trigger thermocline deepening and mixing events that can distribute hypolimnetic water enriched from internal P loading into the surface mixed layer (Carey et al. 2012), along with a lengthening of the ice-free period and temperature increases that extend the cyanobacterial growing season.

Further, episodes of dry and hot weather and quiescent periods, possibly strengthened by atmospheric stilling, can occasionally stratify shallow, usually mixed, lakes, resulting in anoxia at the sediment–water interface (Figure 2.8B and Section 4.2.3). These periods of stratification are often associated with hypoxia and cyanobacteria proliferation (Favot et al. 2024; Wilhelm and Adrian 2008).

Contributions of internal P load to cyanobacteria growth are especially apparent when blooms occur despite external nutrient restrictions (Bormans et al. 1999). Such unexplained growth was observed during episodes of weakened stratification (e.g., late summer thermocline erosion) that allow the entrainment of nutrient-rich water from the bottom layers in otherwise stratified lakes and during diurnal mixing of near-sediment –water with overlaying nutrient-poor water in shallow systems (Bormans et al. 1999).

Differing reactions to nutrients and climate-related variables (temperature and stability) of mixed, polymictic compared to stratified, dimictic lakes were determined in long-term observations of five, mostly small, eutrophic lake basins in Alberta, Canada (Taranu et al. 2012). As expected, the polymictic lakes produced a high cyanobacteria biomass related to their P content that increased throughout the summer from internal loading. In these mixed lakes, biomass was not controlled by stability (consistently low) nor by the almost constant temperature. Conversely, the two dimictic lake basins produced high cyanobacteria biomass mainly coincident with elevated water temperature and stability during the summer stratification. Here, internal loading effects were delayed until fall turnover when TP concentration increased, but low temperature and light seemed to have prevented further cyanobacteria proliferation.

Further morphometric differentiation of unstratified waters can explain complicated observations, including unexpected cyanobacteria proliferation. Intermittent mixing in very shallow lakes and discontinuous mixing in slightly deeper lakes can bring sediment-released P into the photogenic zone and into proximity with cyanobacteria at different rates, depending on fetch, sheltering, and transparency (Holgerson et al. 2022). Hypoxia from sediment oxygen demand in intermittently stratified lakes is pronounced at higher trophic states and with current or previous nutrient enrichment (legacy load) of P and organic matter. For example, in hyper-eutrophic shallow (5 m) Danish Lake Ormstrup, anoxia followed events of stratification leading to increasing phosphate, ammonia, and methane levels within 2–3 m of the sediments and inducing cyanobacteria blooms of *Dolichospermum* after dynamic and unpredictable mixing (Søndergaard et al. 2023b).

These observations can be further complicated by the fact that intermittent stratification is not always easily detected. For example, intermittent stratification was observed by field sensor and hydrodynamic modelling even in an apparently constantly mixed shallow bay. The high variability of short-term stratification between locations in large, shallow Hamilton Bay of Lake Ontario was related to basin shape and bathymetry (Flood et al. 2021). Variability was high at sites along the mildly sloping, narrow, upwind end of the basin compared to those at a similar depth at the steeper, broad, downwind end. These conditions indicated vast variability in anoxia extending over a large area at shallow sloping sites, thus enabling redox-related P release at various times and locations. The authors conclude that such conditions would easily be missed under routine monitoring of the deep location over typical sampling intervals. The authors further conclude that such conditions likely exist in other large shallow freshwater systems with variable stratification, including Lake Erie and bays around the Great Lakes in general. This study supports the probable redox-related P release in similar lakes without available observations of anoxia (e.g., Lake Balaton, Hungary; Lake Winnipeg, Canada; and the main polymictic basin of Lake Simcoe, Canada).

Another study (Missisquoi Bay of Lake Champlain, Vermont (Isles et al. 2017a)) further found that annual differences of water column stability were related to variability in bloom intensity (measured as cyanobacteria pigment concentration). While cyanobacteria biomass increased in early- to mid-July, peaked in August, and declined through early September in three study years, the extent of bloom intensity differed between years. Blooms were minimal when there was no stratification in summers with high water flow and lower air temperature and were largest when surface temperature was high and flow was low, so that periods of stratification occurred. The authors concluded that the bottom water and internal P load likely fueled the blooms, which were perhaps also larger in response to favourable growth conditions such as a warmer and brighter environment.

Internal load from shallow, unstratified areas enhances cyanobacteria proliferation already affected by deep-area P release. In Finnish and neighbouring lakes, total internal load (including sediment P release from shallow areas) was significantly correlated with chlorophyll concentration (derived mostly from cyanobacteria), but internal load just from anoxic sediments in stratified lakes showed no relationship (Tammeorg et al. 2017). This was explained by the relatively short growing period but long stratification in these northern lakes. These factors delay entrainment into the photogenic epilimnion and therefore reduce the influence of hypolimnetic P on summer epilimnetic phytoplankton. Further, these results indicate the probable involvement of shallow areas overlain by mixed, oxygen-rich water in P release, especially in productive lakes (Figure 2.8).

Very deep lakes (those with a high morphometric ratio, Equation 2.3) with increasing periods of more pronounced stratification due to climate change could experience a decrease or delay in nutrient diffusion from the hypolimnion. For example, nutrient-poor, deep dimictic lakes on the Precambrian Shield may not turn over until late fall (Figure 3.9), depending on their stability as indicated by summer hypolimnetic temperature, mean depth, and latitude (Nürnberg 1988a).

This would delay P enrichment to a period of less optimal growing conditions for cyanobacteria (less available light and lower temperature).

Increasing stability has also been described in perialpine lakes, leading to more frequent events of vernal meromixis. In eutrophic, anoxic Lake Lugano, internal P loading has been occurring in both basins, the deep meromictic North Basin and the holomictic South Basin (Lepori et al. 2018). The TP of the South Basin's mixed layer consistently declined with external load reduction (from 120 to 35 μg/L, 1983–2014, despite consistent internal load contribution) but remained higher than that of the North Basin (variable 30–100 μg/L, because of inconsistent internal load contribution). The North Basin TP concentration was dependent on the (spring) mixing depth that produced variable intrusion from the accumulated internal load of its monimolimnion. In years of shallow mixing, when only the upper mixolimnion circulates, external load controlled the TP concentration in the mixed layer. Conversely, in years with deeper mixing, contributions from internal load were substantial but decoupled from external load because they had accumulated over decades (Lepori et al. 2018). These variable TP concentrations from years with different deep mixings (and thus different contributions of internal load) were correlated with variable primary productivity in the North Basin (Lepori and Capelli 2021). Because the decrease in external TP load was followed by a consistent decrease in mixed-layer TP concentration in the holomictic and shallower South Basin, primary productivity also decreased consistently (Lepori and Capelli 2021).

Increased lake stability can also support migrating low-light adapted cyanobacteria (trait R, Table 3.1) by creating unfavourable conditions for other phytoplankton, especially diatoms, which prefer mixed conditions in the spring. Climate-change-related lengthening of the spring stratified period induced vernal meromixis in several deep (>200 m) alpine and perialpine lakes, as described next. Meromixis appears to favour migrating cyanobacteria despite decreased epilimnetic productivity leading to oligotrophication by interrupting internal P load distribution.

For example, in monomictic, mesotrophic Lake Zürich (Section 4.4.3), increased stratification due to climate warming reduced and finally prevented vernal mixing, which interrupted the annual fertilization of the epilimnion with upwelling hypolimnetic phosphate and the mixing and aeration of the hypoxic hypolimnion by down-welling of aerated mixed layers (Yankova et al. 2017). The low-nutrient conditions in the spring prevented spring diatom blooms and decreased productivity leading to re-oligotrophication, even though the hypolimnetic phosphate mass was large enough to sustain spring diatom blooms if mixed into the surface layers (Yankova et al. 2017). Meanwhile, *Planktothrix rubescens* (Section 3.5.3) prevailed. It had been routinely observed in Lake Zürich long before the increase in stratification (since 1975), taking advantage of internal-load-derived hypolimnetic P accumulation while able to withstand high water pressure, migrate vertically, and thrive under poor light conditions. Decreasing light competition in the spring (caused by reduced diatom blooms during limited periods of hypolimnetic P intrusion and oligotrophication) and increasing availability of nitrates from external inputs favoured metalimnetic *P. rubescens* that then is dispensed throughout the mixed layer during fall–winter mixing. It appeared that increased stratification over the last four decades led to changes in ecosystem dynamics which favoured *P. rubescens* despite generally there being less P in the surface water, because it had continued access to hypolimnetic P, while there were fewer deep mixing events that could destroy its vacuoles needed for migration (Knapp et al. 2021).

The climate-change-related trend to increased stability leads to an increase in the duration of anoxia and redox-dependent internal P load in all lakes. The resulting impact of changes in stability on cyanobacteria depends on the morphometric characteristics of the lake. For example, in facultative meromictic lakes (e.g., deep perialpine lakes), increased stability delays nutrient mixing into the surface mixed layer (epilimnion) possibly delaying or hindering cyanobacteria surface blooms but supporting migrating low-light-adapted cyanobacteria. Conversely, in most other lakes including reservoirs and shallow lakes, increasingly frequent episodes of stratification followed by destabilization provide access to the nutrients released from the oxygen-depleted sediment–water interface.

5.1.4 Hypoxia and cyanobacteria relationships

The effects of increasing sediment and hypolimnetic anoxia on cyanobacteria related to climate change (Molot et al. 2021b) are partially triggered by internal P load. This is because much of internal load consists of redox-related P release from lake bed sediments (Section 2.4.2). This type of internal load can be modelled as the product of the spread of anoxia in time and space (quantified as anoxic factor, *AF*, Equation 2.2, or predicted as active sediment release area, *AA*, Equation 2.3) and the P release rate per active area ($L_{int\_3} = AF \times RR$, Equation 2.10). Anoxia in turn greatly depends on lake-mixing state or stability (e.g., as expressed by the morphometric ratio, Equation 2.1), which affects its manifestation differently in stratified and polymictic lakes (Figure 2.8) and significantly improves anoxia prediction.

Evidence of connected stratification, hypoxia, and cyanobacteria proliferation has been described even in polymictic lakes more recently (Section 5.1.3). For example, in Lake Müggelsee, increasing frequency and duration of stratification events have led to bottom hypoxia with pulses of internal-P-load-promoting cyanobacterial growth despite constant external nutrient load (Wilhelm and Adrian 2008). Similarly, bottom water anoxia (*AF*) coincided with various cyanobacterial measures in the stratified Central Basin and the mixed Western Basin of Lake Erie (shown for L_{int_3} with a constant *RR* in Figure 5.3 indicating similar relationships with *AF*). As is observed frequently in relatively shallow but stratified lakes, decreasing hypolimnetic oxygen concentration was associated with increasing hypolimnetic TP concentration, two measures of internal load (L_{int-1} calculated from in-situ summer increases, Equation 2.5, and $L_{int\_3} = AF \times RR$, Equation 2.10), decreasing Secchi transparency, and elevated chlorophyll in mesotrophic Lake Nairne, Charlevoix, Quebec (Labrecque et al. 2012).

Modelling hypoxia in shallow, polymictic, hardwater Lake Balaton, Hungary, suggested an increase in redox-related internal P loading. Here, blooms (*Dolichspermium sp.* and *Aphanizomenon flos-aquae* and the non-cyanobacteria, *Ceratium furcoides*) developed in a recent year (2019, >300 µg/L chlorophyll compared to previous maxima of <80 µg/L) despite a decreased external P load that was insufficient to support the blooms (Istvánovics et al. 2022). Using an ocean-based model (General Ocean Turbulence Model) to determine low DO concentrations in deep water, the authors demonstrated that redox-related P release was the most likely cause of the bloom, likely triggered by climate change-induced stratification and warming. (These changes may also have caused a shift in phytoplankton composition in 2013 from *Cylindrospermopsis raciborskii*, Section 4.5.3, to *Aphanizomenon*.) Therefore, management actions specific to internal P loading (Section 6.2) were proposed.

Positive relationships between annual hypoxia and cyanobacterial biomass were also observed in several large lakes, similar to relationships described for internal P load (Section 5.1.1). For example, decreased anoxia (anoxic factor) coincided with decreased cyanobacteria biomass in the deep basin of Lake Simcoe (Figure 5.2). However, using observed hypoxia to predict internal load in mixed conditions is problematic. These complexities were examined in the deep versus shallow mixed basins of Lake Simcoe (Nürnberg 2020c). The active sediment release area in the shallow basin was best modelled (Equation 2.3) rather than estimated from oxygen concentration in the water column, because predicting internal load using the hypoxic factor based on monitored DO underestimated observed (in-situ) internal load. Underestimation can be expected under mixed conditions because water layers are aerated while sediment surfaces can be anoxic (Flood et al. 2021) (Figure 2.8B), partially explaining Hupfer and Lewandowski's (2008) claim that any correlation between oxygen depletion and P release can be misleading.

The cyanobacteria proliferation in Lake of the Woods (7–46 μg/L TP) detected by paleolimnological analysis of sediment since the early 1980s (Pilon et al. 2019) may be caused by sediment-derived P, because up to 36% of the total load was determined as internal while external P load has been decreasing (Alam et al. 2020). P release rates were highly correlated with temperature, and most were redox-dependent with anoxic release rates up to ten times greater than oxic rates (James 2017b). Variable water column stability and temperature with intermittent oxygen depletion at the sediment–water interface explained some of the cyanobacteria variation in the various basins (Binding et al. 2023).

In another example, the predicted anoxic factor (Equation 2.3) in the relatively shallow (mean depth 2.5 m) basin of Lake Peipsi (Lämmijärv) was significantly and positively correlated with cyanobacteria biomass over several decades (Figure 5.10). In addition, water TP and SRP concentrations as well as cyanobacteria biomass were significantly correlated to internal load and less significantly to external load in three basins of Lake Peipsi (Tammeorg et al. 2016).

Besides planktonic cyanobacteria, benthic cyanobacteria are proposed to be increasing due to warming-induced decreases in sediment redox potentials, even in relatively remote high-elevation lakes (Hampton et al. 2023).

In conclusion, hypoxia enhances the internal loading effect on planktonic (and benthic) cyanobacteria proliferation and is itself considerably affected by the mixing status of specific water systems. Much of the unexplained variation in the relationship between internal load and cyanobacteria, beyond that from direct nutrient enrichment, can be attributed to the difference

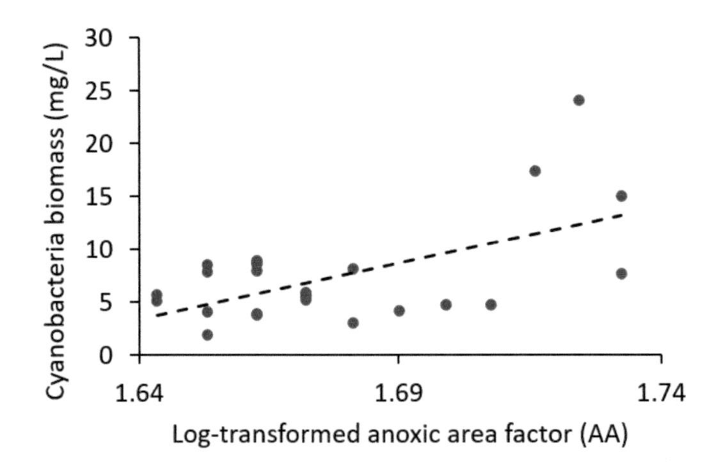

Figure 5.10 Cyanobacteria biomass correlation with anoxia, expressed as the logarithmically transformed anoxic area factor in the Lämmijärv basin (maximum depth 15.3 m) of Estonian/Russian Lake Peipsi, 1997–2021. *Figure 8B from Tammeorg et al. (2024b), open access.*

in the P exchange between sediments and upper waters caused by dissimilar lake shape and stability, which affect the manifestation of hypoxia in stratified and polymictic lakes (Figure 2.8).

5.1.5 Interaction between phosphorus and nitrogen release from sediment organic matter

Nitrogen plays an important role with respect to cyanobacteria proliferation (e.g., Sections 2.5 and 3.3). Under P-replete conditions, when internal P loading provides ample bioavailable P, any additional N will further support the growth of cyanobacteria. When organic matter accumulates in the sediments in highly eutrophic systems and is mineralized (Xu et al. 2021; Yindong et al. 2021), highly bioavailable ammonium can increase as internal N loading (Wang et al. 2008) supporting cyanobacteria, which prefer NH_4 to NO_x (Andersen et al. 2020; Hoffman et al. 2022). At the same time, phosphate is mineralized from organic P compounds in enriched sediments, perhaps from settled cyanobacteria cells (Section 5.4.2), in addition to redox-dependent release (Section 2.4.2).

Changes between bioavailable nutrient forms can lead to changes in cyanobacteria communities. For example, in hyper-eutrophic Lake Chaohu, China (Zhang et al. 2020), cyanobacteria changed from *Dolichospermum* to *Microcystis* as a result of these complex interactions between P and N sediment release.

> In spring, P limitation and N selective assimilation of *Dolichospermum* facilitated nitrate accumulation in surface water, which provided enough N source for the initiation of *Microcystis* bloom. In summer, the accumulated organic N in *Dolichospermum* cells during its bloom was re-mineralized as ammonium to replenish N source for the sustainable development of *Microcystis* bloom. Furthermore, SRP continuous release led to the replacement of *Dolichospermum* by *Microcystis* with the advantage of P quick utilization, transport and storage. Taken together, the succession from Dolichospermum to *Microcystis* was due to both the different forms of N and P in water column mediated by different regeneration and transformation pathways as well as release potential, and algal N and P utilization strategies.

5.1.6 Synergistic effects of temperature and internal load on cyanobacteria

Cases of increasing cyanobacteria biomass with climate warming can often be ultimately explained by climate-induced increases in sediment-derived P as internal loading (Section 4.3.2), even where internal load is not explicitly described (Section 4.4.1). However, in addition to these examples, there are more recent studies which explicitly include internal load.

In that context, a meta-analysis of over 1 260 US lakes showed that the synergistic cyanobacteria-enhancing effect of increasing temperature and P was explained by internal P loading, so that Ho and Michalak (2020) concluded (Section 4.4):

> The linkage between temperature and nutrient concentration . . . implies that the effect of temperature on nutrient availability also needs to be taken into consideration. As the importance of internal phosphorus loading is likely to increase with rising temperatures, this moreover indicates a need to understand the role of internal phosphorus loading in lake nutrient budgets.

Synergistic effects of temperature and nutrients were also investigated by mesocosm studies with water from 39 mesotrophic to hyper-eutrophic Dutch lakes (Lürling et al. 2018) (Section 3.5.1). A nutrient pulse of N and P combined (to yield nutrient saturation) increased cyanobacterial chlorophyll by 900%, compared to an only 18% increase with rising water temperatures (20–25°C) alone. Similarly, nutrients were more important than temperature increases in increasing cyanobacteria, chlorophyll, and microcystins in other experiments using urban Dutch lake water (Lürling et al. 2017). These results suggest that under natural conditions, internal load increases from rising lake temperature are more important than direct temperature effects on cyanobacteria.

In a recent study on shallow lake nutrient cycles, Jeppesen et al. (2020) explored the role of internal P loading on increasing cyanobacteria during climate warming. They projected future cyanobacteria increases in northern Danish lakes from observations of southern Floridian lakes in a space-for-time approach. Data from 1 656 shallow lakes (mean depth <3 m) in north-temperate Denmark and subtropical Florida (FL) were compared to identify responses to increasing temperatures and associated climate variables in these regions of different latitudes and warmth. The differences included the seasonality of internal P loading, where higher temperatures enabled sediment P release during the winter in the southern lakes, while P release mainly prevailed in the summer in the northern lakes. Similarly, oxygen depletion was higher in the southern lakes, and hypoxia prevailed at the sediment–water interface when FL lakes occasionally stratified despite their shallowness. Overall, the summer phytoplankton biovolume and proportion of cyanobacteria were higher in the FL lakes even when compared for similar TP concentrations, indicating the growth-enhancing effect of warmer temperatures and higher irradiation. Nitrogen did not seem to have a direct effect on biomass, as total nitrogen concentration was similar in the summer but far higher in the winter in the northern lakes.

Finally, in all temperature-dependent phytoplankton responses, autecological considerations (e.g., species-dependent temperature optima) as well as the effect of acclimatization (Layden et al. 2022) have to be considered (Section 3.5.1). For example, although high air temperature facilitated the invasion of *Cylindrospermopsis raciborskii* in large, shallow Lake Balaton, Hungary, it was replaced by *Aphanizomenon* due to increased internal P loading from increasingly stratified, hot, and dry summers (Section 4.5.3) (Istvánovics et al. 2022).

5.1.7 Winter and cold temperatures

Despite the generally acknowledged and observed preference of cyanobacteria for higher temperatures (Section 3.5.1), blooms have been observed in north-temperate lakes below 15°C. A literature review on cold-water blooms determined three modes of development for these blooms (Reinl et al. 2023): (1) surface cyanobacterial blooms initiated in cold water temperatures; (2) cold-water cyanobacterial blooms originating from the metalimnion, and (3) blooms initiated in warmer water temperatures and prevailing into colder temperatures.

All modes indicate the potential for blooms to be supported by internal P loading. According to Reinl et al. (2023), cold-water surface blooms (mode 1) occur during upwelling and mixing events that distribute hypolimnetic nutrients across the water column. These events occur typically in the fall when thermocline erosion and fall turnover mix nutrient-rich, cold hypolimnetic layers into surface water (involved species include *Aphanizomenon flos-aquae)*. Metalimnetic originating blooms (mode 2) originate in the P-enriched hypolimnion and the metalimnion (e.g., species often involve *Planktothrix*, trait R, Table 3.1, but also *Dolichospermum* and *Aphanizomenon*). Blooms left over from warm water periods (mode 3) would have had similar potential support of summer and fall internal load as blooms of mode 1.

In addition to these modes of developing cold-water blooms, warming winters have occurred (Section 4.2.2) and been blamed for increased blooms in the subsequent seasons. As warmer winter temperatures are expected to increase oxygen depletion (especially under ice) and sediment P release rates (Figure 2.12), winter internal load can be associated with these blooms. This effect may be offset by shorter duration of ice cover.

For example, Lake Taihu winter–spring temperature has increased since 2011 (during 2007–2017) with an extremely warm winter in 2016 followed by a bloom in 2017 (Qin et al. 2019). Similarly, the annual variability of Lake Winnipeg TP under ice was correlated with subsequent July–October chlorophyll concentration (Nürnberg and LaZerte 2016b) (Figure 5.11). It is conceivable that the large (0.044–0.100 mg/L) fluctuation in mean winter TP concentration was partially caused by internal P load, because when the mean winter (January–March) air temperature was highest (in 2006), winter TP concentration was also high, possibly due to anoxia under ice. Such an event was recorded in shallow Finnish Littoistenjärvi in 1 out of 22 winters (1999 during 1992–2023), when anoxia under ice produced an extremely high TP concentration (Sarvala and Helminen 2023).

In contrast, earlier spring runoff related to climate change appears to have a negative effect on summer phytoplankton, possibly influenced by the amount and duration of external nutrient input. A large collaborative study on 41 lakes in Europe and North America found that earlier spring runoff, associated with earlier snowmelt, corresponded with lower chlorophyll concentration in the subsequent summer (Hrycik et al. 2021).

The seemingly perplexing increase of cyanobacteria when lakes are cold can be explained in many cases by winter internal P loading, sometimes under ice. Further explorations during this under-studied season will help to clarify the most important causal relationships. The

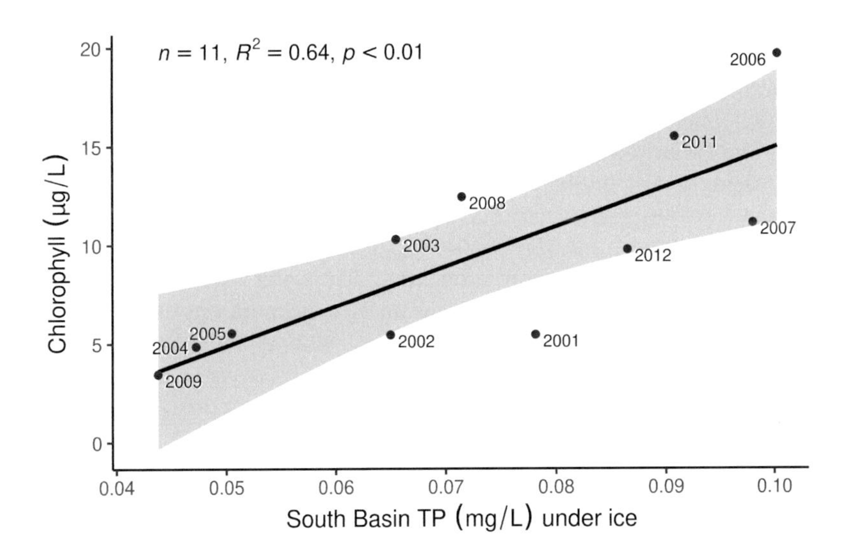

Figure 5.11 Lake Winnipeg mean South Basin July–October chlorophyll concentration versus TP concentration under ice in the previous winter (January–April). *Redrawn from Figure 8 of Nürnberg and LaZerte (2016b), with permission.*

following section (Section 5.1.8) includes further explorations with respect to colder temperature by including lakes in cold-temperate and arctic regions.

5.1.8 Nutrient-poor, formerly acid, and coloured lakes

Historically, increased cyanobacteria and anoxia have been taken as an indication of advanced productivity and trophic state (Chorus et al. 2021; Pick 2016). However, anoxia does not always indicate eutrophic conditions, because it is widespread in oligo- and mesotrophic lakes that are coloured by humic and fulvic acids or lakes that are deep but have small areas (large morphometric ratio, Equation 2.1) (Nürnberg 2019, 1997). When little exposed to anthropogenic influences, the low content of mobile sediment P including the reductant-soluble P fraction limits internal loading (Nürnberg 1988b; Nürnberg et al. 1986) (Figure 2.1, Figure 2.12D). Typically, these lakes have no obvious cyanobacteria proliferation. However, in times and regions of anthropogenic exposure, cyanobacteria blooms have become increasingly frequent. They are partially fuelled by increased P fluxes from the sediments as internal P loading related to legacy external P inputs enhanced by climate change, de-acidification, and lake browning that increases hypoxia.

A detailed review of almost 50 oligotrophic global lakes (TP mostly <10 µg/L) examined the potential causes of increasing cyanobacterial bloom frequency in such lakes (Reinl et al. 2021). Some explanations included the characteristics of buoyancy control, migration, and competitive nutrient uptake in combination with enhanced upwelling of hypolimnetic P-enriched waters due to climate change as well as the intrusion of typically metalimnetic cyanobacteria layers to form surface blooms.

In 29 alpine lakes in the Cascade Mountains, Oregon, USA, hypoxia contributed most to the variation of cyanobacteria in 2019, mainly *Dolichospermum*, despite low nutrient concentrations (Jansen et al. 2024). Internal P loading was considered one of the most likely causes, especially since the TP concentration was on average 60% bioavailable SRP, which is high for low-nutrient lakes without point sources.

Cyanobacteria blooms in previously unfamiliar situations include those in nutrient-poor, mostly oligotrophic, softwater lakes on the Precambrian Shield in Ontario, Canada (Pick 2016). Since phytoplankton composition was monitored in response to public reports starting in 1994 by the Ontario Ministry of the Environment, Conservation and Parks, blooms have been increasingly reported by the public (Favot et al. 2023; Winter et al. 2011). *Dolichospermum* and (less so) *Aphanizomenon* were the most common taxa in waterbodies on the Precambrian Shield, whereas *Microcystis* was most common in the Mixedwood Plains Ecozone in southern Ontario with more productive, hardwater lakes. The waterbodies with confirmed cyanobacteria blooms had a mean spring TP concentrations of 13.0 µg/L ($n = 135$) compared to lakes with no confirmed cyanobacterial blooms and a mean TP of 9.8 µg/L ($n = 918$) (Favot et al. 2023). The small TP concentration difference may indicate internal P loading that has been reported previously in such lakes (Nürnberg et al. 1986; Nürnberg and LaZerte 2004).

A paleolimnological investigation linked increased anoxia and stability with a *Dolichospermum* bloom in a remote, oligotrophic lake in Algonquin Provincial Park on Ontario's Precambrian Shield (Favot et al. 2019). Inferred declining oxygen concentrations coincided with the highest levels of sedimentary chlorophyll *a* and cyanobacterial akinetes. Both were significantly correlated with climate-change-related variables of warming air temperature, atmospheric stilling, and lengthening of the ice-free period (by two weeks). The rapid onset of stratification after a late ice-out in the bloom year 2014 was considered to be caused by increased stability in the spring that prevented mixing (vernal meromixis). These stagnant conditions triggered summer hypolimnetic anoxia and internal P load thus providing the nutrients for cyanobacteria proliferation. Because the lake is remote and generally unaffected by anthropogenic interference, external P loads could unlikely have provided the required nutrients for the cyanobacteria bloom years (including 2014 and 2015). Therefore, climate change was concluded to be the original cause (Section 4.4).

Deep oligotrophic Lake Manitou on an island in Lake Huron, Ontario, Canada, supports cold-adapted lake trout (*Salvelinus namaycush*). A paleolimnological analysis of the sediments determined that the lake has become increasingly more productive and hypoxic, threatening lake trout habitat (Nelligan et al. 2019a). In addition to brownification and changes in catchment usage, the seasonal pattern of observed epilimnetic TP increase in the fall (typical for internal load, Section 2.4.2), as well as the paleolimnological records of increased chlorophyll, hypoxia, and thermal stability, suggests sediment P release to support the eutrophication trend.

Even in the large oligotrophic Lake Superior, sediment-derived P has been suggested as potentially fuelling cyanobacteria in regions with occasional blooms. Two years with surface cyanobacteria blooms in the nearshore areas of the western arm coincided with high precipitation-related river input in one year but vertical movement of nutrients from the sediment to a P-starved population of *Dolichospermum* in another year (Sterner et al. 2020).

Similarly, in the deep oligotrophic northern lake, Great Slave Lake, Northwest Territories, Canada, blooms (*Dolichospermum lemmermannii*) were reported in late summer during warm stratified conditions (2013–2015). These blooms were observed in Yellowknife Bay, which is influenced by the City of Yellowknife, only founded in 1934 (Alambo 2016; Pick 2016). Historically, *D. lemmermannii* has been identified in the Lake's water but never gathered in surface blooms.

In the deep, formerly oligotrophic Lake Stechlin, Brandenburg, Germany, TP concentration increased over a decade from below 5 μg/L to above 40 μg/L accompanied by decreasing Secchi transparency (from above 10 m to below 5 m) and increasing biomass and cyanobacteria blooms not present previously (*Aphanizomenon flos-aquae*, *Dolichospermum spec.*, *Planktothrix rubescens* (Kröger et al. 2023)). The authors concluded that these patterns were likely caused by increasing stability and positive feedback between cyanobacteria biomass and internal P load.

Periods of acidification of lakes and watersheds from high sulfur emissions caused by mining operations prevented internal loading as P release from bottom sediments. Before recuperation from acidity, oligotrophication was observed in two Swedish lakes during the period of industrial acidification (1950–2001), where a substantial portion (up to 76%) of the total external P load was converted to P adsorbed by aluminum (NaOH-P, Table 2.3) and buried in the sediment (Huser and Rydin 2005).

Other formerly oligotrophic and ultra-oligotrophic lakes recovering from acid deposition show signs of increasing productivity. These trends are associated with cyanobacteria blooms and reduced P retention related to increasing internal P load caused by geochemical changes related to increased alkalinity in the lake and its catchment (Kopáček et al. 2015; Nürnberg et al. 2018) (Box 5.1, Figure 5.12). Despite stable atmospheric P inputs, long-term studies in central Europe (Czech Republic) similarly determined increasing P and chlorophyll concentrations in alpine lakes recovering from atmospheric acidification after increased input from the catchment basin and decreased P retention (Kopáček et al. 2015).

Box 5.1 The recovery of acid lakes leading to eutrophication

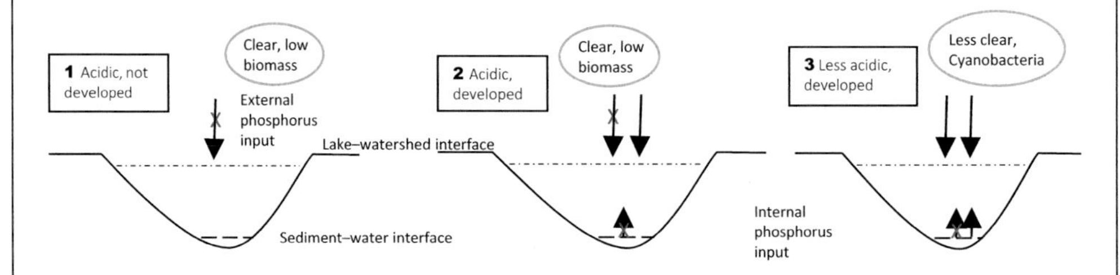

Figure 5.12 Hypothetical phases during the recovery of mining-induced acid lakes leading to eutrophication and cyanobacteria growth. Aluminum compounds in the catchment soils generally diminish external P loading; Al in the sediment prevents internal load as sediment P release (x drawn through arrows). *Redrawn from Nürnberg et al. (2018).*

Phase 1. High emissions of acidifying pollutants during active mining operations: Less developed lakes during mining have little external P input because (a) aluminum (Al) keeps P bound in the catchment soils, and (b) there is not much anthropogenic input. They also have only limited amounts of releasable sediment P. Lakes are ultra-oligotrophic without internal P loading.

Phase 2. High emissions of acidifying pollutants during active mining with increasing development: In anthropogenically disturbed lakes during mining, Al keeps P bound in the catchment soils except for direct developmental P input and land-use change (increased external load). Al hydroxides adsorb P in the water column and settle P to the lake bottom sediment, where it accumulates. Lakes are still oligotrophic without discernible internal loading.

Phase 3. Recovery during acidity-related restoration with reduced acid emissions and continuing development: Declining ionic Al leaching in catchment soils does not prevent P export (anymore). Decreased Al hydroxides in the water column (caused by decreased Al leaching from catchment soils) permit only limited binding of lake P. Newly settled P includes organic P forms that are diagenetically transformed and accumulate as P adsorbed onto iron oxyhydroxides in the surface sediment, causing some internal loading during hypolimnetic hypoxia. Sediment-released P can support cyanobacterial blooms. In this phase, lakes are on a trajectory of becoming meso- and then eutrophic with increasing cyanobacteria proliferation.

Besides getting increasingly productive, formerly acid-stressed lakes, often located at higher latitudes, are also getting browner with increased colour and DOC concentration (brownification, Sections 2.6 and 4.4.4), possibly recuperating to pre-industrial levels (Meyer-Jacob et al. 2020, 2019). For example, much of the recent browning can be traced back to the recovery from the human impact of acid rain, leading to the more natural conditions typical before anthropogenic interference (Myrstener et al. 2021), besides climate change effects (Section 4.4.4).

DOC increases lead ultimately to increased colour that is at least partially caused by iron export (Knorr 2012). Iron concentrations associated with browning are on the rise at an increasing rate in boreal freshwaters in Northern Europe and in North America (Björnerås et al. 2017). Several independent studies (Björnerås et al. 2019, 2017) attributed changes in surface water Fe to decreasing redox conditions in lakes (Ekström et al. 2016), besides increased coniferous forest cover in the catchment (Kritzberg 2017). Further, warming winters and raising groundwater tables in wetlands produced anoxia and were associated with Fe increases in Scandinavia (Blanchet et al. 2022).

In catchment soils of three Swedish rivers, increases in anoxia and decreases in redox were determined as the cause of the mobilization of iron across the terrestrial–aquatic interface, in combination with higher temperature and moisture (Ekström et al. 2016), and not, as also hypothesized, from reduced sulfur deposition from diminishing acid rain (Björnerås et al. 2019).

Loss of transparency from both, eutrophication and brownification, was determined in 30–85% of all 205 reviewed temperate and boreal lakes (Bartosiewicz et al. 2019). The resultant reduction in epilimnetic thickness and earlier onset of thermal stratification with potentially higher temperature gradients and colder near-bottom temperature were described as *thermal shielding*, and it occurs when high phytoplankton biomass or coloured dissolved organic matter increases the radiant heating of surface waters. This increase in water column stability leads to longer thermal stratification and hypoxia (Knoll et al. 2018) as is also observed with climate change. Hence, the effect of browning from DOM increase, itself potentially intensified by climate variables, further amplifies climate-warming effects on lakes (Sections 4.2.3 and 4.3.1).

Increased browning also affects light-dependant benthic oxygen production. The combined effects of increased water levels and flooding of adjacent wetlands doubled or quadrupled DOC, browning, and phytoplankton in a small German lake (Brothers et al. 2014). The resultant diminished light at the sediment surface eliminated benthic primary production thus promoting hypolimnetic anoxia, inducing extensive internal P loading.

Coloured lakes are often relatively productive as compared to clear lakes (Section 2.6), which may explain the increasing cyanobacteria proliferation in browning lakes. While oligotrophic, deep, and clear lakes were dominated by diatoms, lakes with intermediate levels of browning (4–8 mg/L DOC) were associated with cyanobacteria in 71 study lakes on the Ontario Precambrian Shield in 2015 and 2016 (Senar et al. 2021). The coloured lakes had relatively high nutrient concentrations partially derived from redox-related internal P sources. At higher levels of lake browning (DOC 8–12 mg/L), mixotrophic species including bacteria and protozoa were most prevalent.

In general, many of the oligotrophic, brown-water, and acid-recovering lakes are located in the north-temperate and northern regions and can be assumed to support cold-acclimated phytoplankton. In the laboratory, cold-acclimated cyanobacteria grew faster and produced more toxins at warming temperatures than ambient temperature-adjusted and hot-acclimated organisms at low and high nutrient concentrations (Layden et al. 2022). Therefore, the existing cold acclimation of these ecosystems could provide an advantage to cyanobacteria growth and thus further explain the recent increase in cyanobacteria and their toxicity in northern low-nutrient lakes (Section 3.5.1).

The interactions between internal P loading, cyanobacteria, and climate change are complex in northern or nutrient-poor lakes. Oligotrophic lakes experience increasing cyanobacteria could be explained by increasing internal P load as some of these lakes have consistent hypolimnetic anoxia that, combined with recently increasing external P load, would lead to increasing legacy sediment P and internal P loading. Recuperation from atmospheric acid deposition and increased browning decrease the P-adsorption capacity in catchment soils and P retention in lakes also leading to increased internal P loading. While increased stratification and associated anoxia during brownification increase the duration of sediment P release, area-normalized release rates may be negatively affected by organic matter in the sediments (Section 2.6). Differing sources, nutrient concentration, and consistency of the organic material result in varying effects, so that the evidence of lake browning promoting internal P loading to fuel cyanobacteria still needs further exploration (Creed et al. 2018; Tammeorg et al. 2022b).

5.1.9 Effects of drought and decreased flows

Changing hydrology represents one of the most global and significant impacts of climate change on freshwater ecosystems. Specifically, hydrological intensification (Section 4.4.2), where dry areas become drier and wet areas become wetter (Reid et al. 2019), has large but variable influences. Under drought conditions, external inputs are minimal, and internal P loading from the sediment is especially large and obvious (Nürnberg 2009). While elevated temperature, quiescent conditions, and increased stability associated with droughts are directly favoured by cyanobacteria (Section 3.5.3), these conditions also increase internal P load, therefore leading to even more cyanobacterial proliferation.

In a review, Mosley (2015) identified three key processes driving water quality during drought: (1) hydrological drivers and mass balance, (2) temperature and stratification, and (3) increased influence of internal processes and resuspension.

Next, the effects of drought on water quality are explored as related to increased stratification associated with hypoxia and harmful algal blooms.

Drought conditions cause water level decrease (i.e., shallowing of lakes) leading to:

- Greater influence of air temperature on water temperature (elevated summer water temperature)
- Closer proximity of phytoplankton to sediments
- An increased access to nutrient-rich sediments and bottom water
- An increased temporal and spatial spread of anoxia and redox-related P release
- An increased sediment resuspension depending on wind shear and fetch, leading to an increased distribution of nutrient-rich sediment porewater and cyanobacteria cells into the overlying water (discussed in Section 5.1.10)

Much of the impact of decreased flushing and droughts on cyanobacteria (Section 3.5.3) may result from internal P loading. Several incidences of years with high internal loading and high cyanobacteria biomass in North America (Lake Erie, year 2012, Section 5.1.1) (Figure 5.3) and Europe (Müggelsee, year 2006, Section 3.5.3; Lake Nieuwe Meer, year 2003, Section 4.4.1) relate hot and dry weather to the intensification of hypoxia, internal P load, and cyanobacteria blooms.

The impact of increasing meteorological drought on global freshwater quality can be substantial (Mosley 2015). Phytoplankton (mainly determined as chlorophyll and cyanobacteria), nutrients, salinity, organic carbon, and temperature increased in lakes, reservoirs, and rivers, while dissolved oxygen decreased. The authors found that "changed internal processing has been identified as contributing to some of the nutrient trends during drought in lakes and reservoirs" and that "shallowing of lakes can also lead to more turnover of nutrient-laden bottom water".

Based on mass balances over 20 years for two shallow Mediterranean lakes, Coppens et al. (2016) concluded that "internal loading was important during dry years in the absence of external loading". Specifically, when water level and flushing were low, there was little net retention of TP and SRP, and internal P load and chlorophyll concentrations were large. Similarly, in Lake Apopka, Florida, increased drought and decreased flushing were accompanied by increased TP (from internal sources) and chlorophyll, and decreased Secchi transparency (Ji and Havens 2019).

A study of 40 eutrophic and hyper-eutrophic constructed lakes in warm semi-arid north-eastern Brazil determined that drought-induced water-level reduction at the end of the dry seasons favours cyanobacteria blooms in comparison to wet seasons (Brasil et al. 2016). Significant differences included decreased volume, depth, clarity (Secchi), and bottom dissolved oxygen concentration and increased pH, conductivity, NH_4, and water column stability, while SRP and TP were elevated but not significantly so. Chlorophyll concentration, total phytoplankton, and specific cyanobacteria biovolume were significantly elevated, as was the contribution of cyanobacteria to total phytoplankton biomass, while the contribution from diatoms was reduced in dry compared to wet conditions. Cyanobacteria contributed the most to total phytoplankton biovolume under both climate conditions (dry 72.4%; wet 53.8%), possibly because of extreme nutrient enrichment preventing nutrient shortages. Nostocales species (filamentous, potentially N_2-fixing cyanobacteria, especially *Cylindrospermopsis raciborskii*) were the dominant forms year-round, contributing 62% and 42% to cyanobacteria biovolume in dry and wet conditions, respectively. Oscillatoriales (filamentous non-N_2-fixing) contributed 27% and 32% to cyanobacteria biovolume in dry and wet conditions, respectively. A multiple regression model could predict cyanobacteria biovolume best by positive relationships with water column stability, pH, and TN concentration, and negative relationships with Secchi disk and inorganic suspended solids ($R^2 = 0.40$, $n = 80$). The authors concluded that "the high temperature in the sediment, intense bacterial activity, and sediment deoxygenation increase the release of phosphorus from sediments to water column", thus supporting the cyanobacterial population.

In 20 years of water quality observations of a run-of-the-river Texan reservoir (Bellinger et al. 2018), drought-induced discharge reductions promoted earlier onset and longer duration of stable stratification combined with increased hypolimnetic anoxia, higher epilimnetic temperature, low nitrate, and large TP and ammonia concentrations in late summer (Figure 5.13). By linking flushing rates to reservoir stratification and potential phytoplankton bloom magnitude, the authors enabled managers to predict summer water quality conditions from hydrological information.

Water level variation and drought also affect cyanobacteria toxicity, migration, and buoyancy control (Section 3.5.3) and can support the fertilizing effect of internal loading. For example, elevated bottom temperature at low water level benefits the survival and germination of akinetes and vegetative cells (Padisák 1998, Section 4.4.1). Wood et al. (2017) found that an extended drought in shallow eutrophic Lake Rotorua, New Zealand, led to prolonged stratification with increased SRP, decreased nitrate, and high ammonium concentrations, which interacted with temperature to positively influence microcystin cell quotas despite a decrease in *Microcystis* cells.

Low-flow conditions also affect large rivers by inducing internal P loading (Section 5.7.2) and potential cyanobacterial proliferation and toxicity. During low discharge in the Maumee River, Lake Erie watershed, chlorophyll ranged from 50 to 300 µg/L with a low proportion of less than 20% cyanobacteria in the riverine phytoplankton (Laiveling et al. 2022). But a substantial amount of cyanotoxins (38.7% microcystins and 16.7% saxitoxins) and up to 60% cyanotoxin genes indicated a potential for cyanotoxin development under the predicted dryer climate in this Great Lakes region.

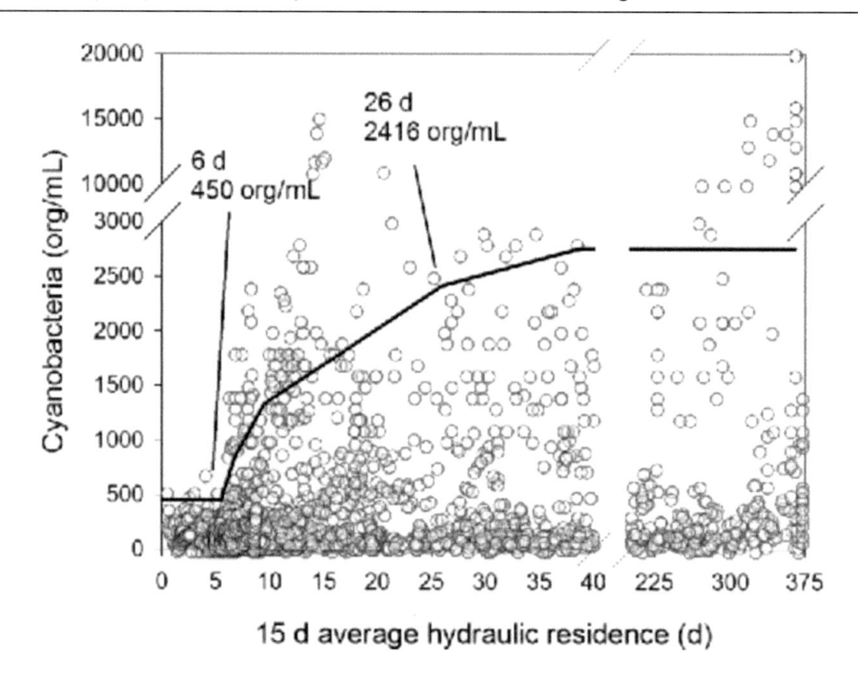

Figure 5.13 Relationship between the 15-day moving average hydraulic residence time of the Lake Austin reservoir and the abundance of cyanobacteria (organisms per mL) during the growing season (May to November) across all years (1992–2014). The solid black line depicts the 95th quantile regression of algal abundance with hydraulic residence time. *Figure 9D from Bellinger et al. (2018), open access.*

5.1.10 Resuspension by bottom shear and bioturbation

Sediment disturbances occur as a result of physical and biological drivers. Shear stress is enhanced by wind and wave action that can be related to anthropogenic disturbances (e.g., boating) and weather events (predicted to be more frequent and extreme because of climate change) and is increased at low lake levels during drought (Mosley 2015). Biological drivers include resuspension by bottom-feeding fish (e.g., carp) and bioirrigation by macrobenthos (e.g., dreissenid mussels, chironomid larvae, burrowing mayflies, and oligochaetes). The effect of sediment disturbances on water quality by any action depends largely on sediment properties, geochemical background (P adsorption capacity, pH, oxygen, and temperature), distribution and quantity of mobile sediment P, lake trophic state, water depth, and water column stability, besides the specific action.

The resuspension of sediment particles caused by shear stress after wind and wave events distributes nutrient-rich sediment porewater, sediment particles, as well as cyanobacterial live and resting cells into the overlaying water. This process especially benefits low-light-adapted cyanobacteria. For example, intensive resuspension coincided with the presence of cyanobacteria in hyper-eutrophic shallow Finnish lake, Kirkkojärvi, when the external TP load was almost non-existent (Niemistö et al. 2008).

Besides the distribution of particulate sediment P upon wave action, phosphate liberated from iron oxyhydroxides into the porewater of eutrophic systems becomes mixed into the open water where it is accessible to cyanobacteria. For example, redox-related P in the porewater was introduced into the overlaying water column in addition to the mineralized organic sediment P in Lake Peipsi (Tammeorg et al. 2020a). Other resuspension studies found relationships between sediment characteristics and P fluxes and determined that resuspension events can be followed by a large increase in diffusive P fluxes (Tammeorg et al. 2016). Significant negative correlations of SRP in the bottom water, the porewater, and diffusive fluxes with oxygen concentration and positive correlations with water temperature indicated that redox-related P release was involved. Both internal load and external load were correlated with cyanobacteria in three basins of eutrophic Lake Peipsi (Figure 5.14). But in all basins, external P load had decreased (from 1985–1989 to 2001–2005) while internal load increased, possibly due to increased temperature and wave action that released P from the porewater during observed prolonged resuspension in the more recent period.

Resuspension caused by biotic activity, especially benthivorous fish, can drastically increase sediment P release and thus hamper efforts to curtail cyanobacteria proliferation by the addition of chemicals to bind P at the sediment surface. Studied benthivorous fish include relatives of common carp (*Cyprinus carpio*), such as goldfish (*Carassius auratus)* and koi (*Cyprinus rubrofuscus*), as well as bream (*Abramis brama*) and gizzard shad (*Dorosoma cepedianum*). For example, common carp increased the sediment mixing depth by 2.5 times from 5.0 (± 1.2) cm to 13.0 (± 3.7) cm, effectively more than doubling the available P reservoir (Huser et al. 2016a).

Bioturbation by fish can further increase resuspension from physical influences. Bottom-feeding fish can make "craters" in the sediment, which are easily disturbed by shear stress from wave and wind action in sediments that would otherwise (without fish) consolidate and resist resuspension (Scheffer et al. 2003). The importance of this additional effect of sediment disturbance is demonstrated by the removal (biomanipulation) of bottom-feeding fish resulting in a reduction of cyanobacteria blooms (Godwin et al. 2011; Huser et al. 2022; Søndergaard et al. 2008) (Section 6.5).

A special case of bioturbation is aquaculture with bottom feeders like silver and bighead carp in nutrient-rich Asian lakes. Fish-mediated sediment nutrient release, resuspension, excretion, and changes in the food web were deemed responsible for a 44% decrease in Secchi transparency

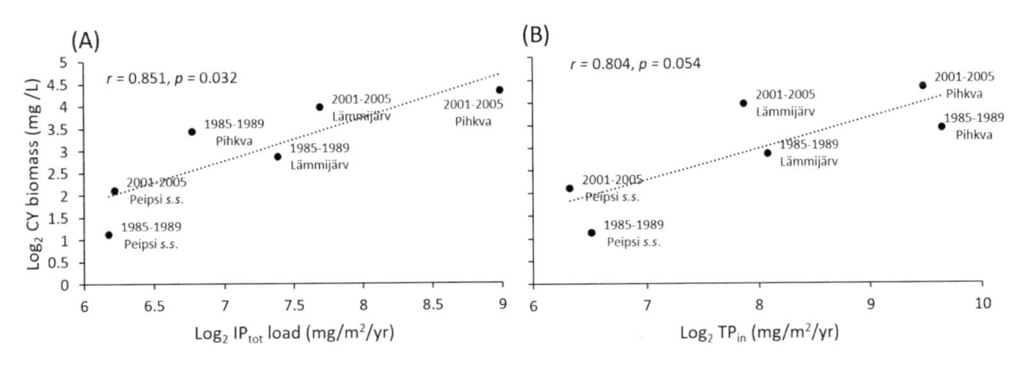

Figure 5.14 Correlations between cyanobacteria biomass (CY) in August and the internal P load (IP_{tot}; A), and the external P load (TP_{in}; B) for Lake Peipsi, Estonia/Russia. Values for periods 1985–1989 and 2001–2005 for three basins of Lake Peipsi are plotted after log2 transformation. *Figure 4 from Tammeorg et al. (2016), with permission.*

in 155 lakes of the Yangtze River Basin, China, which accommodates 54.7% of the total inland aquaculture production (Su et al. 2023).

Resuspension and bioirrigation caused by macrozoobenthos, such as chironomid larvae and oligochaetes, can have differing effects, depending on the nutrient enrichment state and chemical conditions that control equilibrium states (Andersson et al. 1988; Steinman and Spears 2020a). In eutrophic systems, bioturbation and bioirrigation enhance the mixing of sediment mobile P into the water column, but when chemical conditions favour P adsorption, bioturbation can improve sediment P retention. Chironomids and oligochaetes in mesocosms containing sediments rich in P, iron, and organic matter prevented low redox potentials in their burrow tubes so that P precipitated along the oxidized walls, decreasing redox-controlled P release (Lewandowski and Hupfer 2005). Such aeration by intermittent pumping by macrobenthos contributed to the conclusion that phytoplankton biomass can decrease in the presence of chironomids, despite sediment disturbances (Hölker et al. 2015). In contrast, burrowing mayfly nymphs, *Hexagenia spp*., increased benthic P fluxes (Chaffin and Kane 2010). Microcosm experiments with sediment and water from western Lake Erie estimated a substantial SRP flux as internal loading (1.03 mg/m^2/day) when nymphs are present.

5.1.11 Invasive dreissenid mussels

Dreissenid mussels coincide with cyanobacteria blooms in low-productive lakes during climate change (Section 3.6). Changes in the nutrient cycle, especially of P, suggest that sediment P release is facilitated by short-lived redox potential changes from calming conditions surrounding the mussel habitat (Ackerman et al. 2001) and benthic enrichment in nutrients and organic matter. The connection of dreissenid mussels with cyanobacteria (Bahlai et al. 2021) therefore likely includes the bottom-up influences of P release from bottom sediments by increased oxygen demand from calming conditions around the mussel habitat and from the increased organic material of living and decaying mussels.

Changes in nutrient and phytoplankton dependencies related to the recent (since 1980s) mussel invasion in North America are manifested as a decoupling of phytoplankton from nutrients and deviations from established TP–chlorophyll empirical relationships (e.g., Figure 2.4) (Cha et al. 2013, 2011). There may also be possible top-down effects, including changes in grazing pressure (Rowe et al. 2015). Further, greater movement has been suggested of particulate P from the nearshore water column of large lakes to local sediments and from there directly to discharge or offshore bottom sediments (summarized as "nearshore shunt" (Hecky et al. 2004)). According to this theory, P retention is increased, and P and chlorophyll concentrations are decreased in nearshore waters of the Great Laurentian Lakes where dreissenid mussels grow. However, in the shallower areas of several bays of the Great Lakes and other lakes invaded by the mussels, increased P retention was associated with increased cyanobacterial blooms (and chlorophyll) and decaying benthic algae (Cha et al. 2011).

The presence of dreissenid mussels appears to influence the partitioning of P compounds. While P retention from amplified settling of particulate P is increased, small increases of (dissolved) phosphate sustain *Microcystis*. The accumulation of organic material from mussel excretion in combination with settled P can decrease the redox potential at the sediment–water interface, increase redox-dependent and organic releasable P (mobile P), and thus support P release as internal P loading. In addition, flow chambers demonstrated a greater exchange between the benthic layer and the near-bottom water resulting from mussel-exuded jet streams (Nishizaki and Ackerman 2017).

Further evidence of enhanced internal P loading from mussels stems from a model study of Lake Michigan that determined increased dissolved P throughout the year, especially in late summer and fall (when internal load is expected to be largest) (Shen et al. 2020). Most importantly, observed diffusive P transport from the sediment was largely enhanced by the mussels so that the hypolimnetic ratio of dissolved to particulate P was large and the particulate P flux toward offshore reduced, decreasing the "shunt" effect.

Similar to the shallow regions in the Great Lakes, a study recorded decreased TP and chlorophyll concentration after zebra mussel invasion in 25 shallow lakes in the United States (Cha et al. 2013). In deep lakes, effects were less predictable, which could be a consequence of different oxygen states in their hypolimnion. Chlorophyll declined by 40–45% in all studied lakes, but TP only declined in the stratified lakes (–16%) also demonstrating an increased P retention and the uncoupling effect between chlorophyll and P described above.

Some of these observations were reproduced in experimental studies. Microcosms including sediment and water amended with zebra mussels showed a significant increase in the abundance of *Microcystis* and a decrease in green algae (Bykova et al. 2006). These community changes coincided with changes in nutrient fluxes, as nitrate retention increased but P retention decreased.

As anticipated, mussel-related internal P load effects are most important in nutrient-poor conditions so that dreissenid invasion effects are especially strong in oligotrophic lakes. In both oligotrophic and in experimentally created mesotrophic (25 μg/L TP) water from Gull Lake, Michigan (Kellogg Biological Station Long-Term Ecological Research [LTER] site), no or only small P pre-additions helped increase *Microcystis* in response to dreissenid colonization (Sarnelle et al. 2012). In contrast, P-addition had less effect in artificially made eutrophic water (40 μg/L).

These trophic state relationships are supported by a volunteer-based survey of about 168 New York State, USA, lakes (Matthews et al. 2021). In 15 oligotrophic and 41 mesotrophic lakes, the presence of dreissenid mussels nearly doubled the occurrence of cyanobacterial blooms and increased cyanotoxins nearly by 20-fold (Section 3.6, Figure 3.13). In contrast, the presence of mussels did not change the probability of blooms in eutrophic lakes, which was generally high (>75%). Similar results were obtained by a survey that compared the cyanotoxin microcystin with TP in 77 Michigan, USA, lakes (Sarnelle et al. 2010). The largest boost of microcystin by dreissenid mussels occurred at the low TP concentration of 5–10 μg/L compared to almost no impact at 10–26 μg/L.

To conclude, the increase of cyanobacteria in low-productive lakes with dreissenid mussels can be explained by the fact that additional P from the sediment has a greater effect in TP-poor conditions. If the sediment, where the mussels settled, had not been exposed to low redox conditions before, P release could be substantial because newly reduced and enriched sediments can release more phosphate as compared to sediments already experiencing P release. These bottom-up influences of P release from lake bed sediments upon increased oxygen demand due to enhanced calming conditions surrounding the mussel habitat likely increase with climate change (Sections 4.2 and 4.3).

5.1.12 Pre-cultural signs of internal P load effects on cyanobacteria

There are signs that relationships between climate change, toxigenic cyanobacteria, and internal P loading have also occurred in the geologic past so that the dependencies established here may not be unique to the Anthropocene. The study of past environments (i.e., paleolimnology) can provide insight to past, current, and future lake conditions under changing environmental conditions including climate change (Section 4).

Cyanotoxins from cyanobacteria may have contributed to delaying extinction-recovery during climate warming after the End-Permian extinction (EPE; 252.2 million years ago) (Mays et al. 2021). Based on fossil, sedimentary, and geochemical data, microbial blooms occurred in concentrations similar to modern blooms in low-oxygen (inducing internal P load) toxic lakes and rivers during warming events over the past 3 million years. "Comparisons to global deep-time records indicate that microbial blooms are persistent freshwater ecological stressors during warming-driven extinction events". Also, sediment records of a small deep Swiss lake revealed that redox-related internal P loading may have caused cyanobacteria proliferation already 15 000 years before cultural eutrophication, indicating a possible long-term natural eutrophication of deep stratified lakes (Tu et al. 2021).

Other paleolimnological studies on more recent sediments of global lakes discovered internal P loading effects on cyanobacteria by comparing climate data with cyanobacterial records and lake data specific to internal P load, including previous nutrient status (trophic state), hypoxia, thermal stability, and mixing status (throughout Sections 4 and 5).

5.1.13 Conclusions with respect to internal load influences on cyanobacteria

Interactions between internal P load and cyanobacteria in the context of climate change trajectories are numerous and complicated (Figure 5.15).

The lake-mixing state, either polymictic in a shallow lake or stratified in a deep lake including monomixis, dimixis, and meromixis (Section 2.2), and gradations between these extremes (quantifiable as morphometric ratio, Equation 2.1) regulate hypoxia and the influence of internal load on cyanobacteria (Section 5.1.3). Climate change (warmer and dryer and wind stilling, Figure 5.15, blue variables) increases lake temperature and hypoxia, which is related to

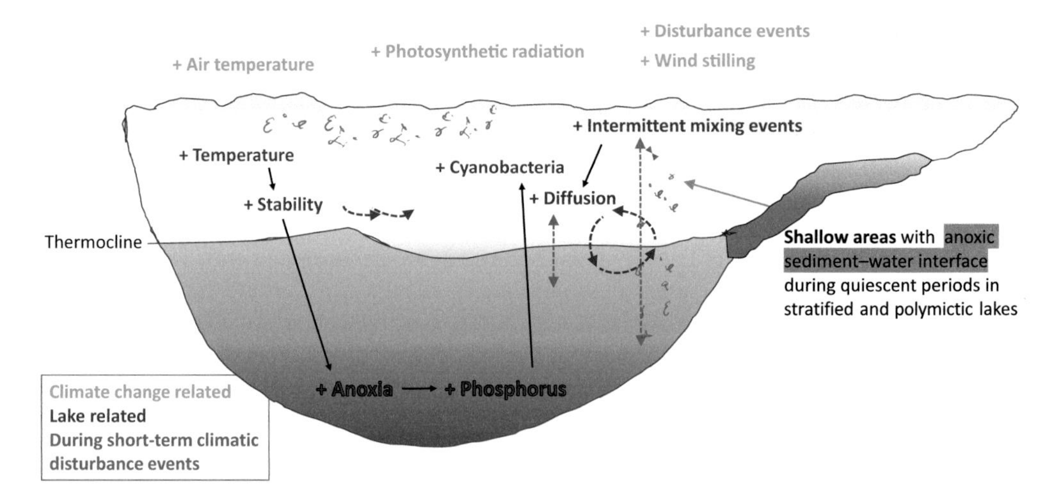

Figure 5.15 Positive interactions between climate change trajectories (blue), internal P load, and cyanobacteria. Increased stability with related lake variables (red) is interrupted by increased frequency of climatic disturbance events (green). Arrows indicate potential influences that can be bidirectional. Potential internal loading by Fe and N is not shown.

increases in both internal P load and cyanobacteria biomass (perhaps except in very deep lakes) (Figure 5.15, red variables). Examples in Section 4.3.1 (e.g., Figure 4.10) indicate an increase in hypoxia in lakes and reservoirs that are related to hotter and dryer weather and decreased flows. Climatic disturbance events are also predicted to increase, which may further fuel cyanobacteria by increasing the diffusion of hypolimnetic nutrient and mineral-rich water into the surface layer (Section 4.3.2, Figure 5.15, green events).

This means that catchment basin, lake morphometry, trophic state, productivity, and historic anthropogenic influences severely control the importance and effect of internal load on cyanobacteria proliferation under climate change in specific water systems. In addition, previous conditions, acclimation, and genetic diversity influence phytoplankton characteristics and their response to internal P loading.

The direct relationships between internal load estimates, sediment fractions, or anoxia with cyanobacteria biomass (Sections 5.1.1, 5.1.2, and 5.1.4), synergistic effects of temperature and internal load on cyanobacteria (Section 5.1.6), and the supporting role of winter internal load (Section 5.1.7) indicate the importance of internal loading under various chemical and physical conditions. Further, influences of lake stability or mixing state on cyanobacteria (Section 5.1.3), drought (Section 5.1.9), and resuspension (Section 5.1.10) can at least partially be explained by internal P loading effects. Even dreissenid mussel enhancement of cyanobacteria (Section 5.1.11) and precultural eutrophication periods (Section 5.1.12) can partially be explained by internal P loading.

While cyanobacteria are often studied in highly productive lakes with especially problematic blooms, cyanobacteria proliferate in a wide range of lakes, that is, lakes that are nutrient-poor, that are recovering from acidity, or that are rich in coloured organic acids (Section 5.1.8). Much of the deviation between temporal trends (e.g., lag periods) for internal load and cyanobacteria can be attributed to differing lake characteristics and trophic states.

5.2 Success of cyanobacteria control by internal load reduction

In most cases, cyanobacteria blooms only fully subside when sediment P release is successfully managed. This is evident from examples where external P input is reduced, but cyanobacterial blooms persist until internal load is controlled as well (Chorus et al. 2020; Cooke et al. 2005; Rönicke et al. 2021; Spears et al. 2022). These observations underline the importance of detecting and accounting for internal P loading in lake management.

Reviews of products and treatment efficiencies to address cyanobacteria highlight those which reduce lake P concentrations by controlling internal loading in case studies (Bormans et al. 2016; Lürling et al. 2016; Lürling and Mucci 2020; Nürnberg 2017; Zamparas and Zacharias 2014) and in reports to governmental committees (Huser et al. 2023; Jones et al. 2020). Such treatments include the withdrawal of nutrient-rich hypolimnetic water among other physical approaches (Section 6.3) and chemical approaches that increase the adsorption (e.g., aluminum- and iron-based chemicals) and binding of P (e.g., lanthanum-modified clay, such as Phoslock) at the sediment surfaces (Section 6.4).

For example, in a small German excavation lake, external and internal P load were successfully controlled for three decades. Despite continued external load control, re-eutrophication with planktonic cyanobacteria blooms occurred as the internal load treatment became ineffective 30 years after an alum treatment (caused by a loss of P adsorption capacity due to chemical aging), and P release commenced again (Dadi et al. 2023). In another example under experimental conditions, a lasting suppression of cyanobacteria was only accomplished when sediment P release was interrupted by chemical adsorption or binding, so that P limitation curtailed cyanobacteria growth (Kang et al. 2022a). In these experiments, lanthanum and aluminum salts, which target P release from sediments, strongly reduced SRP thus inducing nutrient limitation that restricted cyanobacteria proliferation. In contrast, algicides, chitosan, and hydrogen peroxide that attack cyanobacteria directly had only short-lived success in reducing surface-blooming *Microcystis aeruginosa* biomass, sometimes lasting less than seven days.

Re-eutrophication events caused by external and re-established internal load further emphasize the possible control of cyanobacteria by decreased internal P load. In one case, TP concentration and cyanobacteria biomass decreased for two years after a treatment with lanthanum-modified bentonite, but increased again in the third and fourth year, likely because of residual external load and perhaps re-established sediment P release during periods of reduced flushing with increased stability and temperature (Lang et al. 2016). Similarly, the trend toward oligotrophication in an urban Canadian, former gravel pit was interrupted by the external nutrient input from an increasing population of waterfowl (mostly Canada geese, *Branta canadensis*) that overwhelmed the P-binding capacity of lanthanum-modified bentonite in the third post-treatment year (Nürnberg 2017; Nürnberg and LaZerte 2016a).

Internal P load supports the growth of metalimnetic, low-light-adapted cyanobacteria (e.g., trait R, Table 3.1), thus controlling internal load can help control these species. This is particularly evident in the success of the lake restoration method, hypolimnetic water withdrawal. This technique preferentially increases the outflow of nutrient-rich hypolimnetic water relative to surface outflow. During hypolimnetic withdrawal treatments, chlorophyll concentration, *Planktothrix rubescens* biomass, and surface-dwelling cyanobacteria species decreased (Nürnberg 2020b) (Section 6.3.2).

Sediment-dwelling, vertically migrating cyanobacteria also respond to internal load treatments. For example, upward migration and recruitment of *Gloeotrichia echinulate* into the water column decreased after a treatment to bind P at the sediment (alum treatment) in Green Lake, Washington, USA (Perakis et al. 1996).

A detailed analysis of treatment options targeting internal P loading and their success in preventing and restricting cyanobacteria and their blooms is presented in Section 6.

5.3 Modelling cyanobacteria occurrence

Models are often used to predict potential scenarios in situations where environmental conditions are changing. To accomplish such predictions, models are built, validated, and tested using existing data sets, and they are then applied to new situations. In general, models are expected to be more successful in predicting cyanobacteria when they include internal P loading in situations where P is important (without other major limiting variables) but in short supply, than if they do not include internal P load at all.

While ecosystem models often predict TP and phytoplankton biomass as chlorophyll concentration, there are not many models specifically designed to determine the effect of internal P load on cyanobacteria proliferation, either as chlorophyll concentration or biomass (Mooij et al. 2010), even though P upward diffusion from the sediments can be included in dynamic models (e.g., PCLake, LEEDS, LakeMab, reviewed in Nürnberg (2020)). As models do in general, internal P loading models follow two different model approaches: (1) detailed, process-based, aquatic ecosystem models (i.e., mechanistic models) often developed for individual lakes and (2) empirical, cross-system, data-driven models, developed on data of many lakes over a wide range of conditions (Nürnberg 2020) or on single lakes (Rousso et al. 2020). While process-based models require data on many different lake parameters (i.e., are data-hungry) (Robson 2014), the empirical approaches follow the principle of parsimony; they are powerful because they can predict in situations with limited data availability and hence involve lower effort and cost. Nürnberg (2020) concluded that model choice depends on data availability, financial and technical resources, and the application history of specific models. Testability and verifiability combined with the principle of parsimony were considered most important in choosing a model.

Although rare, there are some model studies which explore the importance of internal load on cyanobacteria directly including one on eutrophic Lake Rotorua, New Zealand (Burger et al. 2008), one on a eutrophic tropical Malaysian lake (Tay et al. 2022), and one on the German River Havel system (Lindim et al. 2015) (Section 5.7.2). In these studies, internal load abatement was deemed to be the most effective restoration measure.

A systematic literature review of 123 peer-reviewed modelling studies (Rousso et al. 2020) determined that the main predictor variables for cyanobacteria proliferation were water temperature and nutrients (P and N), followed by meteorology, hydraulics, biological, land use, and chemical water quality (including anoxia in 2% of the studies, Figure 5.16). Similarly, a

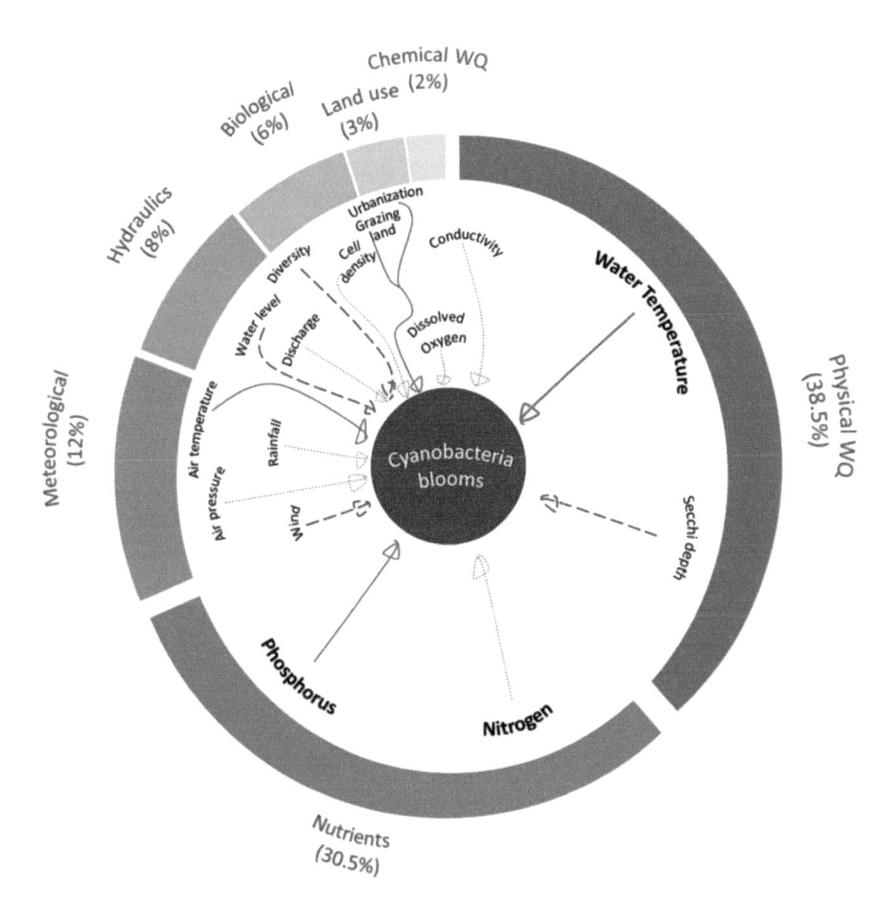

Figure 5.16 Relative frequency for input variables being identified as the main predictor in models of cyanobacteria blooms and their relationship with bloom occurrences. Continuous arrows represent positive relationships; dashed arrows represent negative relationships, and light-grey dotted arrows represent equivocal relationships. *Figure 7 from Rousso et al. (2020), with permission.*

review on predictive cyanobacterial models (using taxonomic biomass data) for 149 lakes in temperate regions across Canada found important predictors to include nutrients, temperature, water column stability, and forms of inorganic N (Beaulieu et al. 2014). According to these reviews, internal loading was not explicitly considered in the modelling studies, only water column TP or SRP and occasional bottom sediment P (Rousso et al. 2020). Nonetheless, changes in often-used physical predictor variables, temperature (increase), stratification (increase), and water level (decrease) indicate that conditions benefitting internal loading have increased alongside cyanobacteria (Section 5.1). Therefore, adding internal P load estimates (derived by various approaches and models, Box 2.3) is likely to improve the modelling of cyanobacteria and related eutrophication.

Internal load may be indirectly implied when, consistent with the large influence of temperature on models of cyanobacteria, a simple model is used that accurately predicts temperature and water level changes following operational or climate influences. For example, these variables (predicted by the open-access one-dimensional hydrodynamic–ecological model, GLM-AED) successfully simulated the biomass of the cyanobacteria species, *Microcystis aeruginosa* and *Chrysosporum (Aphanizomenon) ovalisporum*, in Lebanon's largest freshwater body, Karaoun Reservoir (Fadel et al. 2019).

Various degrees of buoyancy control can provide specific cyanobacteria access to P in the sediment or in the bottom waters (Section 3.5.3). Therefore, differentiation between functional groups (Table 3.1) may further inform models on how internal P load can support cyanobacteria. Suggested groups include: (a) trait R that includes the metalimnetic low-light- and low-temperature-adjusted group, e.g., *Planktothrix rubescens*; (b) trait H1 that can inhabit the metalimnion and occasionally spreads throughout the water column, forming surface blooms, for example, *Aphanizomenon, Dolichospermum*; (c) trait SN in warm and mixed waters, for example, *Raphidiopsis*, *Cylindrospermopsis*; (d) other traits that represent mainly surface-blooming species favoured by stratified conditions, for example, *Microcystis* (M); and (e) trait S1 that prevails in shallow lakes during mixed conditions, for example, *Planktothrix agardhii*. However, there is increased realization that even surface bloomers can be in contact with bottom sediments at shallower sites or closer to the shoreline or in specialized states (as akinetes or vegetative cells, Section 3.5.3).

In deep perialpine lakes like Lake Zürich, *Planktothrix rubescens* was successfully modelled during fall mixing from its buoyancy response to light and irradiance, while nutrients were deemed non-limiting and thus not considered (Walsby et al. 2006). Such evaluation indirectly points to internal loading (distributed throughout the water column during fall mixing) as a precondition for enabling such blooms.

In a historically unproductive lake, Lake Auburn, Maine, USA, modelling the multiple algal blooms that occurred since 2011 indicated that water quality was susceptible to both internal and external P loading (Messina et al. 2020). Several models that connected catchment with in-lake events, simulated temperature, dissolved oxygen, and P (which relates to internal P loading) "proved to be an effective means for predicting the loss of water quality under changing land-use and climate scenarios."

In the deep, eutrophic, monomictic lake Rostherne Mere, the UK, phytoplankton blooms are dominated by diatoms in the spring and by cyanobacteria (*Aphanizomenon* and *Microcystis*) in the summer and fall (Radbourne et al. 2019) (Table 4.1). Modelling (with PROTECH, a lake biophysical model) concluded that internal P load, by increasing SRP at fall turnover, has the largest effect on chlorophyll in future climate change scenarios, more than both external P load and temperature. A process-based integrated assessment model examining synergistic relationships between climate change and release of legacy sediment P similarly concluded that climate-induced increases in internal load outweigh even extreme watershed P load reductions in Lake Champlain, in Vermont and Quebec (Zia et al. 2022).

Increasing water temperature and decreasing flushing positively affected cyanobacteria biomass in Esthwaite Water (in a PROTECH model (Elliott 2010)), one of the first lakes where redox-related internal P loading was studied (Mortimer 1941), indicating the potential effect of internal load. Similar observations were made in a review of ten lake studies (Elliott 2012) (Section 4.4.1). Here, a subsequently developed model showed also an increase in relative cyanobacteria abundance with increasing water temperature, decreased flushing rate, and increased nutrients by internal loading.

Other models are based on the prediction of nutrient increases in the water by internal load (Table 2.5) and the predicted effect on chlorophyll concentration or cyanobacteria biomass (Figure 3.3) as a series of simple regression equations (Nürnberg 2001). Also, the ambitious "roadmap" on "How to model algal blooms in any lake on earth" (Janssen et al. 2019) includes a legacy effect in its lake ecosystem component, which considers sediment P retention and release (i.e., internal load).

When choosing models, those that include potential sediment P release are likely most appropriate for increasing eutrophication and different climate change scenarios. For example, the original model MyLake does not include sediment release when the water column is oxic as in the stratified boreal Lake Vansjø, Norway (Couture et al. 2018), based on the long misunderstanding that water-dissolved oxygen concentration controls sediment anoxia (Section 5.1.3). Only when a sediment diagenesis sub-routine module was incorporated that includes sediment P release, the water quality in Lake Vansjø could be adequately predicted for climate change conditions (Markelov et al. 2019). The consideration of internal P loading had severe management implications, because the results suggested that internal P loading could sustain primary productivity in the lake for several centuries (Markelov et al. 2019).

In conclusion, it seems that successful models for predicting water quality variables (including water TP, chlorophyll concentrations, and cyanobacteria proliferation) under climate change scenarios must contain internal load (e.g., Burger et al. 2008; Tay et al. 2022).

5.4 Enhancement of internal P loading by cyanobacteria

Cyanobacteria themselves have been considered as direct contributors to internal P loading by moving P from the sediments into the pelagic open water upon upward migration (Section 5.4.1) or by creating conditions conducive to sediment P release (Section 5.4.2).

5.4.1 Translocation from sediments to open water

Cyanobacteria display specific abilities of vertical migration, which are possible by buoyancy control, vegetative cell hibernation (*Microcystis*), and the production of seed banks (akinetes, e.g., *Dolichospermum*, *Aphanizomenon*, *Cylindrospermopsis*, *Gloeotrichia*, and *Nodularia*, Table 3.1) (Section 3.5.3). While these mechanisms can move P from the sediment to the water column, their manifestation is short-lived and occurs when shallow sediments are exposed to light and elevated temperature (Head et al. 1999). Therefore, this internal load is usually small as compared to P released from sediments by the reduction of iron oxides or the mineralization of organic compounds.

Estimates on the extent of this P translocation from the sediment to the water are variable. *Gloeotrichia* is a taxon that is primarily dependent on stored P taken up from (shallow) sediments before recruitment and is the most likely cyanobacteria taxon to contribute quantitatively to internal P loading. *Gloeotrichia echinulata* colonies in the sediments of large (23.7 km^2) but well-mixed (morphometric ratio 1.85 m/km) Lake Erken, Sweden, were estimated to have introduced 2.5–3.8 mg $P/m^2/d$ during summer (late July through August 1991) to contribute a total of 40 $mg/m^2/yr$ as internal P loading (Istvánovics et al. 1993; Pettersson et al. 1993). This pulse of *Gloeotrichia's* upward migration slightly increased the pelagic TP concentration (20 µg/L) by less than 10% (<2 µg/L). This is a small amount compared to internal loading from sediment diagenesis throughout the summer in Lake Erken (Section 2.4.2) and was not deemed necessary to be included in lake Erken's sediment P model (Malmaeus and Rydin 2006).

Conversely, *G. echinulate* contributed a considerable internal load in relatively deep (10 m mean depth, 15 µg/L TP) Lake Sunapee, New Hampshire, USA (Carey et al. 2008). Recruitment peaked in mid-September and was 200 times larger from shallow (2 m) than from deeper (5 or 8 m) sites. The sediment surface became enriched in P just before maximum migration, which was attributed to redox-related P release that thus contributed to the P in the migrating cells. Recruitment rate was highly correlated ($R = 0.89$–0.96) to mean surficial sediment P at 7, 11, and 25 days prior. The dependence of recruitment on P pulses was confirmed by laboratory P additions that significantly reduced the time to, and increased the rate of, akinete germination. Lastly, 40–45% of the pelagic *Gloeotrichia* cells were attributed to recruitment from 2-m deep sediments over 14 days.

The importance of sediment recruitment of five different taxa was evaluated after a treatment with aluminum sulfate to control sediment P release in Green Lake, Washington, USA (Perakis et al. 1996). *Gloeotrichia* recruitment was most dependent on adequate light and temperature conditions at the sediment, while *Dolichospermum*, *Aphanizomenon*, and *Coelosphaerium* were independent of sediment conditions once the water column was inoculated. P transfer from the sediments to the water column was only important for large recruitment events of *Gloeotrichia*. Despite no obvious change in recruitment from the sediments, all planktonic cyanobacteria numbers and bloom frequencies decreased in the years after the alum treatment (Jacoby et al. 1994). This also indicates the importance of sediment P release (intercepted by the alum treatment in Green Lake) to sustain cyanobacteria blooms (Section 6.4.1).

Another study determined that most *Dolichospermum lemmermannii* was recruited from the littoral rather than the deep sediment of Lake Erken during high light conditions and sediment mixing (Rengefors et al. 2004). The contribution of the inoculum to the pelagic population was only 0.003% to 0.05% (Karlsson-Elfgren and Brunberg 2004) so that P transfer was assumed negligible. Contrarily, an estimated 90% of *Dolichospermum* in the water column of Bugach Reservoir, Bulgaria (0.32 km^2 area), was said to have originated from overwintering akinetes following germination at the sediments, contributing 10% of total P input as internal P (Hellweger et al. 2008). Akinetes of *Dolichospermum* and *Aphanizomenon* accumulated preferentially (11 fold on average) at the deep lacustrine sites close to the dam in six Spanish meso- and eutrophic reservoirs compared to upstream sites, probably due to the process of settling and subsequent sediment focusing (Cirés et al. 2013). Still, P transfer was likely small because only 0.02–0.5% of the cells transferred into the open bottom water by resuspension in the winter and spring.

Microcystis in sediments and over the winter is encountered globally in lakes. *Microcystis* colonies throughout the sediment layer in the eutrophic Grangent Reservoir, France, were numerous (250 colonies per ml sediment at the surface, a maximum of 2 300 at 25–35 cm, and

600 colonies at 70 cm), complete, and healthy down to at least 35 cm, indicating a large survival and inoculation potential (Latour et al. 2007).

Microcystis populations, which overwinter in a vegetative (resting) state, are less dependent on periodically high inputs from recruitment (Perakis et al. 1996). However, winter survival can be important for and related to recruitment in the following spring and summer (Calomeni et al. 2023). Survival was found to depend on ice cover and temperature, where increased light by decreased snow cover and increased temperature were unfavourable (Brunberg and Blomqvist 2002). In incubation experiments, light exposure caused supersaturated oxygen conditions and poor survival of *Microcystis*, while dark conditions and low oxygen concentrations enhanced survival (Brunberg and Blomqvist 2002). This means that oxygen depletion, when encountered in north-temperate lakes under winter ice cover, could enhance *Microcystis* proliferation in the following growing period, especially if redox-related sediment P release provides P-replete conditions.

Cell abundance from different strains, including *Microcystis* with toxin-creating genes, and their survival rates were highly variable in the western basin of Lake Erie (Kitchens et al. 2018). About 6% of *Microcystis* upward migration was estimated to contribute to the weekly increase of cells before ambient P would sustain its growth, indicating only a very small transfer of P from the sediments into the water column. Newly recruited cells *Aphanizomenon flos-aquae* had a higher P content than cells in the water column in a hyper-eutrophic shallow lake in Oregon (Barbiero and Kann 1994) indicating the sediment as P source. Similar to other studies, the direct transfer of P via upward migration here was small as compared to P diagenesis (redox-related internal load), where it contributed to internal P loading at a rate of 3.6 mg/m^2/d for just one week.

In conclusion, there is accumulated evidence that sediments are a source of nutrients for translocating cyanobacteria (Hellweger et al. 2008). High porewater P concentration from processes related to redox changes and mineralization stimulate akinete growth (Carey et al. 2008), but internal P loading from cyanobacteria recruitment itself is likely small-to-negligible on an annual basis, compared to direct P diffusion from anoxic sediment surfaces. The proportion of possible P transfer from bottom sediments via cyanobacteria grows as other P inputs decrease and is probably most important in nutrient-poor lakes with only little P release due to redox processes and mineralization.

5.4.2 Cyanobacteria-triggered sediment P release

Cyanobacteria senescence after settling to the bottom can create hypoxia and enrich the sediments with organic matter and nutrients. P release from settled dead cyanobacteria cells occurs after transformation to organic P, so that the mechanism of sediment P release expands with productivity (Section 2.4.2). While oligo- and mesotrophic systems mainly involve reductant-soluble release, internal P loading from bottom sediments can gradually include an additional P release from organic matter of settled cells (mineralization) in algae-dominated shallow eutrophic and hyper-eutrophic lakes (Albright et al. 2022; Yao et al. 2016).

The benthic biomass of *Microcystis* correlated with both the heterotrophic bacterial activity in the sediments and internal P load in hyper-eutrophic Lake Vallentunasjoen, Sweden (Brunberg and Bostrom 1992). Previous extensive *Microcystis* blooms in the lake water were deemed responsible for the large biomass of viable *Microcystis* colonies within the sediments associated with increased microbial activity and sediment P release.

Oxygen produced by cyanobacteria photosynthesis initially prevented P and Fe release in Lake Taihu mesocosm studies, enhancing P and Fe accumulation in the porewater during blooms

(Chen et al. 2018). However, these compounds were then simultaneously released from the porewater during hypoxic events. When the bloom collapsed, P was decoupled from Fe release indicating its mobilization from degraded cells. Similarly, when comparing sediments from less-productive sites with no blooms and covered by macrophytes to eutrophic sites experiencing blooms in Lake Taihu, macrophyte-covered sites experienced mainly redox-related fluxes, while algae-dominated sites experienced fluxes from both, redox-related and organic P fractions, which were almost three times that of the macrophyte-covered sediments (Liu et al. 2022).

The implications of a change from a macrophyte-dominated to a plankton-dominated shallow lake state have been documented before and contribute to the paradigm of an *alternative state* (Scheffer et al. 1993). For a recent critical evaluation and discussion of the concept, see Davidson et al. (2023). Consequences of this shift include a decrease of light penetration by excessive cyanobacteria growth, as well as heat absorption leading to increased water column stability (Kumagai et al. 2000), reinforcing the shift from macrophytes to cyanobacteria in shallow lakes (Sukenik et al. 2015). In addition to mineralization-induced internal load (from settled and senescing cyanobacteria), these changes are conducive to redox-related internal load as noted in various freshwater systems including Brazilian reservoirs (Rocha and Lima Neto 2022).

In fact, oxygen depletion at the sediment–water interface can be a direct effect of decaying settled surface-blooming cyanobacteria, which led to a positive feedback loop of increased P release, resulting in sustained cyanobacteria blooms in Missisquoi Bay, Lake Champlain (Smith et al. 2011). A similar increase in SRP and TP following bloom collapse was documented in shallow, hyper-eutrophic Lake Søbygaard (Søndergaard and Jeppesen 2020) (Figure 5.17).

Oxygen depletion at the sediment–water interface can also be caused by large-celled, bottom dwelling, benthic filamentous cyanobacteria, such as *Microseira (formerly Lyngbya) wollei*. Dense benthic mats create conditions conducive to anoxic sediment P release by shading (preventing photosynthetic oxygen production) and water column stabilization (preventing the influx of oxygen from the water column).

Below the benthic mat, loose filaments of *M. wollei* can be buried in the sediment, where they remain alive and viable (Hudon et al. 2014; Lévesque et al. 2012). Thus, this species can acquire P reserves as luxury uptake, decoupling their growth from nutrient inputs. Even without heterocytes, *M. wollei* can fix N and survive heterotrophically despite anoxia and low light (Watson et al. 2015). In other words, it creates conditions which facilitate the internal P loading that it needs as a nutrient source.

High pH (>10) can mediate and increase P release from sediments and P desorption from aluminum hydroxide and other compounds sometimes used in lake restoration (Kang et al. 2022b; Reitzel et al. 2013). Cyanobacterial blooms can increase the pH (by photosynthesis). For example, in ten North-Brazilian reservoirs ranging from oligo- to hyper-eutrophic, higher pH, eutrophication, stratification, and lower water transparency were observed during cyanobacterial and mixed blooms (Amorim and Moura 2021). Consequently, cyanobacterial blooms can accelerate the flux of dissolved P from the sediment and resuspended sediment particles by elevating the pH in the water (e.g., Holmroos et al. 2009; Xie et al. 2003).

Cyanobacteria's high affinity for phosphate uptake (Solovchenko et al. 2020) enforces upward diffusion from the sediments. The continuous uptake of phosphate despite aerobic conditions caused low SRP concentration in the water of shallow eutrophic Beaver Lake reservoir, Arkansas (McCarty 2019). Experimental P release rates and sediment equilibrium (EPC_0) conditions in sediment cores confirmed that cyanobacteria growth was not P-limited as long as there was access to sediment porewater.

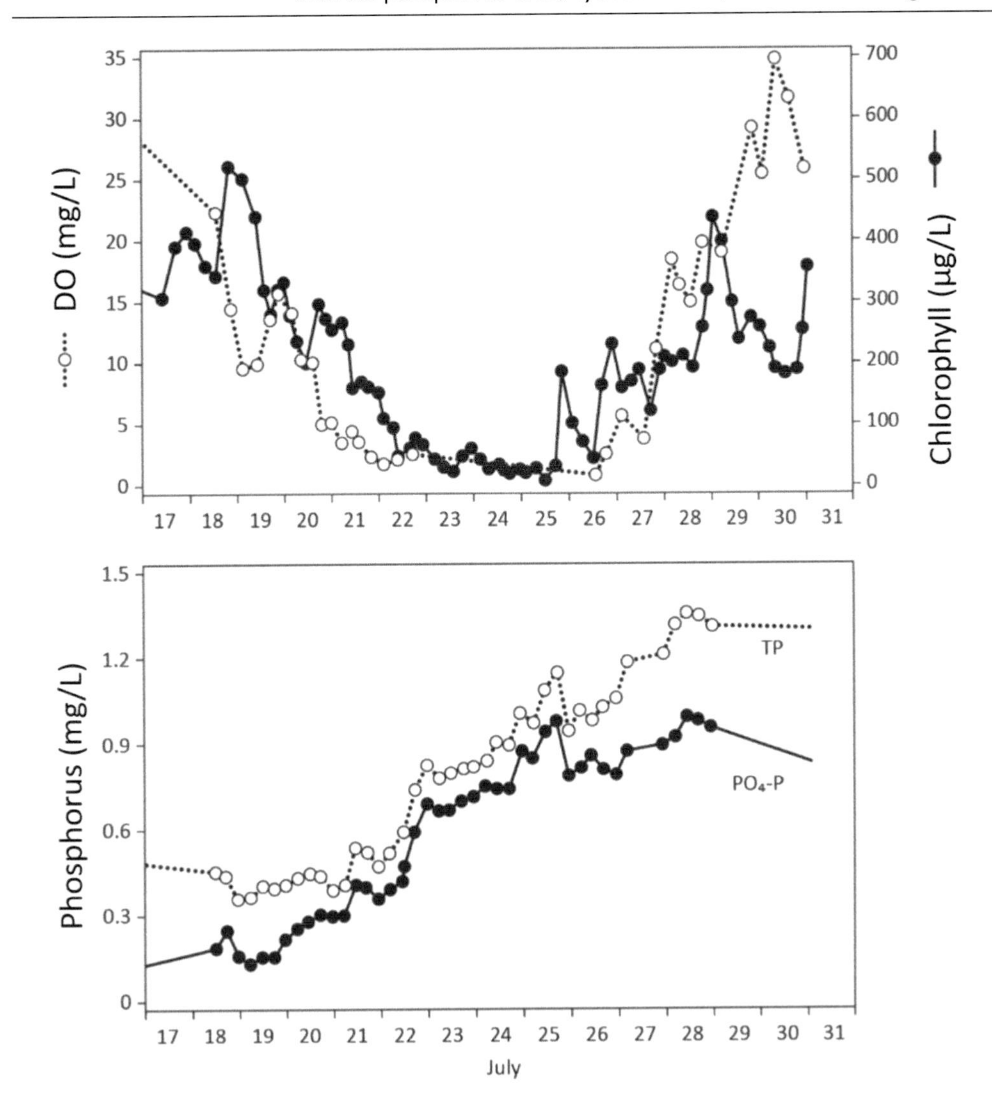

Figure 5.17 Changes in hypertrophic and shallow Lake Søbygaard during a phytoplankton collapse in July. As the phytoplankton collapse occurred, concentrations of dissolved oxygen (DO) and chlorophyll decreased (top panel) simultaneously with an increase in total P (TP) and phosphate (PO_4-P) (bottom panel) in the water column. After several days, DO and chlorophyll returned to pre-collapse concentrations, and the P concentrations stabilized. *Figure 2 from Søndergaard and Jeppesen (2020), with permission.*

Various characteristics of pelagic and bottom cyanobacteria provide conditions that facilitate internal P loading. It is difficult to determine the direction of causal relationships because of positive feedback loops. For example, *Microcystis aeruginosa* increased the release of BD-P, NaOH-P, and organic P in a mesocosm study and then accumulated the released P, especially under slightly mixed conditions (Ren et al. 2022). The investigation of 11 water bodies with variable cyanobacteria blooms from the middle and lower reaches of the Yangtze River, China, determined that a decrease in SRP led to increasing trophic state variables (chlorophyll, TP, and phytoplankton biomass) (Wang et al. 2023b). Organic matter from cyanobacteria simultaneously reduced the sediment equilibrium P concentration and increased P release rates, effectively increasing internal load.

The effects of cyanobacteria on the eutrophication process led to their designation as ecosystem engineers in recently eutrophied Lake Stechlin, Germany (Kröger et al. 2023). Positive feedback loops become a self-proliferating process in which the environment itself is modified and affects resource availability for potential competitors. In conclusion, cyanobacteria intensification during predicted increases in productivity and eutrophication with climate change (e.g., Beaulieu et al. 2019) will also increase the internal load by mineralization directly from organic matter of decaying cells, as well as from Fe-oxyhydroxides after diagenesis and increased spatial and temporal extent of anoxia.

5.5 Challenges in studying effects of internal load on cyanobacteria

To collect evidence for links between cyanobacteria and internal P loading, there are numerous difficulties that have to be overcome, as the complexity of internal P load can hinder simple testing of its effects on cyanobacterial blooms. Reliable data on internal loading and matching cyanobacteria biomass are rare, so that internal P loading influences are not always recognized. Potential idiosyncrasies and difficulties to be considered when determining internal load effects on cyanobacteria are reviewed next, based on discussions in the previous sections.

1) Difficulties concerning the *detection and quantification of internal P load* (Nürnberg 2009):
   a) Undetected internal load – Especially in shallow lakes, P released from sediments mixes throughout the water column and is immediately taken up by phyto- and bacterioplankton. This makes simple determination by P concentration (or hypoxia) monitoring ambiguous.
   b) Controversy about the precise origin of sediment-released P and ambiguity about its chemical form – Most internal P loading originates from dissolved iron oxyhydroxides under reducing conditions (<200 mV), but some P release stems from mineralization processes in organically enriched sediments of highly productive lakes (Section 2.4.2). It is difficult to quantify internal P load and to model lakes if the origin and form of P are not understood.
   c) Unclear definitions and inconsistent units and dimensions – Sometimes, only the portion of internal load that affects epilimnetic phytoplankton is considered, rather than the gross release from bottom sediments. In these cases, internal load estimates are not comparable between studies and may be underestimated.
   d) Failure to distinguish net versus gross estimates of P flows – To obtain a total gross estimate of internal load (as is done for external load), retention by settling must be considered (as an additional term to a net estimate). Failure to do so leads to inadequate quantification and modelling of internal load (by combining downward with upward fluxes).
   e) Inadequate determination of P sedimentation in mass balance models that predict lake TP concentration – P retention is an important yet difficult variable to model that must be tested (if an established model is used), calibrated (if no model is readily available), or otherwise supported. Else, predictions are meaningless.
   f) Lack of long-term quantification of internal loads – Internal P load varies between years as it depends on climate conditions, water level, external load, and other factors. For

quantitative statistical analysis and comparison with potential driving variables, several years of internal load estimates must be available. Means over several years can be compared to means of potential drivers, thus increasing relevance (by statistical corroboration) of results (Nürnberg 2020c).

g) Interpretation of experimental results – Bench, microcosms, and full lake experimental nutrient additions do not always include a sediment component to adequately determine sediment effects with respect to P release and retention. Newly settled material in the experiments may underestimate internal P loading, since legacy P can take a longer time to accumulate. For this and other reasons related to experimental ambiguity, these experiments may not adequately represent natural conditions.

Despite these potential problems, when internal P load is quantified for several years and by several methods, estimates are available to explore quantitative relationships with phytoplankton and cyanobacterial biomass (e.g., Section 5.1.1). Such evidence-based analysis is rarely available for other hypothetical causes of cyanobacterial biomass fluctuations, such as N and Fe internal load (Sections 2.5 and 5.6.2).

Other variables that are quantifiably related to cyanobacteria include the extent of hypoxia, water level, flow, browning, pH, and salinization. These variables may be related to internal P loading because they affect sediment P release directly or indirectly by interactions with sediment redox state, lake stability, stratification, dilution, and light.

2) Lack of *long-term meaningful records of cyanobacteria abundance*:

a) Chlorophyll *a* pigment concentration has regularly been used as a surrogate for cyanobacteria biomass, which can be adequate in late summer blooms when most of the phytoplankton are cyanobacteria. However, cyanobacteria are often temporally, vertically, and horizontally patchy, for example, due to wind-affected transient distributions. In addition, non-standard and incomplete chemical analysis can lead to analytical errors, rendering this variable error-prone. Integrated sensor-based, potentially in-situ, records of chlorophyll *a* concentration (or even better, phycocyanin pigment, which is specific to cyanobacteria) could improve cyanobacterial biomass estimates in the future.
b) Secchi depth transparency is a robust measure of phytoplankton biomass, when the colour of the lake water is low or does not change considerably. Because it integrates phytoplankton over the water column, it is easy to measure and often the only monitored water quality variable available. However, the portion of Secchi transparency that is attributable to cyanobacteria also depends on colour (Nürnberg 1996) and on the biomass of other phytoplankton, which are rarely measured at the same time. Thus, Secchi transparency values are not often used in large meta-analyses to reflect cyanobacteria.
c) Cyanobacteria biomass evaluation including species identification could increase the predictability of internal load influences on specific cyanobacteria. For example, metalimnetic *Oscillatores*, while supported by internal load, may only become a visible (surface) problem when physical conditions allow for increased mixing (Nürnberg et al. 2003). However, long-term observations that include species identification are rare.

3) Incorrect or unidentified *characterization of the water system:*

a) Misunderstanding the effect of *mixing state* – If a lake is deep, stratified, and without oxygen depletion in the hypolimnion (bottom water DO > 3 mg/L) like Lake Washington, USA, internal load can be assumed to be negligible. External load abatement, especially that of raw sewage, is likely to (and did, in the case of Lake Washington) improve water quality (decrease TP and cyanobacteria) (Edmondson and Lehman 1981; Hampton et al. 2006).

Conversely, if a lake is productive and mixed, its sediment is enriched with P fractions that are released under low redox conditions at the sediment surface despite oxygenated water over the sediment. The lack of apparent anoxia in the water column does not imply the absence of redox-related P release (Smith et al. 2011). Under such mixed conditions, sediment-released P is taken up by benthic and pelagic phytoplankton and cyanobacteria and is difficult to trace back to its sedimentary origin. In highly productive systems, this redox-related release is accompanied by mineralized P from recently deposited organic P (Liu et al. 2022). In this situation, external and internal load, both, have to be evaluated and compared to ensure most efficient management.

b) Effect of *productivity and trophic state* – Not all sediments in lakes with hypolimnetic anoxia (<2 mg/L) release P, because in low-productivity lakes, there is not much P in the mobile P sediment fraction that is releasable. Consequently, many low-productivity lakes fail to experience cyanobacteria blooms despite hypolimnetic anoxia (Nürnberg 1997).
c) Importance of *anthropogenic history and legacy P* – Sediment P release is especially enhanced in lakes with a history of high external P input. Therefore, past lake productivity and current mobile sediment P fractions may be important co-predictors for the influence of climate change on internal load. For example, the influence of bottom water temperature changes on internal loading mass depends on the sediment characteristics, especially the degree of mobile P enrichment from legacy P, which dictates the areal P release rate.
d) Effect of *coloured organic acids* – The presence of humic and fulvic acids (DOC) in coloured (stained) lakes affects sediment P release and P availability in the overlying waters in specific ways (Section 2.6). Consequently, quantitative relationships between cyanobacteria and trophic state variables are often different in stained versus clear lakes. Colour should therefore be considered in lake evaluations and data analysis (Havens and Nürnberg 2004; Nürnberg 1996, 1995a; Tammeorg et al. 2022b).

4) *Complex responses to climate warming* prevent simple and clear correlations:

a) *Air temperature increases* may actually lower bottom water temperature due to an earlier onset of stratification. Cooler sediment temperature can then potentially reduce sediment P release rates. However, this effect is offset by an increase in the duration of ice-free conditions, stratification, and hypolimnetic anoxia (Section 4.2).
b) *Lake browning and loss of transparency* can decrease bottom temperature by increasing water column stability and enlarging the hypolimnion, potentially lengthening the stratified period. For example, the estimated increase in methane genesis and release suggests that the period of a redox potential low enough for P release would be similarly increased (Bartosiewicz et al. 2019) (Section 4.3.1).

5) *Other often hidden causes* of increased cyanobacteria (also see Section 5.6):

When P is in abundant supply and not limiting, other conditions related to climate change and anthropogenic development can positively affect cyanobacteria and support their blooms so that internal P load does not appear to have any additional enhancing effect. Nonetheless, internal load can ensure that the P supply remains sufficient. Further, the limiting nutrient often changes over the season (Box 3.2) and can vary spatially (e.g., in stratified versus shallow regions). Cyanobacterial metabolism, species, traits, and genetics, as well as previous exposure to nutrients and other chemical and biological factors are often unknown. Accordingly, simple summaries, comparisons, and correlations can miss the influence of internal P load on cyanobacteria in many situations, leading to ineffective lake management approaches.

5.6 Alternate or auxiliary hypotheses for causes of cyanobacteria increases

The presentation above (Sections 5.1, 5.2, and 5.3) provides support for the hypothesis that internal P load is enhanced by climate change effects and thus causes the increasing cyanobacteria proliferation in many cases. This section critically evaluates observations that could lead to alternate hypotheses, where causes for recent bloom increases would be unrelated to increasing internal P load.

5.6.1 *External P input*

In less studied systems, some external sources may not be obvious; in others, changes in the various sources and chemical forms may have occurred. For example, changes in atmospheric inputs may be overlooked. Increased frequency of storm events and large fires could lead to increased unmonitored atmospheric deposition reaching remote, formerly low-productive lakes (Hamilton et al. 2022).

In general, external load is most important where it occurs as large quantities of bioavailable P (Section 2.4.1; e.g., Figure 5.2). For example, an early model study determined the March–July bioavailable P loading from Lake Erie's largest P contributing stream, the Maumee River, as best predictor of the following season's cyanobacteria biomass in the western basin (Stumpf et al. 2016).

Due to changes in agricultural practices and best management practices, external input may have changed from less available and fast settling particles to soluble and bioavailable phosphate. Especially, changes in agricultural practices including increased tile drainage that retains particulate P can decrease TP concentration without decreasing SRP (Jarvie et al. 2017). Even changes in herbicide usage was proposed as an increasing source of SRP (Spiese et al. 2023) (Section 5.6.3).

However, agricultural intensification was considered less likely to be the cause of the recently increasing SRP concentration across the Great Lakes Basin, studied from 2003 to 2019 (Singh et al. 2023). Over 370 watersheds across the Great Lakes Basin, a widespread increase in SRP concentrations (in 83% of watersheds, with 46% showing significant increase) was reported without changes in TP. Because small, forested watersheds at higher latitudes were identified as areas experiencing the largest relative increases of SRP, changes in agricultural practices were not considered the cause. In all watersheds, winter temperatures were key drivers of winter concentration trends. Therefore, it is possible that increased anoxia and associated P export from internal P loading, especially during warmer winters with more frequent snowmelt episodes in northern wetlands, may be the cause (Section 5.7.3).

5.6.2 *Iron release from anoxic sediments (iron internal loading)*

Because cyanobacteria require iron for N assimilation and the reduction of nitrate to ammonium (Section 3.4.2), their apparent proliferation during anoxia has been attributed to redox-related

Fe release (Molot et al. 2021b, 2014). In this hypothesis, internal loading from reduced ferrous iron (Fe^{2+}), which can be estimates from iron mass balances (Nürnberg and Dillon 1993), is the main reason for cyanobacteria proliferation. Cyanobacteria could reach sediment-released hypolimnetic Fe by downward migration, after which they produce strong ferric iron (Fe^{3+}) chelators or siderophores. In contrast, other phytoplankton do not produce siderophores.

In general, it is difficult to distinguish Fe from P internal loading effects on cyanobacteria, because the redox-controlled release occurs to both elements from the same compounds of iron hydroxides and often at the same time unless P is released via mineralization (Section 2.4.2). For example, Fe concentration increases that have been observed in European and North American north-temperate lakes were partially attributed to redox-related internal loading that includes both P and Fe and is possibly associated with organic matter by increased browning (Björnerås et al. 2017) (Section 5.1.8).

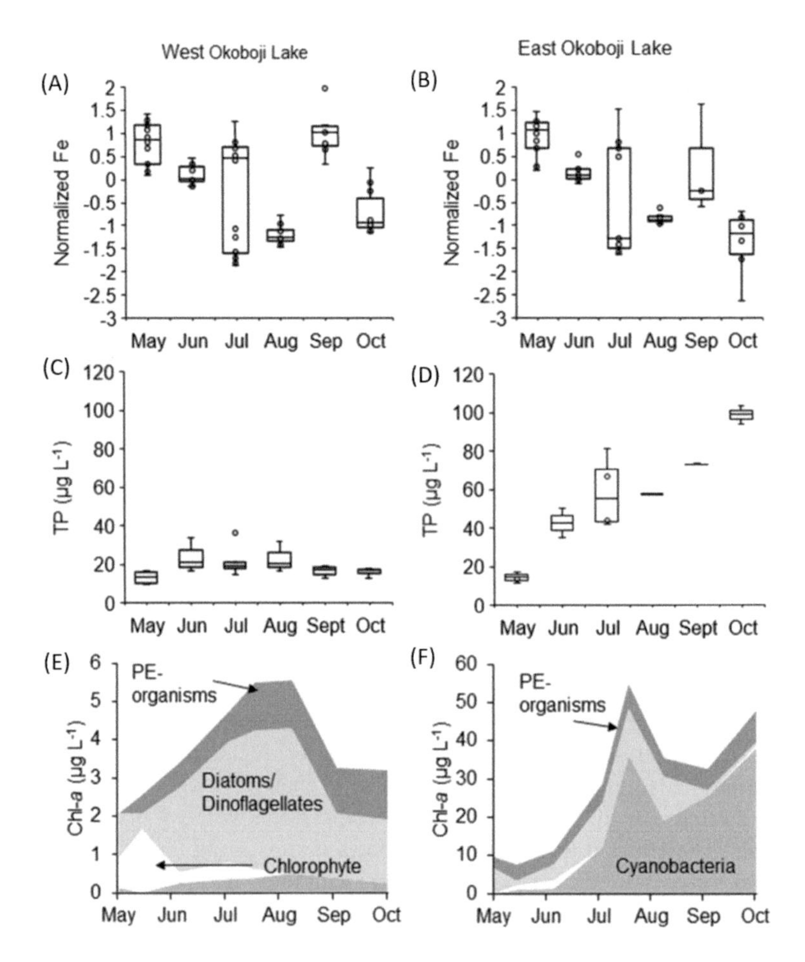

Figure 5.18 High internal P load coincides with high cyanobacteria biomass in unstratified eutrophic East Okoboji Lake, Idaho (right column) in summer 2017, while dissolved iron concentration tends to decrease. The Fe pattern is similar in stratified mesotrophic West Okoboji (left column), while TP and cyanobacteria remain low. *Figure 2 from Leung et al. (2021), with permission.*

While in nutrient-replete systems, added iron may increase cyanobacteria (Section 3.4.2), an increase in Fe should not necessarily cause an increase in cyanobacteria, because they can thrive at extremely low Fe concentrations. Iron-binding chelators improve Fe availability to cyanobacteria in freshwater lakes, especially under low Fe^{3+} concentration in oxic, hardwater eutrophic systems (Du et al. 2019; Sorichetti et al. 2016).

Observations in two connected hardwater and eutrophic prairie lakes (Idaho, USA) do not support the hypothesis that Fe internal load is the major driver of increased cyanobacteria blooms. There was no effect of dissolved Fe enhancement, while cyanobacteria only thrived on elevated TP from internal loading in one of the lakes (Leung et al. 2021) (Figure 5.18). Increased P from internal sources provided the nutrients needed by cyanobacteria in unstratified eutrophic East Okoboji Lake, where high internal P load coincided with high cyanobacteria biomass in summer blooms of 2017. Conversely, cyanobacteria did not bloom in stratified mesotrophic West Okoboji without apparent internal P loading. Dissolved iron concentration tended to decrease over the summer in these hardwater prairie lakes, perhaps due to Fe uptake by cyanobacteria according to the authors. But iron concentration was similar in the two basins, so that there seemed to be no Fe internal load necessary to sustain the increased cyanobacteria biomass in East Okoboji.

Iron internal load as major cyanobacteria driver is also unlikely for theoretical reasons. In highly eutrophic, sulfur-rich systems, hydrogen sulfide binds iron in the porewater to form solid iron sulfide thus preventing Fe release without interrupting P release from redox-related processes and mineralization (Smolders et al. 2006). Nonetheless, there is no evidence of Fe limitation on cyanobacteria proliferation in such systems. Further, cyanobacteria do not accelerate growth upon iron additions in attempted lake restoration treatments, as long as the sediment–water interface is kept oxic to prevent P release, sulfate concentrations are low, and the dosage is sufficient (Bakker et al. 2015) (Section 6.4.3).

5.6.3 *Increase in herbicides and fungicides*

Herbicides from agricultural activities may improve the growth conditions for cyanobacteria by decreasing the eukaryotic algae, thereby removing this competitor (Beaulieu et al. 2014, French review, cited in 2020 p. 22 *Guidelines for Canadian Drinking Water Quality: Guideline Technical Document*).

While the increased use of P fertilizers has often been acknowledged as a source of increasing P input globally, the increased use of P-containing herbicides is less often considered. Any input of bioavailable P in runoff and input streams can support cyanobacteria growth. Recent investigations in the Lake Erie watershed observed that SRP concentrations increased in the two most eutrophic rivers (Maumee and Sandusky) during the period of decreased and plateaued fertilizer use in these watersheds since the mid-nineties (Spiese et al. 2023). Instead, glyphosate use had increased almost parallelly with the SRP concentrations, and its inclusion in the model yielded closer predictions of observed SRP (which were still higher than predicted) in a mass balance analysis. While glyphosate does not analyze as SRP, its degradation in terrestrial and aquatic systems can convert glyphosate-P to SRP. According to Spiese et al. (2023), the global increase in the use of glyphosate may contribute to the globally observed increase in SRP and hence could also contribute to increases in cyanobacterial blooms.

Fungicides at environmentally relevant concentrations can promote the proliferation of toxic bloom-forming cyanobacteria by inhibiting natural fungal parasite epidemics that would impair cyanobacteria growth (Ortiz-Cañavate et al. 2019). Using genetic differentiation, a diverse, spatially and temporally varying fungal community was detected in western Lake Erie and Saginaw Bay of Lake Huron during *Microcystis* blooms, including potential parasites and decomposers of various phytoplankton species, including cyanobacteria (Marino et al. 2022).

These examples of increased herbicide and fungicide use show that there are anthropogenic disturbances apparently unrelated to direct nutrient additions that could support the proliferation of cyanobacteria and deserve consideration.

5.6.4 Road salt in lakes and salinization

An increase in salinity attributed to anthropogenic activities and climate change has been documented in many freshwater ecosystems. Drought (Brasil et al. 2016), hydrologic alterations, land use, and intrusion of brackish water because of sea-level increases are predicted to increase salinization and thus favour the development and expansion of freshwater cyanobacteria versus other freshwater phytoplankton (Georges des Aulnois et al. 2019). The problem, identified as Freshwater Salinization Syndrome (FSS), has been recognized throughout freshwater systems including lakes, reservoirs, and rivers (Kaushal et al. 2023, 2018; Kinsman-Costello et al. 2022). Freshwater salinization is counted as 1 of 12 emerging threats to freshwater diversity (Reid et al. 2019).

Salinization increases both nutrients N and P, as hypothesized from multiple reviewed studies on urban wetlands and coastal saltwater (Kinsman-Costello et al. 2022). Direct effects on P include decreased nutrient removal by microbial processes and increases in sulfate, pH, and alkalinity, which reduce phosphate adsorption in bottom sediments. Direct effects on N include changes in plant–microbe competition and less denitrification because of the conversion of nitrate to ammonium rather than gaseous N_2 under low redox conditions by promoting dissimilatory NO_3 reduction to NH_4 (Kinsman-Costello et al. 2022).

Indirect salinization effects include increased lake stability and stratification which enhance oxygen depletion, possibly leading to sulfidation, thus supporting redox-related sediment P release. Based on sensor data and modelling of two north-temperate Wisconsin lakes, Mendota and Monona, Ladwig et al. (2021) speculated that winter salt use could delay spring turnover, prolong summer stratification, and increase stability in all seasons but fall. But the stabilization effect from FSS can also decrease the internal load effect on epilimnetic phytoplankton by slowing upward P diffusion through the summer thermocline. In lakes with a high morphometric ratio (Equation 2.1), increased stability can even lead to vernal meromixis preventing spring P input from the hypolimnion but potentially prolonging the anoxic period (Section 5.1.3).

The increase of cyanobacteria under FSS can partially be attributed to higher salt tolerance by cyanobacteria compared to other phytoplankton, based on their ability to cope with osmotic pressure changes. But such tolerance seems specific to species and taxa. For example, only strains of *Microcystis aeruginosa* from brackish regions accumulated sucrose but strains collected in freshwater habitats did not (Georges des Aulnois et al. 2019).

Another suggested effect of FSS includes changes in the ecological network including predator–prey relationships due to increasing zooplankton mortality (Hintz et al. 2022). Possible eutrophication effects by increased salinity and associated lake level decreases due to water abstraction were explained by biological influences of macrophytes, fish, and zooplankton, in addition to increased internal load (Jeppesen et al. 2015).

In conclusion, there are several potential factors that facilitate cyanobacteria proliferation in saline waters during FSS, and internal P loading can be a contributing factor.

5.7 Freshwater systems besides lakes

The influence of internal P loading on cyanobacteria and their proliferation can be affected by the type of freshwater system. The discussion in other sections is mostly based on comparatively natural, freshwater lakes, while here we explore other systems such as reservoirs, rivers, wetlands, and estuaries.

5.7.1 Reservoirs

In general, reservoirs respond with cyanobacterial blooms to climate change and related stressors just like lakes. But specific characteristics render reservoirs especially vulnerable. Their water volume management means that water level, volume, and associated morphometric and hydrological features vary in response to anthropological requirements. Because of their manipulation- and weather-related short- and long-term variability, directional changes from climate change can be more severe with more frequent extremes. With respect to P, reservoirs have a higher retention than lakes with similar morphometric–hydrological conditions (Hejzlar et al. 2006; Nürnberg 2009; Straskraba 1998), so that sediments accumulate P faster than lakes. Consequently, there is more sedimentary P that can turn into mobile P to be released as internal load.

Climate-related droughts creating very low reservoir levels, enhanced stratification, and increased internal P loading and cyanobacteria biomass (*Microcystis* and *Dolichospermum*) are reported globally (Baldwin et al. 2008; Messager et al. 2021; Naselli-Flores 2003) (Section 5.1.9). In their review on mitigation strategies to combat potential impacts of climate change on reservoir services, both internal and external factors were deemed important (Yasarer and Sturm 2016). Elevated temperature increases evaporation which decreases water level (in addition to a direct positive effect on phytoplankton), followed by increased water residence time, stability, anoxia, and the potential for internal P loading. In addition, increased climate extremes such as storm events lead to increased disturbance of watershed processes and greater sediment and nutrient loads to a water body. Expected and experienced droughts exacerbate climate-related stressors on reservoirs that are supposed to deliver water and impede reservoir services by decreasing withdrawal water volume and quality (Yasarer and Sturm 2016).

Low water levels can also be intentionally caused by drawdown in reservoir management, creating similar conditions as from natural climate variation and change. Observations from such reservoir management can thus inform potential climate change consequences (e.g., Section 4.4.2). Reservoirs that experience low water level (Section 5.1.9), either due to natural climate impacts or by management have increased anoxia (Figure 4.10), methane, and redox-related metal concentration in the bottom water, combined with redox-related P as internal load (Deemer and Harrison 2019; Haggard et al. 2023; Nürnberg 2002). Water level drawdown increased chlorophyll concentration and cyanobacteria biomass of *Microcystis* in the northern, semi-arid regional Lake Diefenbaker, Saskatchewan (Abirhire et al. 2023, 2019). It was hypothesized that the high abundance of cyanobacteria during drawdown was associated with thermocline deepening and subsequent internal loading of nutrients, as it was correlated with reduced water level, increased air temperature, and moderate wind speed.

Experimental drawdown in a small dimictic eutrophic reservoir in Virginia, USA, increased stratification strength and surface phosphate concentrations leading to cascading surface water quality deterioration (Lewis et al. 2024a). After a phytoplankton bloom connected to the drawdown, bloom degradation was associated with increased DOC, dissolved CO_2, and NH_4

concentrations, and decreased DO (down to 41% saturation), while total iron and manganese concentrations increased, indicating associated P release.

Effects of drought on internal P load and cyanobacteria have been extensively studied in Brazil (Section 4.4.2, 5.1.9). The location at lower latitude dictates a higher temperature that affects anoxia (anoxic factors are higher for the same TP concentration compared to the temperate lake model, Box 2.1 (Carneiro et al. 2023)) and sediment P release rates, so that internal loading in the southern reservoirs is especially high (Rocha and Lima Neto 2022). In 30 Brazilian water supply reservoirs (Rocha and Lima Neto 2022), internal P loading rates during the dry period were significantly correlated with chlorophyll concentration and Secchi transparency (Figure 5.19) and increased with trophic state, being ten-fold greater for hyper-eutrophic than oligotrophic conditions. Internal P load was controlled by climate variables and correlated positively with water temperature and wind speed, and negatively with water level indicating resuspension mechanisms.

An important feature of reservoirs that influences internal load is their shape. An in-stream (run-of-the-river) reservoir consists of a shallow, riverine inflow, a transition zone with advanced settling, and a deep stratified lacustrine section (Lehner 2024). As discussed, for sediment-released P to cause cyanobacteria biomass to increase, the conditions have to be right, and the system has to be P or P and N co-limited (Section 3.3). For example, in the Snake River reservoir, Brownlee Lake, the western United States, internal P loading increased phytoplankton biomass only in the shallow and transition zone (where chlorophyll was predictable by TP

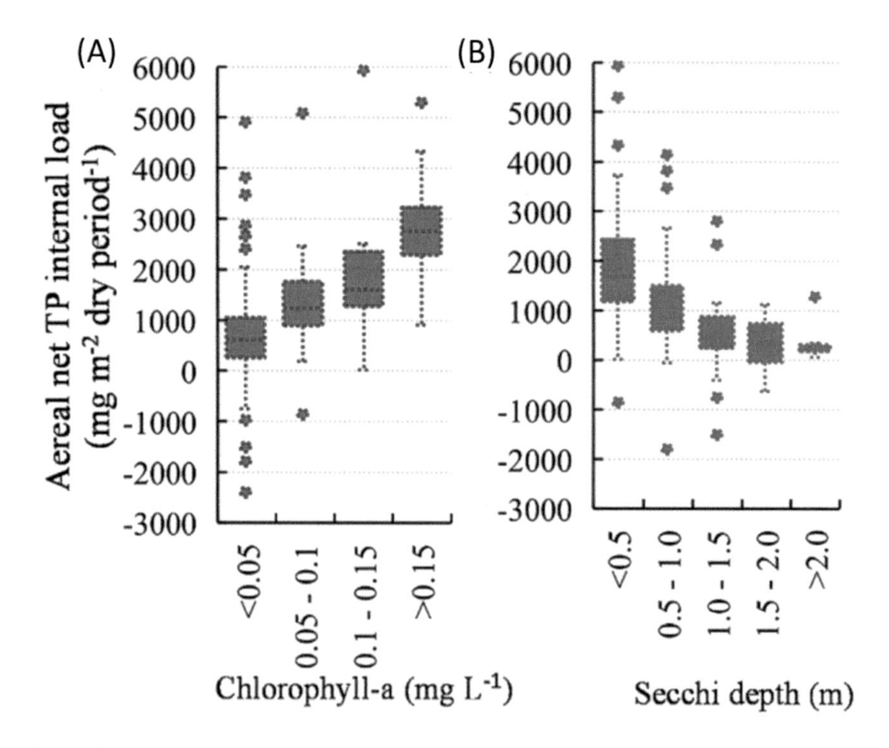

Figure 5.19 Trends of the seasonal net internal P load with (A) chlorophyll and (B) Secchi transparency in 30 Brazilian reservoirs ($p < 0.01$). *Figure 5 from Rocha and Lima Neto (2022), with permission.*

concentration), while chlorophyll concentration remained low (only about 50% of prediction based on TP) in the light-limited deep section at the dam despite large amounts of TP and SRP (Figure 5.20A). Light limitation (Section 3.5.2) occurs frequently in the lacustrine section of reservoirs especially when they contain a meta- or hypolimnetic outlet that generates a large and deep mixed epilimnetic layer. When algal cells are exposed to such deep-reaching water, their residence time in the shaded layer increases, which leads to their light limitation. Under such conditions, internal P loading (determined by anoxia and P depth profiles) remains in the highly bioavailable form of SRP. This elevated SRP (despite decreased TP due to settling) only marginally affects the phytoplankton in the deep reservoir section where it is released from the sediments but fertilizes downstream water systems (Figure 5.20B, C).

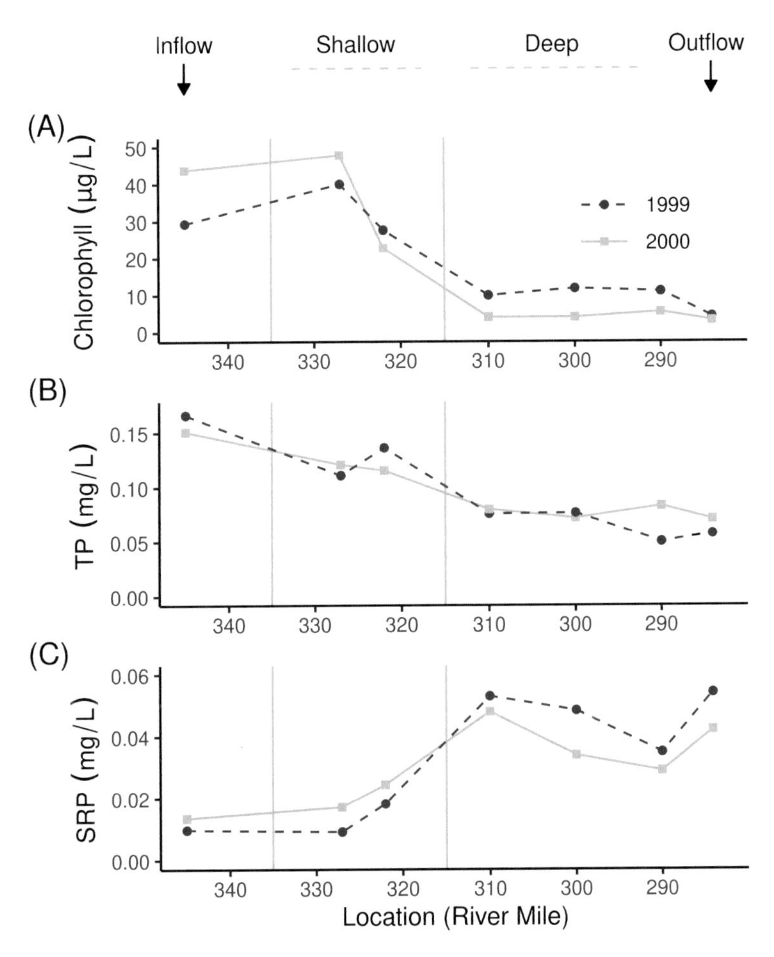

Figure 5.20 Spatial changes along the run-of-the-river reservoir, Brownlee Lake, Idaho/Colorado (Snake River, Hells Canyon), in the mixed surface layer for two growing periods (May to September). Chlorophyll (mainly of cyanobacteria) decreases sharply in the deep lacustrine section close to the dam due to light limitation (A). TP gradually decreases due to settling from inflow to the deep section (B), while SRP sharply increases in the lacustrine basin, indicating a lack of P limitation (C). *Nürnberg unpublished data.*

Because almost all external load originates as riverine input in reservoirs, where inflow TP settles along the reservoir bottom (e.g., Figure 5.20C), long-term decreases in inflow concentrations of TP and chlorophyll are expected to also eventually decrease legacy sediment P and internal load, especially if it corresponds to a decreased anoxic volume as was observed in Brownlee Reservoir, 1995 to 2021 (Naymik et al. 2023).

While reservoirs have special characteristics related to their hydrology and morphometry, internal load effects on cyanobacteria are still similar as for natural lakes. However, climate change effects can be more pronounced and societally important because many reservoirs are in naturally dry regions, increasing their vulnerability in a warmer and dryer climate to both water shortage and eutrophication.

5.7.2 Rivers

Internal loading processes in slow-moving rivers resemble those of reservoirs and lakes, where low flow increases water residence time and supports oxygen depletion and anoxia at the sediment–water interface. For example, when nitrate was used up in the Thames River, Ontario, in the Great Lakes Basin, which occurred during increasing N assimilation and hypoxia, cyanobacteria proliferated in slow-moving sections (Box 3.3, Nürnberg (2007a)), conceivably supported by redox-related sediment P release. Similarly, sediment P release in streams of three east German catchments mimicked a point source although domestic point sources were absent (Dupas et al. 2018). Instead, SRP concentration peaked from summer anoxic SRP release in wetland soils enriched by diffuse agricultural sources and was controlled by temperature, flow rates, and NO_3 inflow.

Modelling the effects of reduced nutrient loads (of both N and P) in the River Havel, a tributary to the River Elbe, Germany, by including sediment mobile P, Lindim et al. (2015) concluded that phytoplankton growth (mainly of cyanobacteria) depended on internal load and the accumulated sediment P in late summer and fall. The reduction of both external P load to combat spring blooms and of internal P load to combat summer and fall cyanobacteria blooms was deemed necessary to improve water quality in this system, while N abatement did not have any effect.

A meta-analysis of 45 studies showed that the potential for P exchange is moderated by sediment and stream characteristics including sorption affinity, stream pH, exchangeable P concentration, and particle size (Simpson et al. 2021). Sediments in streams function as P buffers and often have the potential to either remove or release P to the water. For example, fractionation and P-exchange experiments with stream sediments of mostly agricultural land from north-central Ireland determined large, variable amounts of the reductant-soluble P fraction, BD-P (Table 2.3) (Li et al. 2023). This highly bioavailable P was released into the ditch and stream water during sediment disturbances from storm events, especially following calm-weather periods.

In general, river TP and chlorophyll concentrations are positively and significantly correlated (Figure 5.21) like those of lakes and reservoirs (Figure 2.4), although with different parameters (indicating generally less chlorophyll per unit of TP) and different trophic state boundaries (Dodds et al. 1998). Rivers, especially large ones, can also have similar cyanobacteria taxa. A genetic survey of 12 large US rivers determined mainly *Microcystis* and *Planktothrix* in various quantities with about half contributing to cyanotoxicity (microcystin synthetase gene, mcyE) (Graham et al. 2020; Linz et al. 2023).

Internal loading effects on cyanobacteria can be different in shallow streams and rivers where illuminated water encourages benthic cyanobacterial biofilms or mats (Bouma-Gregson et al.

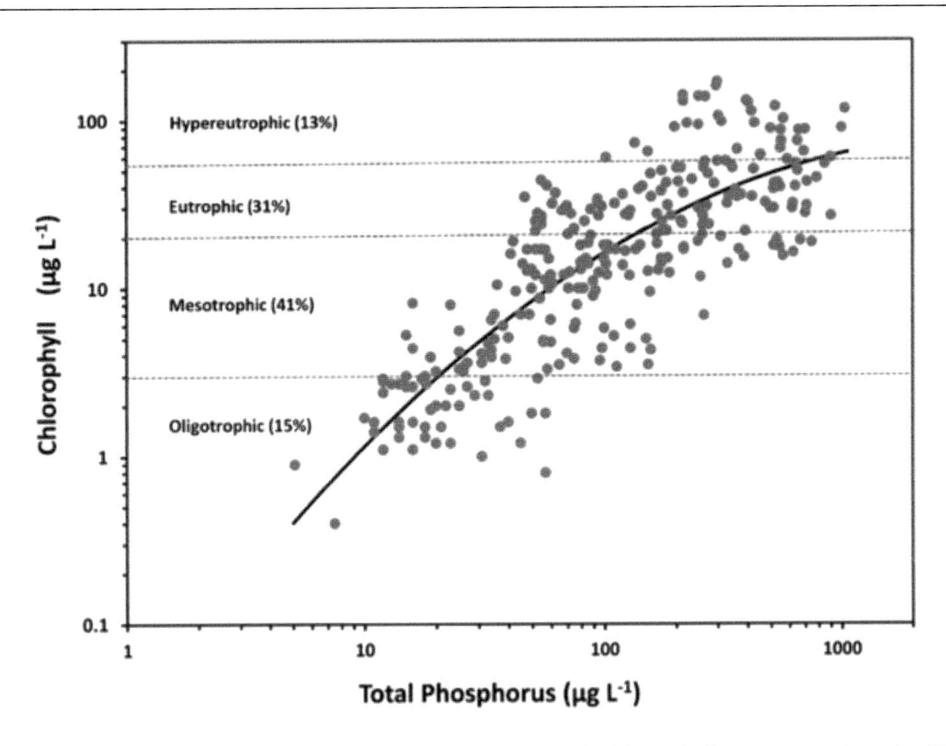

Figure 5.21 Relationship between total phosphorus and chlorophyll concentrations in 292 streams and rivers in North America (79%), Europe (16%), and elsewhere. Dotted lines show suggested boundaries between oligotrophic, mesotrophic, and eutrophic river conditions. *Figure 3 from Wurtsbaugh et al. (2019), with* permission.

2018; Chorus and Welker 2021; Poirier-Larabie et al. 2020). These mat-forming benthic cyanobacteria demonstrate a direct pathway of sediment-derived P to cyanobacteria. The mats can be dangerous to live stock, pets, and humans because of their potentially high toxicity.

A genetic study on New Zealand river biofilms summarizes the known characteristics of benthic cyanobacteria (Tee et al. 2020):

> In freshwater systems, these benthic proliferations form visible mats that coat the beds of ponds, lakes, and rivers. . . . These mats can be readily ingested by animals, and globally many cases of canine death have been linked to the consumption of benthic mats containing anatoxin-producing cyanobacteria . . ., such as *Microcoleus autumnalis and favosus (formerly Phormidium autumnale and favosus*).

As these benthic cyanobacteria species thrive in oligotrophic, low P systems, they appear to obtain their phosphate from the sediments as internal P loading, in addition to organic P (Tee et al. 2020).

> Despite low external phosphorus concentrations, well-established mats are highly enriched in DRP [SRP]. . . . This internal pool of bioavailable phosphorus is thought to be partly derived from large quantities of fine sediment trapped within the mats. Phosphorus release

> from this material likely occurs due to high daytime pHs (>9) generated by photosynthesis, and also due to low nighttime dissolved oxygen concentrations (<4 mg/L), resulting from respiration. . . . In addition, alkaline phosphatase activity in *Microcoleus autumnalis* cultures suggests they have the capability to mineralize organic phosphorus.

In conclusion, there are indications that internal P loading in rivers supports pelagic cyanobacteria in ways similar to those in lakes and reservoirs, but shallower rivers are especially recognized for their benthic, mat-forming cyanobacteria that obtain P from the sediment porewater and can be highly toxic. Climate change likely influences riverine cyanobacteria development in various ways, depending on river flow rates, water levels, nutrient inputs, and the location within its watercourse.

5.7.3 Wetlands

Wetlands as defined here comprise areas where water covers the soil or is close to the soil surface most of the year. That means that they are shallow, but depending on the plant community, which can be emergent and dense (e.g., *Typha* spp. in a cattail Typha marsh), under the water surface, or absent, they can stratify, have enhanced particle and P sedimentation, and can be hypoxic during stagnant conditions in the summer at warm temperatures and in the winter under ice. While natural wetlands are more or less unchanged by anthropogenic influences, constructed wetlands are newly created or transformed from natural wetlands to increase nutrient retention and fauna and flora diversity. All wetlands are exposed to climate change, and the interactions of sediment P release with cyanobacteria are much dependent on hydrological conditions and surface water warming, similar to very shallow lakes.

Natural wetlands are often mentioned in the context of browning from organic acid and Fe-stained DOC increases (Sections 2.6 and 4.4.4). Increased mobilization of redox-related nutrients in wetlands (as internal load to these shallow systems) occurs globally and is especially important in the P delivery to larger lakes and reservoirs, including the Great Laurentian Lakes. Increased frequency of cyanobacteria blooms in extended wetland areas as in the Everglades, Florida (Reddy et al. 2011), the Southern Balkan wetlands (Berthold et al. 2018; Karstens et al. 2015), and boreal freshwaters in Northern Europe and in North America (Björnerås et al. 2019, 2017) with associated Fe increase is, at least partly, attributed to increased redox-related sediment P release (Ekström et al. 2016).

The seasonal pattern of increased SRP export in winter and during snowmelt, with decreased SRP in summer due to uptake by phytoplankton, is well-known in northern natural wetlands and has been observed for a long time (Devito et al. 1989), indicating the importance of redox-related internal P loading in these systems (more details in Section 4.3.2). Further, in a recent experiment with enriched wetland soils of the Maumee River, Lake Erie watershed, P retention increased in warmer and aerated treatments, likely due to P uptake by wetland organisms (Hurst et al. 2022). In contrast, P retention decreased in cooler and anaerobic treatment conditions, indicating redox-related P release. Similarly, spikes of SRP in German streams were attributed to anoxic SRP release in wetland soils (Dupas et al. 2018) (Section 5.7.2).

Increased mobilization of redox-related nutrients in wetlands (as sediment-derived internal load in these shallow systems) around large lakes may also partially explain the high loading of SRP to large lakes and the resulting impact on cyanobacteria proliferation (Larson et al. 2013). For example, April through September SRP fluxes were positive and substantial in two river mouths of Green Bay, Lake Michigan (Larson et al. 2020). SRP concentrations also increased in recent years in the Great Lakes watershed especially from northern forested areas (Singh et al. 2023), and spring tributary SRP load was found to be more important than TP load in predicting cyanobacterial blooms in Lake Erie (Stumpf et al. 2016).

5.7.4 Estuaries and enclosed marine systems

Estuaries and enclosed marine systems also experience cyanobacteria blooms likely related to internal P load with spreading hypoxia. Here, we consider enclosed marine systems or inland seas and include marine areas with less salinity than open oceans that are sometimes affected by anthropogenic enrichment, similar to many estuaries.

The location of estuaries between typically P-limited freshwater and N-limited marine water renders estuaries P and N co-limited to a various extent along their saline gradient. Increased eutrophication by enhanced riverine input of P and N has led to increased hypoxia and cyanobacteria proliferation (Diaz and Rosenberg 2008).

N is usually more limiting than P in estuaries because P is more easily released from clay as salinity increases, there is less planktonic N fixation, and rivers tend to bring relatively P-rich waters. The abundance of P, partially because of release from P-enriched riverine sediments, means that any internal P loading in estuaries due to sediment anoxia can be expected to be less correlated with the proliferation of cyanobacteria than in lakes, while it still fertilizes and supports phytoplankton biomass.

Estuaries encounter climate change effects in a manner similar to freshwater. Heatwaves (ranging up to 10 events per year, and lasting up to 44 days with mean duration of 8 days) have also been increasing in frequency, duration, and intensity as in freshwater systems (Section 4.2.1) and co-occurred with hypoxia and low pH events between 1996 and 2019 at 17 water quality monitoring stations in 12 US protected areas (Tassone et al. 2022). Although co-occurring with increased heatwaves, only changes in external load were considered as an explanation for changes in annual hypoxia and low pH events over the years. The veiled character of internal P loading in systems that are not P limited may be the reason why this mechanism was not considered earlier. Both an early review seeking to determine causes of estuarine harmful algal blooms (Anderson et al. 2002) and a later paper (Anderson et al. 2012) only mentioned groundwater, in addition to known external inputs, as stimulating phytoplankton blooms in estuaries.

In contrast, the importance of internal P loading in the Baltic Sea has been recognized for several decades. Here, a correlation between the extent of hypoxia (annual mean area with <2 mg/L oxygen occupying 5% to 27% of the total bottom area over the time period 1961–2000) and P accumulation was recognized as a contributor to the cyanobacteria blooms and the general state of severe eutrophication (Conley et al. 2009b, 2002). Consequently, a recent review involves internal P load in the causal mechanisms for increasing cyanobacteria in the Baltic Sea by the interaction with climate-change-induced increases in temperature, stratification, and decreases in redox state (Berthold and Campbell 2021).

Chesapeake Bay, the largest estuary in the United States (surface area: 11 600 km^2), has been studied because of its continuing eutrophication for decades. Hypoxia consistently develops in the deeper main stem portion of the Bay (z_{max} = 12 m), while shallower regions are subjected to fluctuating flows and hypoxia. Internal P loading from anoxic sediments and from mineralization of organic sediment has been noted repeatedly as an important cause of increased productivity (Li et al. 2017). Nonetheless, recent models depicting P (and N) loads, retention, and exports did not include explicit estimates of internal loading (Shen et al. 2022) despite the available values of annual hypoxic volumes (Bever et al. 2018) which could be used to compute redox-dependent internal P loads (similar to Approach 3 described in Box 2.3, Section 2.4.2).

Hypoxia and related eutrophication since the 1950s have also been documented in the northern Gulf of Mexico where the Mississippi River discharges into the ocean (Rabalais and Turner 2019). The hypoxic area can reach 23 000 km^2, and relevant internal nutrient loading from sediments has been acknowledged, but not quantified so far.

The importance of redox-related sediment P was determined in a modelling study of a coastal bay in Queensland, Australia (Hamilton et al. 2007). The authors determined that the proliferation of the cyanobacteria *Lyngbya majuscule* was best predicted when including redox state of the sediments as a predictor variable, besides nutrient pool, water temperature, current velocity, and light. They concluded that increased availability of phosphate caused the *Lyngbya* population to grow.

More comprehensive global surveys tend to support individual case studies on estuaries. Mapping of daily marine coastal algal blooms between 2003 and 2020 using global satellite observations at 1-km spatial resolution calculated a global increase in bloom area (13.2%) and frequency (59.2%) but a decrease in tropical and subtropical bloom areas of the Northern Hemisphere (Dai et al. 2023). These authors also found that significantly more blooms occurred in estuaries compared to the open oceans.

An analysis of a global data set (from the Harmful Algae Event Database, HAEDAT, and Ocean Biodiversity Information System for the period 1985–2018) did not find any increase in the frequency and distribution of marine harmful algal blooms globally (Hallegraeff et al. 2021). Instead, they suggested that an increased monitoring of aquaculture production results in a perceived increase in harmful algae events. In particular, aquaculture production increased 16-fold from a global total of 11.35 M tonnes in 1985 up to 178.5 M tonnes in 2018. The number of HAEDAT bloom events over time was significantly correlated with aquaculture production, with all regions exhibiting more HAEDAT events as aquaculture expanded. Aquaculture can enrich sediments with nutrients and organic matter and increase hypoxia in freshwater systems, leading to internal P loading (Li et al. 2020), which can be expected in marine systems as well.

Studies on estuaries and enclosed marine bays suggest: (1) cyanobacteria proliferation caused by potential increases in internal P loading is expected when N ceases to be limiting (demonstrated by N supply increases across the freshwater to marine continuum, Wurtsbaugh et al. (2019)). (2) Aquaculture increases not only nutrients (both N and P) in the surrounding areas but also P (and organic matter enriched with N and P) in the sediments, which then is released back into the open water under coinciding increased hypoxia.

5.8 Importance of internal P load to cyanobacteria proliferation during climate change (Conclusion)

The ultimate objective of this analysis is to increase the understanding of the relationships between cyanobacteria and internal P load in the wake of climate change. Such an understanding then can inform the development of data sets, models, and management strategies for the prevention and mitigation of cyanobacteria.

Theory and case studies support the hypothesis that sediment-derived internal P load can drive cyanobacteria proliferation. Thus, conditions that favour internal P loading and increase its spatial and temporal availability can promote cyanobacterial growth. P limitation occurs frequently in freshwater and brackish systems and is often prevalent during the periods favourable to cyanobacteria growth (Box 3.2). Even when other factors are or become growth-limiting over the growing season, additional quantities of P can be expected to further promote cyanobacteria. However, growth limitation is a construct that varies under natural conditions (Glibert 2020). Detailed observations of physical, chemical, and biological interactions in combination with weather patterns are needed to provide quantifiable evidence of the specific impact of internal load on cyanobacteria proliferation during climate change (Figure 5.15).

Evidence that internal P load supports cyanobacteria is presented throughout this text and includes the following:

1) Elevated P in the anoxic hypolimnion and sediments is highly bioavailable phosphate (Section 2.4.2) and can fertilize P-deficient plankton and cyanobacteria when it is distributed close to the phytoplankton.
    a) In stratified lakes, fertilization of the mixed layer occurs during fluctuations and erosion of the thermocline and during complete turnover (fall and spring).
    b) In polymictic shallow lakes, fertilization occurs throughout the summer and fall; it is most apparent in more productive lakes (eutrophic and hyper-eutrophic lakes).

    As a result:

    c) Chlorophyll concentration and cyanobacteria biomass increase (correlated to TP concentration) throughout summer and fall in shallow lakes and, later, during thermocline erosion and fall turnover, in stratified lakes.
    d) Chlorophyll concentration and cyanobacteria biomass can increase even when external P load is small (i.e., without being correlated to external load).
2) Correlated hypoxia: when sediment P is released from redox-sensitive iron-P compounds, cyanobacteria proliferation is correlated with hypoxia.
    a) Cyanobacteria can be correlated with hypoxia even for low-productive conditions (oligo-, mesotrophic lakes).
    b) Cyanobacteria are not always correlated with hypoxia under interfering chemical conditions, that is, in brown-water or acid-sensitive lakes with P-adsorbing humic acids or aluminum from acid deposition.
3) Cyanobacteria proliferation is correlated with decreasing lake volume and lake level during periods of water withdrawal or drought, mainly because of increased exposure to sediment-derived P. Reduced water volume can also increase the concentration derived from internal P loading by amplified hypoxia and P release because of increased temperature and the concentrating effect of evaporation.
4) Cyanobacteria blooms are reduced after a successful abatement of internal P load.
5) Cyanobacteria proliferate in nutrient-poor lake water with accumulated legacy sediment P.
6) Models that successfully predict water quality variables, including water TP and chlorophyll concentrations in climate change scenarios, include internal load (Section 5.3).

Many current publications are in general agreement that internal P loading associated with anoxia is augmented by climate change and can cause increases in eutrophication and cyanobacteria (Table 4.1 and Table 4.2). Even when the predominant influence of internal loading is not always realized, redox-sensitive P pools are acknowledged as important factors when comparing synergistic effects of urbanization, agriculture, climate change, and industrialization on cyanobacteria bloom formation (Scholz et al. 2017).

The start of internal P loading in a previously unaffected lake may trigger rapid water quality deterioration. Potential resistance and resilience with respect to increasing internal P load from climate change are based on several pre-conditions that likely affect specific lakes in different ways. In this context, *resistance* is the tendency of a system to oppose change, and *resilience* is the level to which a system recovers after a disturbance (Thayne et al. 2022).

- *Amount of mobile P in the sediment.* The amount of mobile P influences the potential P release. If there is little releasable sediment P, climate change will not effectively increase internal load; but where former or current anthropogenic influences provide legacy P accumulations, the potential for P release is high. Geochemical changes in P adsorption properties of lake sediments and catchment soils can influence P release, for example, in areas affected by acid deposition.
- *Spatial and temporal variability in anoxia at the sediment–water interface.* Increasing anoxia is expected from climate change and will affect shallow poly- and oligomictic lakes the most. Increasing anoxia is expected to be slower in very stable and deep lakes.
- *Spatial and temporal contact between P resources and cyanobacteria.* The movement of sediment-released P into cyanobacteria habitat (i.e., surface waters with suitable solar radiation, temperature, secondary nutrients, or metalimnion for deep-water trait R taxa) is affected by global location and altitude. In shallow mixed lakes, contact between cyanobacterial habitat and P-releasing sediments is constant throughout the optimal growth season, and climate change is predicted to extend this season. In deep stratified lakes, thermocline erosion may be delayed by climatic warming, creating fewer cyanobacteria blooms and shifting them to later in the season. In alpine lakes, light is plentiful so that increased warming and atmospheric stilling can extend the optimal habitat and growth season.
- *Sources of cyanobacteria.* Resistance to the proliferation of cyanobacteria is weakened by the ample availability of suitable species of cyanobacteria. Upstream sources and inflows carry cyanobacteria from neighbouring water systems, and vegetative resting cells and akinetes are introduced from the lake bottom to inoculate surface waters as has been described for many systems (Cottingham et al. 2021) (Section 3.5.3).

The concept of regime shifts and tipping points has been used in ecology and limnology to describe system changes that are extreme and not easily reversible (Scheffer et al. 2009) and likely apply to climate change (Woolway et al. 2022b). It seems plausible that an increase in internal P loading in response to climate change could create such tipping points. For example, elevated external loading, followed by external load abatement and by eutrophication from internal sources, was described as a regime shift in the large shallow mixed Lake Balaton in Hungary (Istvánovics et al. 2022).

Early warning signals for such tipping points and critical transitions (Scheffer et al. 2009) are needed to provide guidance for lake management. Their detection requires quantitative analysis of large populations of lakes using high-frequency monitoring combined with the investigation of past changes using paleolimnological records (Spears et al. 2016a). Generally, early warning signals include increased variability and fluctuations in variables such as lake water TP concentration, thus providing an indicator of imminent eutrophication (Carpenter and Brock 2006). Early signals pertinent to internal P loading include:

- Increased water column stability, especially during the potential cyanobacteria-growing period.

- Increased anoxia at the sediment–water interface and above.
- Increasing trends in variables related to climate change that can enhance internal P load, including increased air temperature and drought, decreased precipitation and ice cover, increased frequency of extreme events, and trends in specific climatic indices.

Overall, both limnological and climate variables determine the potential importance of internal P loading on eutrophication and on cyanobacteria proliferation in any specific freshwater system.

Chapter 6

Treatment options

Abatement and prevention of cyanobacterial proliferation

Some treatment options address cyanobacteria characteristics specifically. For example, artificial mixing of lake surface layers can suppress buoyant cyanobacteria (Visser et al. 2016). Deepening of the mixed surface layer can induce light limitation, thus limiting cyanobacterial bloom development. Treatments including algicides and other approaches aiming to directly affect cyanobacteria are available as well. However, such methods treat the symptom and not the cause. Algicides often provide only short-term benefits or, worse, increase toxin availability by cell lysis, because they can promote the release of otherwise intracellular cyanotoxins and move them from particulate into dissolved forms (Lürling and Mucci 2020; Mucci et al. 2017; Weenink et al. 2015).

Instead, the cause, for example, increased nutrients, should be best addressed. Considering the involvement of internal P loading with cyanobacteria bloom proliferation and eutrophication increase during climate change (Section 5.8), internal P load detection and quantification, followed by internal load abatement and management, are often needed, in addition to external load abatement and other measures.

The following sections concentrate on ways to diminish cyanobacteria by addressing internal P loading, including legacy internal load by means of external load reduction.

6.1 External nutrient input control

6.1.1 P input reduction

Large amounts of external P load can support cyanobacteria, especially if the input is in a bioavailable form like untreated waste water. For example, Lake Washington, in Seattle, USA, developed cyanobacterial blooms after receiving sewage effluent for over two decades, without having an anoxic hypolimnion and redox-related sediment P release. Lake Washington's water quality improved rapidly in response to a drastic decrease in external P load from sewage diversion (Edmondson 1970). Chlorophyll concentrations and the share of cyanobacteria decreased with the reduced P-loading after a lag of several years (Edmondson and Lehman 1981).

In addition to the possible direct impact on cyanobacteria, P input from any source settles to the sediment after potential recycling within the water column depending on P availability. This P sediment retention can be high (up to 90% of annual input) and is foremost controlled by hydrology (including the annual flushing rate or water load), as realized in P mass balance modelling (Section 2.4). Highly bioavailable P forms, for example, from point sources, can be immediately taken up to support cyanobacteria and can contribute to internal P loading after settling (Fastner et al. 2016) (Section 2.4.1). Consequently, external P load needs to be addressed to prevent and decrease the accumulation of legacy sediment P and future internal P load (Jeppesen et al. 2014).

DOI: 10.1201/9781003301592-6

External load is generally most significant when the area ratio of watershed to lake is large, which provides a quick estimate of the relative importance of the nutrient sources. Because of the dependence of internal load on variables that often vary between years, including stratification, hypoxia, temperature, and other climate-related variables, and do not influence external inputs in the same way, a tight correlation between external and internal loads is not expected. In stratified lakes without anoxia (Nürnberg 1984c) and in lakes formerly impacted by acid deposition (Nürnberg et al. 2018), external load and P retention can be high, without any internal loading. In such lakes, the beginning of sediment P release provides a tipping point (Section 5.8).

The disconnect between external load and lake water TP after a major external load decrease, without any subsequent change in water TP concentration has been described repeatedly since the seventies and was reported as a sign of internal P load (Larsen et al. 1981; Nürnberg et al. 2013b; Sas 1989). It is likely that the diagenetic processes that create a lag between the accumulation of P from external sources and P release as internal load prevent simple correlations of simultaneous load values. Such processes may be responsible for the sudden beginning of P release and associated increase in lake TP concentration after long periods of external load input (Gebremariam et al. 2021; Kröger et al. 2023).

Therefore, while long-term averages of annual internal loads tend to be lower at lower external loads (especially if the external input was high in the past but is declining, e.g., in Lake Simcoe, Ontario (Nürnberg et al. 2013a)), these averages depend on the history of loading and are not always closely correlated (Nürnberg and LaZerte 2001). Similarly, annual changes with time rarely follow a comparable pattern for external and internal loads, as observed for 18 years of load estimates in shallow Lake Winnipeg (Figure 2.13). Nonetheless, Tammeorg et al. (2022a) observed an internal load decrease by about three fold that was significantly correlated to that of external load in large shallow Võrtsjärv in 1985–2020. The change was attributed to an observed decrease of the mobile P pool. On the other hand, a fast release of settled organic P from same-year external input was described in a deep German lake, where recent sediment capping was apparently preventing the release from redox-related P diagenesis (Hupfer and Lewandowski 2005).

Diminished external P input also leads to a decrease of organic material that fuels sediment oxygen demand. But the decrease in anoxia usually lags behind a decrease in sediment P release rates. For example, modelled nutrient cycling and lake metabolism predicted a much slower recuperation from anoxia compared to P concentration and algal biomass after potential external nutrient reduction in eutrophic Lake Mendota, Wisconsin, USA (Hanson et al. 2023). Also, a lag in the recuperation of oxygen was observed until 1990 in Lake Erie after large waste water reduction efforts in the seventies (Charlton et al. 1993). Meanwhile, P and chlorophyll concentration and Secchi transparency reached mesotrophic values already in 1979.

Nonetheless, hypoxia eventually decreases with load reduction measures. For example, decreased inflow concentrations of TP and chlorophyll corresponded slowly to a decreased anoxic volume over a period from 1995 to 2021 in the Brownlee Lake reservoir (Naymik et al. 2023) (Section 5.7.1). This trend indicates that redox-related internal P load likely decreased in response to a decrease in external load.

In conclusion, external P load management promises to help prevent the accumulation of releasable sediment P and of oxygen-depleting organic matter. The prevention of internal P load by various procedures of external P load abatement was considered to be most important in three lakes in Europe, New Zealand, and China (Spears et al. 2022). Addressing P release was determined to delay lake degradation involving cyanobacteria proliferation.

6.1.2 N input reduction

The vast increase in recent N fertilizer production using the Haber Bosch process (Glibert 2020) (Section 2.5) has created a large surplus of legacy N compounds in catchment soils and associated large N loads into freshwater and marine systems. Further, N input may contribute to the worldwide proliferation of cyanobacteria in already highly enriched systems where increased internal load provides sufficient P resources. Increased urea use (60% of all N fertilizer throughout much of the world) has been held responsible for the increase in HABs worldwide (Glibert et al. 2014a). The likelihood of N increase throughout freshwater systems is projected due to increased fertilizer application rates and increases in total precipitation and extreme events (Sinha et al. 2017).

In this context, the benefit of controlling only a singular nutrient (usually P, Section 3.3) to combat cyanobacteria, rather than both P and N, has been challenged (Glibert 2017). As the recent eutrophication with an explosion of cyanobacteria blooms is supported by the increase of internal P loading, nitrogen management may be an additional approach to address and reduce cyanobacteria proliferation and possibly toxin production (since cyanotoxins require a surplus of N, Section 3.3), if the following conditions are met:

- The lake is eutrophic or hyper-eutrophic (Table 2.2).
- There is no P limitation throughout the cyanobacteria growing period.
- Nitrogen levels are high.
- There are few N-assimilating species. (This is not a prerequisite but would accelerate N-management effects.)

When P and other factors (e.g., light, temperature) are not limiting, N limitation can arise so that N becomes an important nutrient to control cyanobacteria (Section 3.3.3). The interplay of trophic state and nutrient limitation is summarized by Qin et al. (2020b) who suggest further investigations since N abatement is so costly.

> The importance of N limitation increases but P limitation decreases with lake trophic status while N and P colimitation occurs primarily (59.4%) in eutrophic lakes. These results demonstrate that phosphorus reduction can mitigate eutrophication in most large lakes but a dual N and P reduction may be needed in eutrophic lakes, especially in shallow ones (or bays). . . . While these results imply that more resources be invested in nitrogen management, given the high costs of nitrogen pollution reduction, more comprehensive results from carefully designed experiments at different scales are needed to further verify this modification of the existing eutrophication mitigation paradigm.

Also, a simple decrease of TN may not decrease eutrophication most effectively because of the variety of N-forms involved (Section 2.5). Redox states of inorganic N-compounds affect sediment anoxia and consequently redox-related P release. Reductive processes during oxygen depletion can convert nitrate to N_2, and N-forms with higher redox state like NO_x are replaced by those with lower redox state like ammonium later in the summer (Box 3.2). Nonetheless, a low spring nitrate concentration would reduce the possible cyanobacteria proliferation later in the summer in systems where very low nitrate concentrations (down to 0.1 mg/L) coincide with cyanobacterial blooms (e.g., in a eutrophic river system in the Lake Erie catchment, Box 3.3, Nürnberg 2007). Therefore, N management must include a lake system analysis to determine which N forms, reduced (ammonia and urea) or oxidized (nitrate and nitrite) compounds should be decreased to best control cyanobacteria blooms (Chen et al. 2019b).

While ammonium is a potent fertilizer and easily incorporated into cyanobacteria, the influence of urea is more complicated (Section 3.3.2). Urea's effect on anoxia and sediment P release depended on the resulting ultimate NO_3 concentration in a mesocosm experiment (Ma et al. 2022). The conversion of urea to NH_4 and then to NO_3 in oxic waters reduced P release by increasing the oxygenated sediment surface layer, but urea first promoted P release by lowering sediment pH and stimulating the production of alkaline phosphatase activity (APA).

Several case studies demonstrate the potential benefit of external N control (Section 3.3.3). Less available N decreased cyanobacteria proliferation in Lough Neagh during the late summer periods of N limitation due to internal P loading (McElarney et al. 2021). While heterocyte development in N-assimilating species increased, they could not make up for N limitation and thrived on ammonium (Andersen et al. 2020, Figure 3.8). An internal N load of NH_4 was not sufficient to support cyanobacteria blooms in late summer in Maumee Bay, western Lake Erie. It was concluded that "processes driving eutrophic systems to NH_4 limitation or colimitation under warm, shallow conditions support the need for dual nutrient (N and P) control" (Gardner et al. 2017).

It was further observed in the Maumee River that the ratio of reduced N to NO_x was positively correlated with chlorophyll and phycocyanin (from 2009 to 2015 (Newell et al. 2019)). This observation was taken "to demonstrate the urgent need to control N loading, in addition to current P load reductions, to western Lake Erie and similar systems impacted by non-N-fixing, toxin-producing cyanobacteria."

Similarly, decreasing external N inputs controlled summer cyanobacteria including N_2-fixing taxa, which did not compensate for the N-deficit in Müggelsee, Germany (Shatwell and Köhler 2019). Despite a correlated decrease of P and N external load, the authors concluded that a "P-only control strategy would not have been as successful."

Most of these case studies involved lakes that were hyper-eutrophic and experienced large internal P loads from former nutrient inputs. Without this added P source, the cyanobacteria biomass would not have been sufficiently high so as to experience N limitation. Future N increases and any additional input from internal P loading are expected to impact lake productivity, leading to the demand for dual nutrient reduction in specific high-productive lakes (Jeppesen et al. 2024a; Paerl et al. 2020).

In conclusion, external N load abatement may be proposed as cyanobacteria control (depending on the size of external P load), when internal P loading is so large that methods directly addressing it (Section 6.2) would not work or are not feasible. Therefore, one could say, when all else fails and when N is high, control N input. However, without a consistent and long-term commitment to P control, the sediments will keep accumulating releasable legacy P that can support cyanobacteria blooms as soon as needs for other nutrients and conditions are met. Further, species can respond to changes in nutrient availability under many conditions so that N-fixing cyanobacteria may develop during periods of temporary N shortage (e.g., Box 3.3, (Nõges et al. 2020)).

6.2 Internal P load abatement in general

While the way to ultimately reduce internal loading is the reduction of P input from external sources to prevent the accumulation of legacy P in the lake bottom sediments, for several reasons, the direct management and reduction of internal P load can be required.

The mechanisms and characteristics of internally derived P are much more capable of supporting cyanobacteria than P from external sources, and this difference is enhanced by climate change trends (Figure 5.1). In many cases, the summer cyanobacteria proliferation is supported by internal sources, while the spring diatom bloom is often triggered by external P sources. Even though the settling diatoms may contribute to the summer internal load, the most direct approach in general is the prevention of summer P release. In this context, the treatment of internal P load rather than of cyanobacteria directly is recommended. Further, external load abatement is not always possible to the extent necessary, could take a very long time, or would not prevent blooms in the presence of internal load (Lehman 2023). Therefore, the direct management of internal sources is needed for the more immediate restoration of many eutrophic systems (Huser et al. 2023; Messina et al. 2020; Osgood 2017; Zamparas and Zacharias 2014). Many case studies support cyanobacterial bloom abatement by treatment approaches that diminish internal loading from the sediment (Section 5.2).

For example, early simulation modelling of Lake Rotorua, New Zealand (Burger et al. 2008), based on the enhancing effect of sediment nutrient (SRP and NH_4) release rates on water nutrient concentrations, indicated that "reductions in sediment nutrient fluxes would be more effective in reducing cyanobacterial biomass than similar proportional reductions in catchment fluxes, due to the coincidence of large sediment nutrient release events with high cyanobacterial biomass."

But, as well, insufficient abatement of external P sources has decreased in-lake restoration effects with time. Restoration attempts were compromised, partially because of the persistently high loads from the catchment during a hypolimnetic withdrawal treatment (Tu et al. 2020), after dredging (Kiani et al. 2020), after alum additions (Nogaro et al. 2020), and after a P-binding lanthanum treatment (Nürnberg and LaZerte 2016a).

Climate change is predicted to increase the challenges to lake restoration. Extreme air temperatures as encountered in predicted heatwaves changed the sediment P exchange processes from a chemical to a more prominent biological base, so that biogeochemical processes were significantly accelerated, involving pH and DOC effects on chemical adsorption and stimulated nitrification (Zhan et al. 2022, 2021). Mesocosm experiments in a hyper-eutrophic urban canal determined that a heatwave likely increased nutrient release (SRP, NO_x) and cyanobacteria in all treatments (Zhan et al. 2022). Consequently, of the treatments described below (Sections 6.3.1, 6.3.4, 6.4.2, and 6.4.3), "dredging and lanthanum modified bentonite exhibited the largest efficacy in reducing phytoplankton and cyanobacteria biomass and improving water clarity, followed by iron-lime sludge, whereas aeration did not show an effect."

Sustainability of the proposed actions should be considered (Tammeorg et al. 2024a) throughout the planning process for internal load abatement. The principal of circular economy would ensure "the use of nutrients, sediments, or biomass that are removed from a lake, in agriculture, as food, or for biogas production". Suggestions for a sustainable approach include lake-specific assessment, prevention of any harm to the target and downstream ecosystems, cost-effectiveness, and the minimization of conflicts in public interests and of the need for repeat interventions.

The trajectory of a lake with an outflow to recover with respect to nutrient concentration and eutrophication depends on the interplay of external input that accumulates on the sediments and internal load that decreases mobile sediment P. To determine whether a specific lake sediment P concentration decreases because of sediment P loss as P release or still increases because of settling P, the principal of *threshold P external load* was developed (Nürnberg 1991). *Threshold external load* is defined as the external P load value at which the flux downward from external sources (e.g., modelled as the product of external load and retention due to settling, Equation 2.7) matches the flux upward from anoxic sediments by also considering sediment focusing (using the ratio of lake area over hypolimnetic area). To prevent sediment P accumulation, external load must be equal or below the threshold value, which will eventually lead to a decrease in internal loading from bottom sediments. This term can be used to quickly determine the trajectory in a lake's development with respect to restoration by external and internal load abatement.

In this context, Horppila (2019) warns that sediment nutrients are often not taken into account in general lake assessment, so that an attempt is made to improve water quality while ignoring the sediment. This addresses the aim of many restoration approaches, including chemical interventions and aeration, which strive to remove nutrients from the water to settle and retain them on or in the sediments. These approaches can delay the long-term recovery of lake ecosystems, when they harbour accumulated harmful sediments and decrease the value of threshold external load.

The time it takes for lakes to recover is an important consideration in the choice of a lake treatment and varies between and within the approaches. In chemical applications, the time depends foremost on the dosage and extent of application. In aeration and oxygenation treatment, time and energy input play a role. In hypolimnetic withdrawal applications, the amount and rate of P export influence the outcome. Dredging requires technical skills that avoid the disturbance of sediments and dispersal of P-enriched porewater into the lake water; success is largely influenced by the depth and quantity of removed sediments and the resultant lake depth.

Models have allowed more precise predictions for the time it takes to achieve lake restoration success after external load reduction while considering internal P load (Carleton and Lee 2023). This study modelled mean lag times of 13.1 and 39.0 years to reach 50 and 75% declines in water column TP concentration for more than 70 000 lentic water bodies in the continental United States. Another study involving Irish Lough Neagh and 23 lakes from the literature estimated 8–20 years for lakes with an internal load to reach their targets (Rippey et al. 2022). Much thought has been put to how to best determine lake response or recovery times expected after interventions by mass balance and mechanistic models (Chapra 1997; Cooke et al. 2005; Sas 1989)

Lake management to combat sediment P release comprises three basically different approaches and addresses (a) the external P source (Section 6.1.1), (b) the internal P source directly, or (c) increased P retention, mainly by chemical treatments, which leads to an accumulation of sedimented P (Figure 6.1, Figure 6.2). The most direct internal P load abatement includes hypolimnetic withdrawal and sediment removal. These two approaches also reduce nitrogen and organic matter as an additional benefit. Other approaches tackle the two mechanisms that lead to internal P load (Equation 2.10), which are the areal P-release rate during favourable conditions (foremost dependent on the mobile sediment P concentration and environmental conditions related to redox potential and temperature) and the spatial and temporal extent of the area that supports P release (e.g., anoxic factor, Box 2.1). All successful treatments of internal load decrease lake productivity and hence are associated with decreases in nitrogen and organic matter.

The various approaches to combat sediment P release shown in Figure 6.1 and Figure 6.2 are based on physical (Section 6.3), chemical (Section 6.4), and combined approaches, including biological control (Section 6.5). In-depth descriptions of internal P load abatement are available in many publications (Cooke et al. 2005; Lürling et al. 2020; Zamparas and Kyriakopoulos 2021), sometimes in combination with treatment suggestions specific to cyanobacteria control (Chorus and Welker 2021; Huisman et al. 2018; Ibelings et al. 2016b; Kang et al. 2022a; Paerl 2017). Here, only an overview of the relevant treatment approaches with their suitability and success is presented.

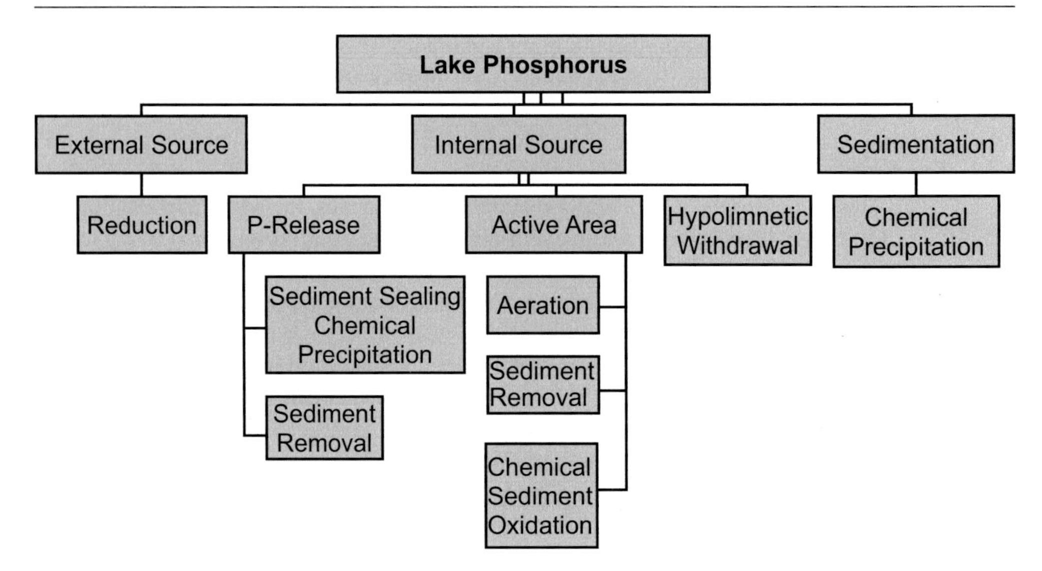

Figure 6.1 Approaches to reduce or prevent legacy P accumulation and current internal P load.

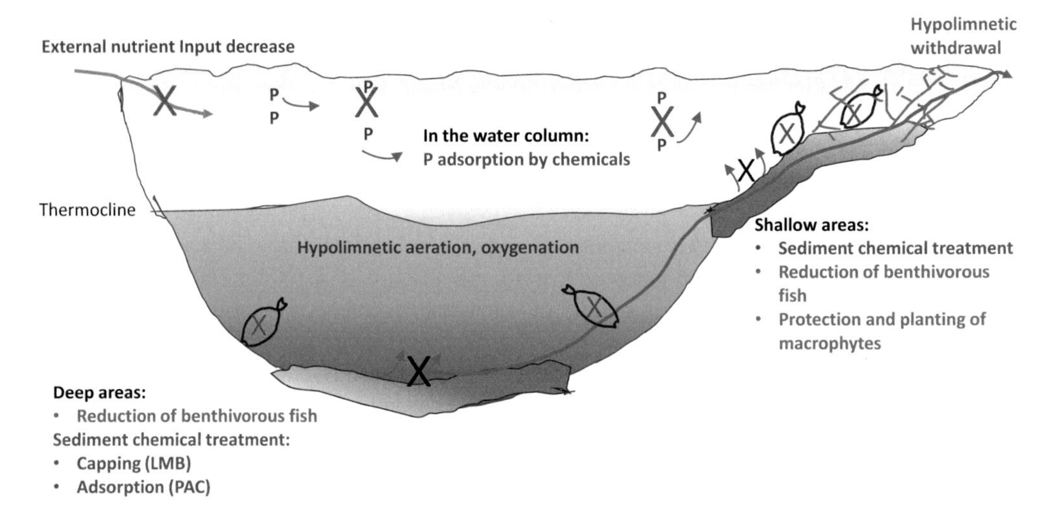

Figure 6.2 Selected procedures to treat internal P load.

An efficient plan to address internal P load would include the following steps:

1. *Detection* of internal P loading (Table 2.4):

   a. Summer: early detection and quantification of P (orthophosphate and TP) and related variables are needed to identify trends that are potentially related to climate change: increasing trends during the warm season in anoxia; and P, Mn, and Fe concentrations (in the hypolimnion in stratified lakes and in the mixed water layer in shallow, polymictic lakes); in stability and stratification; and in low-flow and drought periods.

   b. Winter: increasing trends of stratification under ice; of anoxia; and P, Mn, and Fe concentrations; and winter internal P load. Conditions under ice are useful indicators of conditions in the summer open water especially in shallow polymictic lakes that do not stratify in the open-water season. (If a lake experiences anoxia and P release in the cold, stratified layer under ice, it will experience even higher P release rates and extended anoxia at the sediment–water interface during warmer periods, even if these incidences are not obvious under the summer mixed conditions.)

2. *Quantification* of internal load in comparison to external load:

   a. Internal load estimates have to be comparable to external load estimates. Estimates must cover the same period, need to be a gross estimate and retention by settling must be considered, describe the same compounds, which is usually TP, and be expressed in the same units (Section 5.5).
   b. There exist several approaches to estimate internal P load including (1) *in-situ* water-column TP increases, (2) net and gross estimate from annual P mass balance, (3) gross estimate from P release rate and anoxic factor (Box 2.3), and lake-specific process-based models (Section 2.4.2).

3. *Management* to decrease internal P load:

   a. External P load reduction to decrease the accumulation of sediment P and (often delayed) anoxia (Section 6.1.1)
   b. Internal P load reduction (Sections 5.2, 6.2, 6.3 and 6.4)

4. *Monitoring* and reporting treatment outcome

   a. Comparison of water quality and lake conditions before and after treatment by extensive monitoring efforts.
   b. Evaluation, reporting, and dissemination of the treatment effect to the scientific and societal community.

6.3 Physical internal load control

A review on "controlling internal phosphorus loading in lakes by physical methods to reduce cyanobacterial blooms" implies cyanobacteria abatement through in-lake restoration techniques that directly address internal P load (Bormans et al. 2016). Such physical measures include sediment removal (Section 6.3.1), flow-related treatment (Section 6.3.2), artificial mixing and destratification (Section 6.3.3), and aeration or oxygenation of the whole water column or specific water layers (Section 6.3.4).

6.3.1 Sediment removal: dredging and excavation

Sediment removal addresses both parts of internal loading, the active release area and the P release rate (Figure 6.1). It can be conducted in the lake water without drawdown (dredging) or after the complete removal of the water (excavation). When using this technique for internal load abatement, the exposed sediment layer must have less accessible mobile P after removal, which can be challenging (Moore et al. 1988). Sediment removal is expensive and often requires an extensive permitting process. Removed material has to be examined for toxicity and then deposited away from the lake; any drainage from the dredged material would have to be captured to prevent fertilization of ground and surface water.

Nonetheless, 17 applications were described with variable effects on cyanobacteria and often poorly addressed external P input (Lürling et al. 2020). A successful application in a small shallow Estonian lake decreased mobile sediment P and internal P loading for several years (Kiani et al. 2020). The removed sediment could then be successfully reused in agriculture as fertilizer (Kiani et al. 2023). But the improved lake water quality was later impeded by large nutrient inputs from external sources that overrode these beneficial effects.

Dredging has been used to combat the fertilizing effects of nutrient-enriched sediments surrounding aquaculture. Aquaculture increases organic and nutrient settling leading to hypoxia and sediment P release in freshwater (Bristow et al. 2008; Paterson et al. 2011) and marine systems (Section 5.7.4). Shortly after dredging, SRP diffusion flux and organic matter were reduced at the sediment–water interface but then rebounded in a newly formed deposited layer (Li et al. 2020).

It appears that dredging benefits are often hampered by continuing P loading that recreates an enriched sediment layer. If the external P flux cannot be sufficiently decreased, "additional in situ restoration techniques that improve the oxide layer of surface sediments and reduce sediment suspension" are recommended (Li et al. 2020).

6.3.2 Flow-related treatment: reservoir flow management and hypolimnetic withdrawal

Operational management strategies by altering hydrodynamic conditions are especially useful in reservoirs. Hypolimnetic withdrawals, surface flushing, pulsed inflow, and artificial mixing were reviewed to "increase understanding of the relationships between HAB drivers and reservoir operations" (Summers and Ryder 2023). In addition to the control of internal P loading, flow management can render growth conditions for cyanobacteria less favourable as discussed elsewhere (Section 3.5.3).

Bottom water hypolimnetic withdrawal is a treatment to reduce internal P loading in stratified lakes (Nürnberg 2007b). This treatment withdraws lake water preferentially from the nutrient-rich, low-oxygen hypolimnion instead of the surface and hence increases P export. The decrease of internal P loading resulting from forced P export was significantly correlated to decreases in lake water TP concentration (Figure 6.3). The aim of hypolimnetic withdrawal is the decrease of mobile P in the sediments as was determined in Swiss Burgäschisee (Tu et al. 2020). Hypolimnetic withdrawal is relatively cost-effective and needs only low maintenance efforts and energy input when passive syphoning by gravity is possible. It is especially successful for the treatment of metalimnetic, low-light-adapted cyanobacteria (e.g., trait R, Table 3.1). Chlorophyll and biomass of *Planktothrix rubescens* decreased during hypolimnetic withdrawal, in addition to the decrease of surface-dwelling cyanobacteria species (Nürnberg 2020b, 2007b, 1987; Nürnberg et al. 1987; Sosiak 2022). As in all in-lake restoration activities, insufficient decline in other (external) P sources diminishes success. For example, Burgäschisee remained eutrophic after 40 years of hypolimnetic withdrawal despite decreased P retention and lower mobile P content in the surface sediments due to persistent hypolimnetic anoxia and external P input (Tu et al. 2020).

Related treatment methods include re-release of the withdrawal water into the same lake. With any withdrawal method, the withdrawn water has to be treated before being discharged, possibly in a water treatment plant, best after removal of the nutrients for potential reuse (Nürnberg 2020b; Silvonen et al. 2023, 2022), increasing the potential sustainability of this method (Tammeorg et al. 2024a)

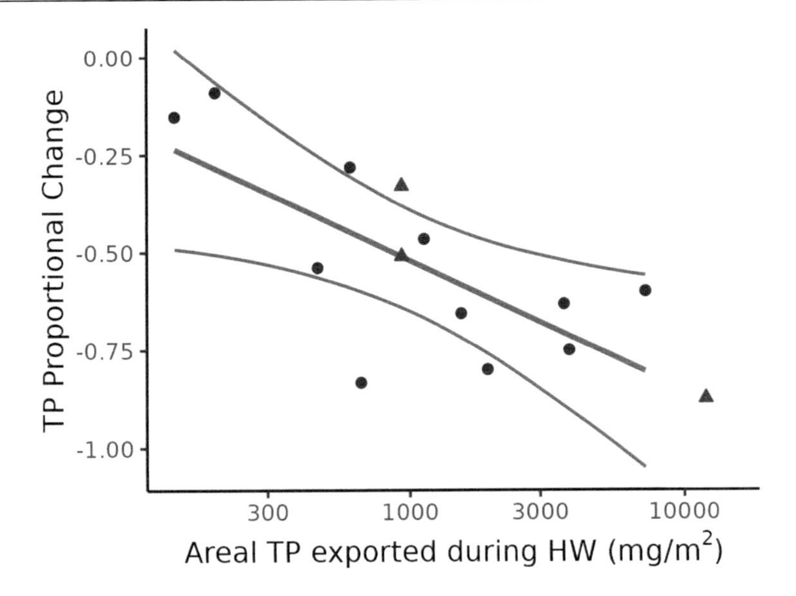

Figure 6.3 Proportional change in epilimnetic summer TP as a function of TP exported during the corresponding period (3–30 years) by hypolimnetic withdrawal (HW). The regression line with the 95% confidence band is shown for 11 long-term operations (filled circles, n = 11, R^2 = 0.50, p < 0.001), together with three independent case studies (triangles) that support the relationship. *Figure 5 from Nürnberg (2020b) with permission.*

Surface flow augmentation that increases the water flushing rate would enhance P export and could dilute surface P concentration when sufficient water with low P is available (Kowalczewska-Madura et al. 2022; Noordhuis et al. 2016; Welch and Jacoby 2009). However, such a treatment is rarely used because clean water sources are usually lacking.

6.3.3 Artificial mixing and destratification

Strong thermal stratification delays the nutrient intrusion from the hypolimnion into the epilimnion thus benefitting surface water quality, while periods of entrainment enhance phytoplankton biomass (Section 5.1.3). Nonetheless, artificial mixing and destratification have been applied usually in an effort to increase oxygen in the whole water column and, consequently, to decrease anoxic sediment P release. However, as artificial summer destratification distributes the nutrients throughout the water column at a time when light and temperature are high, it can benefit cyanobacteria and enhance cyanobacteria blooms; for example, destratification by aeration has decreased water quality by increasing P in a study on 212 Minnesota lakes (Beduhn 1994).

Similarly, destratification for almost 20 years by pumping epilimnetic water into deeper water layers in formerly stratified regions of eutrophic Finnish lake Tuusulanjärvi was accompanied by an increased internal P load, despite increased oxygen concentration (Horppila et al. 2017). These effects were attributed to increased temperature at the sediment–water interface, increased sediment resuspension by treatment-induced turbulence, and accelerated mineralization leading to increased organic P release. Also, artificial destratification exacerbated diffusive P flux from bottom sediments in hyper-eutrophic Cedar Lake, Wisconsin, USA, where the sediment–water interface remained anoxic (James et al. 2015). Here, low-sediment Fe release rates prevented any adsorption of the released P during aerobic entrainment events, so that cyanobacteria blooms prevailed in late summer.

Even if destratification is not complete, the pumping of epilimnetic water to the hypolimnion can affect the stability by increasing the hypolimnetic temperature so that the stratified period is shortened and prevails over a smaller volume of the lake, counteracting potential beneficial effects of aeration. During several years of pumping epilimnetic water to the hypolimnion in a basin of Finnish Vesijärvi, hypolimnetic summer temperature increased by 5–6°C on average, inducing early fall turnover with elevated water temperature and light favourable to cyanobacteria (Horppila et al. 2015; Niemistö et al. 2016).

In general, effects of mixing had to be evaluated against conditions favoured by cyanobacteria in specific systems (Section 3.5.3). For example, artificial mixing can shift the phytoplankton genera from buoyancy-controlled surface-blooming cyanobacteria to green algae, possibly even with unchanged internal P load (Visser et al. 2016). Increased vertical mixing of the epilimnion can also lead to light limitation by increasing the time phytoplankton cells spend in less-illuminated conditions, thus decreasing phytoplankton growth efficiency (Köhler et al. 2018). Mixed conditions, while growth-limiting to surface bloomers, are favoured by the low-light-adapted cyanobacteria of trait R (Table 3.1). For example, (accidental) fall destratification has resulted in a bloom of toxin-producing *Planktothrix rubescens* in a deep kettle lake (Nürnberg et al. 2003).

Negative effects of increased stability and stratification in wake of climate change (Section 4.2.3) are associated with increases in anoxia, internal P loading, and cyanobacteria (Section 4.4.3), apparently indicating a potential benefit of artificial mixing. For example, natural *Microcystis* blooms worsened after each period of experimental stratifications in Nieuwe Meer (Jöhnk et al. 2008, Section 4.4.1). Thus, it appears that with respect to artificial mixing and destratification, the specific goal, lake conditions, and treatment variables have to be considered (Donis et al. 2021).

6.3.4 Aeration and oxygenation

A successful aeration (addition of air) or oxygenation (addition of pure oxygen) treatment needs to keep the surface sediment oxidized, thus reducing the internal P load by decreasing the anoxic area involved in redox-related P release (Figure 6.1). Oxygen conditions in the water column are less important; instead, the oxygen state at the sediment–water interface has to be considered (Gächter and Wehrli 1998; Hupfer and Lewandowski 2008; Tammeorg et al. 2017). When the flux of organic matter exceeds the supply of oxygen, as can happen in more eutrophic systems (Katsev and Dittrich 2013), the added oxygen can increase organic P mineralization, enhancing internal P loading.

Aeration and oxygenation are often applied below the thermocline to the hypolimnion in an attempt to prevent destabilizing the lake and prevent the pitfalls prevalent in destratification treatments (Section 6.3.3). Nonetheless, only in some of these cases was internal load diminished (Ashley 1983; Gerling et al. 2014; Horne and Beutel 2019). In these successful cases, P accumulates in the sediment during aeration and can be released when treatment is discontinued (Kowalczewska-Madura et al. 2020; Tammeorg et al. 2022a).

In many other cases, internal P load was not successfully reduced. Applications did not manage to oxygenate the sediments sufficiently to inhibit sediment P release (Gächter and Wehrli 1998), or release had become independent of the oxygen state during severe eutrophication (Katsev and Dittrich 2013). Oxygenation did not improve mixed layer water quality in four out of five Danish lakes (Liboriussen et al. 2009), but treatment effects are often described more favourably, for example, in a review on hypolimnetic oxygenation (Preece et al. 2019). Observations showed that aeration rarely decreased water TP concentration in about 30 Finnish lakes, leading to the conclusion that it does not appear to successfully inhibit redox-related P release, especially in warm, shallow, and enriched systems (Tammeorg et al. 2020b, 2017). A recent (installed in 2019) aeration treatment of the relatively shallow eutrophic Lake Carmi, Vermont, increased sediment P release and DO consumption rates and resulted in an increased summer surface TP concentration, possibly due to decreased stability and increased bottom temperature (Kirol et al. 2024). An extensive literature review of about 70 references did not reveal any positive effects except that hypolimnetic anoxia was generally prevented where the systems were sized appropriately, and fish habitat was improved (Wagner 2015). But other water quality indicators including TP concentration and cyanobacterial biomass did not noticeably decrease in lakes treated with hypolimnetic aeration or oxygenation.

For internal P load abatement, the treatment by itself remains questionable (Lürling et al. 2020). Only hypolimnetic aeration or oxygenation coupled with a chemical to adsorb P, such as iron and aluminum (Section 6.4), has consistently decreased internal load and delivered positive effects on trophic state (Cooke et al. 2005; Moore et al. 2012).

Also to be considered is the required long-term commitment. These treatments need to operate continuously to be effective, which requires constant energy input, management, and maintenance. If they are interrupted, water quality can become worse than it was before treatment, because oxygenation increases P retention and decreases export, as shown for a reservoir (Gantzer et al. 2009). This accumulated sediment P can lead to enhanced P release once the hypolimnion becomes oxygen depleted again. With respect to practicality, aeration usually requires equipment on the lake surface that is open to vandalism, presents navigational challenges, and is prone to be destroyed by ice build-up (Nürnberg et al. 2003).

Nonetheless, hypolimnetic aeration and oxygenation can be beneficial to treat elevated metal concentrations (Fe, Mn, and Hg), and associated taste and odour problems in drinking water reservoirs (Gantzer et al. 2009; Gerling et al. 2014; Munger et al. 2019), and is useful to improve habitat for fish and macrobenthos (Müller and Stadelmann 2004). State-of-the-art oxygenation designs have improved and can overcome the often-encountered insufficient distribution of oxygen throughout the hypolimnetic sediment–water interface (Gerling et al. 2014; Mobley et al. 2019), but effects on internal P load have not been studied.

A novel technique of oxygenation involves the preparation of natural minerals that can release nano-bubbles of pure oxygen (Waters et al. 2022). In an experimental environment, these minerals increased the oxygen level at the sediment–water interface after deposition on the sediment surface and decreased redox-related P release by 96.4% (Wang et al. 2020), but field tests and further evaluation of the practical, ecological, and social applicability are still needed (Waters et al. 2022).

6.4 Chemical internal load control

Chemical lake treatment includes additions to the water to precipitate or adsorb P and increase retention via settling (Figure 6.1, right) and additions to the sediment to reduce P release by adsorption and binding or to decrease the active release area by chemical sediment oxidation (Figure 6.1, centre left and centre).

Lake surface chemical additions that aim at reducing phytoplankton in the water column often involve Al-based chemicals. Benefits are usually short-lived because the dosage typically only considers the P pool in the water column that is much smaller than in the sediment, so that the chemicals have to be applied repeatedly (Lewandowski et al. 2003). Even though such treatment induces P sedimentation, it is rarely efficient enough for considerable P removal and does not abate internal P loading. In contrast, increased sedimentation may increase mobile sediment P and future P release.

More effectively, chemical treatments aim to improve the sediment by increasing its P adsorption capacity or preventing P release. Hence, treatment efficiency strongly depends on sufficient dosage of the material. Dosage should be calculated considering mainly the mobile P pool of the sediment and, in eutrophic systems, of inorganic P in the water (James 2011; Kasprzak et al. 2018; Meis et al. 2012). Different application techniques seek to further enhance treatment efficiency by considering time (best before or after the growing period) and space (applications aimed at specific depths and over eutrophic sediments) and by planning multiple applications (Schütz et al. 2017).

Many chemicals have been examined to increase P retention in sediments (Lürling et al. 2020). They include foremost compounds with Al for P adsorption and clays with incorporated lanthanum (La) for permanent P binding. Other chemicals are less effective and include Ca, Fe, zeolites, and allophane: adsorption capacity is generally improved by adding Al or La to these substances. As specific conditions of the water body require consideration, restoration outcome can be optimized by the careful choice of chemicals with different characteristics (Kang et al. 2023; Lürling and Mucci 2020). In a recent synthesis of known chemical applications in lake restoration, important details such as potential use as fertilizers and a role in circular economy were also explored (Zamparas and Kyriakopoulos 2021).

6.4.1 Aluminum compounds

Aluminum (Al) compounds to treat internal P loading have been used for several decades and include alum, an aluminum sulfate $Al_2(SO_4)_3$, the compound most often applied in North America, and other compounds, like poly aluminum chloride (PAC), often applied elsewhere.

Alum applications produce a flocculent gelatinous precipitate of aluminum hydroxide, $Al(OH)_3$, which is chemically stable up to a pH of 8.8, even under hypoxic conditions. But at higher pH values that occasionally occur during high phytoplankton biomass, P is easily desorbed from Al precipitates (Reitzel et al. 2013; de Vicente et al. 2008). P-binding efficiency can decrease with time due to aging and polymerization processes (Berkowitz et al. 2006; de Vicente et al. 2008) and is stated as one of the reasons for less-efficient P removal by alum than expected from dosage calculations (James and Bischoff 2020). Therefore, most pertinent lake characteristics to consider for an alum-based treatment are the lake water's buffering capacity and its pH levels. The pH level for most efficient adsorption is relatively narrow, around 6–8.8. Beyond this window, Al desorbs and becomes potentially toxic depending on the fish species and water chemistry (e.g., at Al ionic concentration above 0.1 mg/L in neutral water for lake trout (Gensemer and Playle 1999)). In addition, suffocation due to Al polymerization and subsequent precipitation on fish gills can lead to acute toxicity, even in hardwater lakes (Poléo 1995; Wauer and Teien 2010).

Potential toxicity during application is somewhat diminished in a PAC treatment. PAC is also the preferred option of Al treatments because it does not contain sulfate that is easily turned into hydrogen sulfide under anoxic conditions. Hydrogen sulfide would potentially decrease the P retention capacity of the bottom sediments. For these reasons, PAC is the only permitted Al-based treatment in many European countries. Different types of PAC can be engineered for different water characteristics involving organic content, humic acids, and pH for use in water treatment processes (Yan et al. 2007).

114 lakes were identified as being treated with alum or PAC to reduce internal phosphorus loading (Huser et al. 2016b). Treatment longevity varied for lakes of different morphometry, applied dose and other factors, and averaged at 11 years. Shallow, polymictic lakes had average longevity of about six years. A past application of a large dose of alum in a German recreational lake almost completely prevented cyanobacteria growth for 27 years (1987–2014) (Rönicke et al. 2021). After that time, alum in the sediment finally lost its P adsorption efficiency, which coincided with increased lake water SRP concentration attributable to internal P loading. Similarly, a PAC treatment successfully decreased chlorophyll concentration and cyanobacteria in two Finnish lakes (Sarvala and Helminen 2023) (Figure 5.7).

6.4.2 Lanthanum-modified clays

There are several clay-based P-binding materials of which lanthanum-modified bentonite (LMB) has been the most researched and successfully applied material in lakes. The main commercially available form of this material is called *Phoslock* and was developed by the Commonwealth Scientific and Industrial Research Organization (CSIRO) of Australia and patented in 2002 (Douglas 2002). Phoslock is made by combining bentonite, a naturally occurring clay, with lanthanum (La from $LaCl_3$), a metallic element that forms a solid compound with phosphate.

In LMB applications, the clay settles to the bottom sediments so that the active ingredient, La, can intercept any phosphate released from the sediment over a wide range of pH values, independent of the redox state (Mucci et al. 2018; Ross et al. 2008). Consequently, cyanobacteria blooms are less frequent and spatially and temporally more confined. Phoslock treatment has been applied even in jurisdictions where Al-based applications are not permitted, because of its general lack of toxicity (Lürling and Tolman 2010; Ontario Ministry of Environment 2009; Spears et al. 2013a).

Phoslock has been used in more than several hundreds of applications (Copetti et al. 2016) throughout Australia (Robb et al. 2003), New Zealand (Burns et al. 2009; Hickey and Gibbs 2009), China (Han et al. 2021; Liu et al. 2012; Yin et al. 2018), Europe (Epe et al. 2017; Lürling and van Oosterhout 2013; Meis et al. 2012; Spears et al. 2016b; Yasseri and Epe 2016), the United States (Bishop et al. 2014), and Canada (Nürnberg 2017; Nürnberg and LaZerte 2016a) to treat lakes, rivers, stormwater ponds, and reservoirs.

High P adsorption capacity of La (due to the formation of a bond with phosphate) under varied redox conditions and its low toxicity encouraged the development of additional LMB materials. Novel material has varied La versus clay ratios to be used in different situations and include *Zeofixer* (Zhang et al. 2024) and *Eutrosorb*.

6.4.3 Other compounds

Iron is involved in redox-related P release (Section 2.4.2), and its adsorption of P under aerated conditions has invoked lake managers to use Fe compounds in lakes, especially when lake Fe concentrations are naturally low. But because of the redox characteristics, Fe additions cannot be expected to consistently treat internal P loading, as adsorption requires reliable elevated redox conditions. If natural aeration does not suffice (which is rarely the case), simultaneous aeration or oxygenation is required that is strong enough to prevent redox conditions at the sediment surface to decrease to 200 mV and below.

Therefore, Fe addition treatments do not bind P permanently and have not been recommended (Cooke et al. 1993). Instead, enhanced Fe adsorption and sedimentation in the sediments increase the mobile P fraction so that future deteriorating redox conditions induce sediment P release. Other negative characteristics include the reaction to organic acids. Stained or non-stained organic acids can combine with Fe ions to prevent P adsorption, rendering the chemical inefficient. For example, after a two-year improvement, iron application ($FeCl_3$) resulted in higher surface water P concentrations than before treatment (Münch et al. 2024). Continued redox cycling and enhanced P release under anoxic conditions, in addition to Fe stabilization by organic matter that diminished P adsorption, prevented quantitative internal P load control.

Further, iron addition may enhance cyanobacteria blooms, because of potential iron limitation (Section 3.4.2). Iron chloride created an *Aphanizomenon* bloom in the first year of application, attributed to alleviated iron limitation in a small lake in British Columbia, Canada (Hall et al. 1994). When dosage was increased, internal P load remained lower during consistent aeration but increased when aeration was interrupted.

Nitrate additions can delay redox-related P release by raising or preventing low redox potentials. As long as NO_x is present, redox-dependant internal P load ceases, even when oxygen concentration is low. A decrease of NO_3 by the diversion of wastewater from a wastewater treatment plant was suggested to increase sediment anoxia and thus increase internal P load and productivity including cyanobacteria in hyper-eutrophic east German Lake Jabel (Kleeberg et al. 2000). Increased eutrophication with NO_3 depletion was observed in several highly eutrophic lakes (e.g., Müggelsee, Tegelsee, and Quitzdorf Reservoir, Germany, reviewed in (Beutel et al. 2016)). Low NO_x concentration often coincides with cyanobacteria (Molot et al. 2022; Nürnberg 2007a). In an attempt at lake restoration, nitrate additions have long been used occasionally to prevent redox-related processes including internal P loading (Ripl 1976) and heavy metal and mercury accumulation under hypoxic conditions as reviewed in Beutel et al. (2016). However, other studies were unsuccessful (Noonan 1986) or determined large increases of ammonia concentrations and severe toxicity to zooplankton and macrobenthos as reviewed by Lürling et al. (2020), rendering the benefit of a nitrate addition treatment ambiguous.

It appears that nitrate treatment interrupts the sediment P release only if (1) the addition of nitrate consistently prevents ferric iron reduction at the sediment–water interface (and N remains in an oxidized form) or (2) nitrate is combined with redox-independent adsorbents (Beutel et al. 2008). Similar to an Fe treatment, the added potential fertilization (nitrate is typically limiting during maximum cyanobacteria abundance, Section 3.3.3) can prevent any benefit and rather cause water quality problems from eutrophication. There is also the added risk of nitrate and ammonium toxicity at elevated concentrations in the lake water. Under experimental conditions, N concentration increases were minimized by the application of slow-release compounds (Wauer et al. 2005) or sediment-capping material (Zhan et al. 2019).

Calcium compounds have been applied to mimic the natural occurrence of authigenic calcium carbonate formation and co-precipitation with P that occurs in hardwater lakes during increasing temperature and pH conditions, inducing a turbid stage sometimes referred to as lake whitening (Wetzel 2001). While this occurrence in natural Ca-rich lakes increases P retention, for example, in alpine lakes (Nürnberg 1998), attempts at precipitating P with calcium hydroxide (lime) in Canadian and German lakes have not provided consistent P removal (Cooke et al. 1993; Koschel et al. 2001; Murphy et al. 1988; Prepas et al. 2001).

Clay without amendment is unspecific to P and does not retain it in the sediments (Zamparas et al. 2012). A clay layer on the sediment will only provide a barrier to P release with the amendment of P-adsorbing particles (e.g., Fe, La, Al). A specific clay, *zeolite*, when amended with P-absorbent chemicals also adsorbs N due to its natural NH_4 exchange capacity (Kang et al. 2022b). But clay, especially bentonite, can aggregate with *Microcystis* spp., leading to enhanced sedimentation of the otherwise buoyant cyanobacteria (Verspagen et al. 2006). Calcium-poor, clean soils successfully adsorbed P at the sediment–water interface under experimental and in-situ conditions (Sun et al. 2023a). Porous *geopolymers*, which are amorphous inorganic materials including clays, capture various contaminants including phosphate (Ettahiri et al. 2023).

Industrial waste products from construction and mining, from drinking water treatment sludge (Gao et al. 2020) and as coal ash (Jeong et al. 2022), have been evaluated in various studies and can provide a cost-effective alternative to the more studied material. But many such potential P adsorbents still need specific studies with respect to dose effect and potentially toxic and interfering characteristics regarding the specific receiving systems (Spears et al. 2013b).

6.5 Combined treatments including fish biomanipulation, macrophyte growth, and other methods

Benthivorous, bottom-feeding fish species including goldfish (*Carassius auratus*), carp (*Cyprinus carpio*), and bream (*Abramis bramacan*) can stir up the sediments and increase internal loading. A study on a shallow US lake determined that the sediment-mixing depth increased more than twofold, down to 13 cm, in carp-exposed areas, which increased the amount of releasable P by up to 92% (Huser et al. 2016a). These fish usually thrive in turbid phytoplankton-dominated lakes and ponds. They hinder macrophyte establishment by consuming plant material, reducing light availability, and increasing siltation, thus creating or prolonging a turbid phytoplankton-dominated state (Jeppesen et al. 2024b). Nonetheless, the removal of planktivorous (roach, *Rutilus rutilus*) and benthivorous (bream) fish alone did not appear to have lasting effects beyond 6–10 years on water quality in 36 shallow, eutrophic Danish lakes (Søndergaard et al. 2008).

Sediment disturbance by omni-benthivorous fish can lessen the effectiveness of a chemical sediment treatment such as with LMB or PAC. But biological methods (removal of planktivorous and benthivorous fish) are powerful in combination with such chemical applications.

Biomanipulation by reducing benthivorous fish mass is therefore often combined with chemical treatments (Dondajewska et al. 2019; Mehner et al. 2008). It is recommended before the application of restoration techniques that combat sediment P release with chemicals (Section 5.1.10 (Godwin et al. 2011; Huser et al. 2022; Søndergaard et al. 2008)). The enhancement of a submerged macrophyte community by protected planting can further decrease sediment disturbance. The biological approaches of reducing fish biomass and increasing submerged macrophytes insured that an accompanying chemical treatment targeting sediment P release (LMB) effectively controlled internal P load in mesocosms (Han et al. 2022) and in a lake basin (Liu et al. 2018).

Such a combined approach, macrophyte establishment, benthivorous fish removal, and the interruption of sediment P release may be more cost-effective than single distinct methods because of possible synergistic effects on the physical, chemical, and biological interactions (Jeppesen et al. 2024b).

Other combined approaches include aeration that is often combined with the application of chemicals to ensure even distribution (calcium), oxic conditions (iron), and improved adsorption capacity (alum) (Pereira and Mulligan 2023). Likewise, organic matter and P coagulation in the water column (with PAC) is sometimes combined with sediment capping (with LMB) to treat both the water column and the sediments (Su et al. 2021; Waajen et al. 2016).

A combination of restoration approaches is particularly important in urban lakes, where neglected stormwater inflow, constructed hard banks, introduced bottom-feeding fish (e.g., goldfish), waterfowl and its feeding by the public, and organically enriched sediments under stagnant shallow water can contribute to high internal and external loading (Lürling et al. 2023; Nürnberg and LaZerte 2016a).

Benefits of internal load abatement do not always last (van Oosterhout et al. 2022). Sometimes, continued monitoring and comparison of internal with external load quantities are needed to determine whether repeat applications are necessary and should be planned routinely (Lewandowski et al. 2003) or whether other strategies including more aggressive external load abatement are more promising.

6.6 General considerations and conclusions about the treatment of cyanobacteria by internal P load reduction

The restoration of lakes with respect to the abatement of internal P load and cyanobacteria blooms has become much more urgent because of climate change. Specific applications and restoration success depend on several pre-conditions besides the availability of known treatment options. As technologies improve, it can be expected that novel techniques in combination with proven approaches will provide more options in future lake management.

Non-technical considerations include the societal support and acceptance of envisioned lake restoration; community involvement is a major contributor to ultimate restoration success (Cianci-Gaskill et al. 2024).

While remediation is often needed in eutrophic systems, prevention in less affected systems must also be considered. A meta-analysis of 400 large-scale disturbances impacting terrestrial, marine, and freshwater ecosystems found that recovery is rarely complete (Jones et al. 2018). That study further concluded that passive recovery (simply ending the disturbance) should be considered as a first option, because restoration activities did not generally result in faster or more complete recovery. Hence, "early actions to prevent degradation of lakes currently in good ecological condition are preferable to attempting to restore lakes that have been allowed to degrade" (Spears et al. 2022). Accordingly, the potential impact of internal P load on cyanobacteria should be prevented by a pro-active approach that includes the following attributes.

Determine probable signs of internal P load:

1. Establish a (routine) monitoring network that can identify internal-P-load-related signals.
2. Identify an early warning system that describes and detects signs of internal loading.

If signs of internal load are apparent:

3. Compare the internal with external P load, which requires quantification of both P loads.
4. Determine the climate-related relationships with (annual) internal load estimates.
5. Establish some knowledge on N concentration and loading to evaluate the importance of N in the specific system.
6. Use models to forecast the expected eutrophication based on anticipated anthropological development and climate change.
7. Initiate the process of creating a remediation plan that includes approaches for prevention and restoration.

A *remediation plan* includes the following general steps:

Deciding on most sustainable and long-term option, considering:

- P export versus settling and burying
- Reuse options
- Societal consensus and benefits
- Economic implications

Evaluating available and feasible technical restoration options:

- Increase P export: hypolimnetic withdrawal in stratified lakes or sediment dredging in shallower lakes. While providing an opportunity to recycle nutrients (and support circular economy), both options need extensive treatment of waste products (water or sediment). For flow augmentation and dilution, the water source is rarely available.
- Increase settling, burying, and capping: chemical P precipitation in stratified and mixed lakes and capping of sediment; hypolimnetic aeration or oxygenation in stratified lakes (not consistently effective).

The abatement of cyanobacteria blooms via the treatment of internal P loading requires detailed knowledge of the system to be treated. Important information includes the size and timing of the sediment P release as well as the characteristics and temporal and spatial distribution of potentially affected cyanobacteria. For example, deep cyanobacteria of trait R (Table 3.1) are easily missed by routine monitoring (Erratt et al. 2022b) but are especially sensitive to hypolimnetic withdrawal. In contrast, most surface-blooming cyanobacteria respond to a general decrease of late summer phosphate pulses (i.e., internal P loading) from the sediments and the hypolimnion.

The importance of lake-specific analysis in the context of nutrients was described as one of three elements to control cyanobacteria blooms (Ibelings et al. 2016a): (1) understanding of the key ecological traits of the dominant cyanobacteria taxa; (2) system analysis of the lake, in particular its morphometry and water and nutrient balance; and (3) adequate design and execution of the management methods of choice. In summary: "Control of internal P loading requires comprehensive, site-specific diagnosis (i.e., a lake system analysis) to ensure the measures are appropriate, cost-effective, and have public support" (Lürling et al. 2020).

Glossary and definitions

Active area release factor (AA) Active period (days/summer–fall or days/year) and area (of the size of the lake surface area) that releases P and contributes to internal load (*see summary in* Nürnberg 2019, 2004).

Akinetes Thick-walled resting cells; development triggered by unfavourable growth conditions (e.g., low light, low nutrients, extreme temperature).

Alkaline phosphatase activity (APA) Process of enzymatic actions that convert dissolved organic P to phosphate. The expression of APA is under environmental control and triggered by P-deficient conditions.

Annual areal water load, q_s (m/yr) The annual outflow volume (Q, m^3/yr) per surface area (A, m^2), where $q_s = Q/A$.

Annual water detention time *or* annual water residence time, *tau (yr)* lake volume (V) divided by annual outflow volume (Q), where tau = V/Q.

AF *and* HF, Anoxic *and* hypoxic factor *(days/summer-fall or days/year)* Extent and duration of anoxia (DO <1 or 2 mg/L) or hypoxia (threshold between 3 and 6 mg/L). These factors summarize the information available by DO profiles and hypsographic information (areas corresponding to various depths) into one number per season. These factors represent the number of days in a season or year that a sediment area equal to the lake surface area is anoxic or hypoxic (*see summary in* Nürnberg 2019, 2004).

Bioturbation and bioirrigation The disturbance of sediments and the siphoning of water into sediments by organisms such as dreissenid mussels, chironomid larvae, burrowing mayflies, and oligochaetes.

Brownification, lake browning The phenomenon of increased DOC from humic and fulvic acids creating an increasingly brown water colour in lakes.

Catchment basin *or* watershed area Area around a lake that includes all upstream land and water draining into the lake. (Typically does not include the lake area.)

Chlorophyll a The green pigment in surface water that serves as a measure of *algal* or *phytoplankton* biomass. Chlorophyll *a* is used, when referring to *chlorophyll* without any further description.

Climate indices Based on normalized values of specific climate-related variables:

ENSO – El Niño – Southern Oscillation, a global atmospheric circulation that influences global temperature and precipitation. ENSO is extensively used in weather forecasts.

NAO – North Atlantic Oscillation, based on sea-level pressure differences. (E.g., NAO is related to temperature in European lakes.)

PDO – Pacific Decadal Oscillation, based on North Pacific sea surface temperature.

PNI – Pacific Northwest Index, based on three terrestrial climate variables in the Northwestern United States: the air temperature at San Juan Island, total precipitation at Cedar Lake in the Cascade Mountains, and snowpack depth on Mount Rainier.

Cyanobacteria Also called *bluegreens* or *bluegreen algae*, although they belong to bacteria and are prokaryotic. They can produce toxins that create health effects if ingested in quantity (life stock, pets). Three important groups:

Nostocales For example, *Cylindrospermopsis* or *Raphidiopsi raciborskii, R. curvata, R. mediterranea, Chrysosporum (Aphanizomenon) ovalisporum, Chrysosporum (Anabaena) bergii, Dolichospermum flos-aquae, Aphanizomenon flos-aquae, Aphanizomenon gracile and Anabaena lapponica.*

Chroococcales For example, *Gomphospheria, Merismopedia, Microcystis, Synechococcus, Synechocystis, Woronichinia.*

Oscillatoriales For example, *Limnothrix, Microseira (Lyngbya) wollei, benthic Oscillatoria, Hormoscilla pringsheimii, Planktothrix.*

Detritus Fragments of disintegrated, usually organic, material suspended in water.

Diazotrophic Nitrogen-fixing microorganism compared to *non-diazotrophic* microorganisms that do not fix N.

Dissolved oxygen (DO) Concentration of oxygen gas dissolved in lake water.

Dynamic ratio (km$^{0.5}$/m) Quotient of square root of surface area (km^2) over mean depth (m). Indicates the potential of mixing in a temperate lake. The larger the ratio the more (deeper and longer) mixing occurs. Its inverse is called *morphometric ratio* and indicates the strength of stratification.

El Niño The warm phase of the El Niño – Southern Oscillation; it is associated with a band of warm ocean water that develops in the central and east-central equatorial Pacific.

Epilimnion (epilimnetic) mixed surface layer in lakes and reservoirs during periods of thermal stratification, as opposed to meta- and hypolimnion.

Eukaryotic algae Phytoplankton and "real" algae but not cyanobacteria, which are prokaryotic.

Euphotic layer depth, z_{eu} Determined as the depth to which 1% of the surface light (photosynthetic active radiation, PAR) penetrates. In theory, it represents the depth at which phytoplankton photosynthesis balances respiration on a daily basis.

Export, L_{out} The mass of TP that leaves the lake or reservoir via the outlet stream. Units are similar to loading units – kg/yr or mg per square metre of lake surface area per year (mg/m^2/yr).

External load, L_{ext} The sum of TP inputs from all external sources, that includes streams, non-point and point sources, precipitation, and groundwater. Units are kg/yr or mg per square metre of lake surface area per year (mg/m^2/yr). External load is a gross estimate. Much of its phosphorus is in a chemical form that is not immediately available to algae.

Heterocytes Thick-walled cells where N fixation occurs, a process that transforms atmospheric N_2 via reduction into NH_4 and incorporates it into the cell body. Only certain cyanobacteria possess heterocytes and usually only under conditions of N limitation.

Hypolimnion (hypolimnetic) deep, supposedly stagnant, layer during times of thermal stratification.

Internal load, L_{int} Annual TP inputs from internal sources, especially the bottom sediments. Units are kg/yr or mg per square metre of lake surface area per year (mg/m^2/yr). Gross estimates are usually used, but net estimates, based on mass budgets, can also be calculated. Most of the TP in L_{int} is in a chemical form (phosphate) that is highly bioavailable to phytoplankton and bacteria.

Lake In a strict sense, naturally occurring low points in the landscape that contain standing water year round, predominantly in the form of open-water habitat (Hayes et al. 2017). In a wider sense, reservoirs are included when referring to "lakes", but see *Reservoir* definition.

Light ratio (z_{eu}/z_{mix}) Controls the species distribution of phytoplankton and functional groups (traits) of cyanobacteria.

Limiting nutrient Algae, bacteria, and phytoplankton in lakes are often nutrient limited, so that any addition of the bioavailable form of that nutrient would increase such biomass. The nutrient that elicits the largest response is called the limiting nutrient (usually phosphorus or nitrogen).

Luxury P uptake The ability, especially of cyanobacteria, to take up and store phosphorus in large excess of the current metabolic demand.

Mean depth, z (m) Lake depth computed as the lake volume divided by lake surface area. This value is always smaller than maximum depth – the depth at the deepest location.

Mixed layer depth, z_{mix} (m) Depth of the boundary between the epilimnion and the metalimnion.

Mobile sediment P fractions Sediment P fractions that are released as internal loading: porewater P, iron bound P (BD-P), and labile organic P (non-reactive NaOH-P) (Table 2.3).

Morphometric ratio (m/km$^{0.5}$) Quotient of mean depth (m) over square root of surface area (km^2). Indicates the likelihood of stratification in a temperate lake. The larger the value the stronger (and longer) is the stratification. Also called *Osgood Index* (Osgood 1988). Its inverse is called *dynamic ratio* and indicates the potential of mixing.

Phosphorus retention, R Retention is a proportional value based on the external P load. It can mean two different quantities – it can be a theoretical value and implies sedimentation or a measured value from a mass balance. When retention is measured from an annual TP budget as R_{meas} = (in – out)/in, it is a *net* term that combines downward fluxes of settling and upward fluxes from the sediments (internal load). The proportion of TP load that is retained due to sedimentation, R_{sed}, is a *gross* term and has to be modelled or predicted. The difference between R_{meas} and R_{sed} depends on the internal load and can be used to estimate it.

Photogenic zone Illuminated water layer.

Photosynthetically active radiation (PAR) Radiation in the 400–700 nm waveband.

Phytoplankton Photosynthetic plankton that include algae and cyanobacteria.

Planktonic Characteristic of phytoplankton that typically reside in the water column.

Polymixis, polymictic The mixing state in lakes and reservoirs that describes frequent (daily to weekly in the summer) mixing of the whole water column.

Primary productivity (PP) PP is a measure of phytoplankton productivity in lake water. It can be expressed as volumetric PP (e.g., mg C/m^3/d) or as areal PP (e.g., C/m^2/yr).

Redox potential (Eh) Redox potential is the measurement of the tendency of an environment to oxidize or reduce substrates. It can determine the degree of anoxia in hypoxic lake regions such as in the sediment, the sediment–water interface, and the hypolimnion. Eh is reported in volts or mVolts, against the hydrogen electrode under standard conditions. Eh values relevant to redox-related sediment P release include: Mn= ~400 mV, NO_3= ~300 mV, Fe= ~200 mV, and less for H_2S, methane, and other reduced gases.

Representative Concentration Pathways (RCP) Climate scenarios based on the balance of solar radiation entering the climate system and infrared leaving it, where 2.6 is low, 4.5 and 6 are intermediate, and 8.5 is high.

Reservoir Anthropogenic, constructed water body that is usually located within a river system (Hayes et al. 2017). Reservoir volumes and water levels often fluctuate with usage and season. Many aspects known in lakes apply to reservoirs, but they may be numerically different (e.g., flow rates and catchment basins are typically larger, P retention is higher for the same water load, etc.)

Reynolds Functional Groups, *also called* Codons or Traits Used in phytoplankton classification; for cyanobacteria, see Table 3.1.

Secchi disk transparency The depth at which the round black and white disk disappears is an integrated measure of algal biomass. Because its use is widespread, many relationships with nutrients and chlorophyll concentration from other lakes are available (as regression equations). It is affected by water colour and turbidity.

Sediment focusing The redistribution of fine sediment material toward the deeper areas.

Sediment oxygen demand (SOD) Organically enriched bottom sediment takes up oxygen from the overlaying water as oxygen demand that creates hypoxic conditions.

Seston *also called* suspended particulate matter Particles suspended and settling in the water. Seston consists of organic and inorganic particles including plankton and detritus.

Settling velocity, v *(m/yr)* The average distance that TP settles downward within one year.

Shade index, z/z_sec Used in shallow lakes as a measure of the underwater light climate. Represents the ratio of average lake depth to Secchi depth transparency. Shade is related to light limitation and increases with value.

Siderophores Glycoprotein chelators with a strong affinity for oxidized iron (Fe^{3+}) and can be secreted by microorganisms such as cyanobacteria and fungi when ambient iron concentration is low. The oxygenated iron in these complexes will be transported across the membrane to be reduced to Fe^{2+} before it can be used.

Soluble reactive P (SRP) Soluble fraction of TP that consists mostly of the biologically available (bioavailable) phosphate. Also called dissolved reactive P (DRP).

Stability Water column stability. Often computed from temperature differences between depths as *Schmidt stability* or *relative thermal resistance to mixing* (RTRM).

Sulfuretum Environment in the sediment or water that is anoxic with a low redox potential where hydrogen sulfide is stable.

Thermal stratification Period when a deep lake is warm in the surface mixed layer (*epilimnion*) but remains cold in the bottom layer (*hypolimnion*). Temperature changes rapidly between these layers at the *thermocline* in the *metalimnion*.

Dimictic Spring and fall mixing occurs during spring and fall turnover with a stratified period in the summer and in the winter under ice.

Monomictic Stratified only in the summer but mixed throughout the rest of the year, including in the winter when there is no ice formation.

Top-down versus bottom-up *control* The mechanisms that control the growth and productivity of organisms. Bottom-up refers to chemical (nutrients, micronutrients) and physical (light, turbulence, temperature) conditions, while top-down relates to predator–prey and other relationships from organisms and includes allelopathy, the interaction between organisms by expelled chemicals.

Total phosphorus (TP) All phosphorus (P) that can be analyzed in a water or sediment sample. It includes phosphate (highly bioavailable for phytoplankton), particulate forms (include algae and non-living suspended particles), and P forms, including organic compounds, that cannot be easily used by phytoplankton.

Trophic state The trophic state classification provides a measure of nutrient enrichment and productivity of a water body. In lakes, it is based on the growing season averages for several water quality variables of the mixed surface layer (Table 2.2).

Volatile organic compounds (*VOCs*) VOCs impact taste and odour in drinking water and include *Geosmin* (*trans*-1,10-dimethyl-*trans*-9-decalol) and *2-MIB* (2-methylisoborneol).

Watershed area, also catchment area Area around a lake that includes all upstream land and water draining into the lake (typically does not include the lake area).

Watershed to lake area ratio This ratio (using consistent units) is often used to differentiate between the importance of external versus internal processes and inputs, where smaller values indicate that internal load is more important than external load.

References

Abirhire, O., Davies, J.-M., Imtiazy, N., Hunter, K., Emmons, S., Beadle, J., Hudson, J., 2023. Response of phytoplankton community composition to physicochemical and meteorological factors under different hydrological conditions in Lake Diefenbaker. *Sci. Total Environ.* **856**, 159210. https://doi.org/10.1016/j.scitotenv.2022.159210

Abirhire, O., Hunter, K., Davies, J.-M., Guo, X., de Boer, D., Hudson, J., 2019. An examination of the long-term relationship between hydrologic variables and summer algal biomass in a large Prairie reservoir. *Can. Water Resour. J.* **44**, 79–89. https://doi.org/10.1080/07011784.2018.1531064

Abtew, W., Trimble, P., 2010. El Niño – Southern Oscillation link to south Florida hydrology and water management applications. *Water Resour. Manag.* **24**, 4255–4271. https://doi.org/10.1007/s11269-010-9656-2

Ackerman, J.D., Loewen, M.R., Hamblin, P.F., 2001. Benthic-pelagic coupling over a zebra mussel reef in western Lake Erie. *Limnol. Oceanogr.* **46**, 892–904. https://doi.org/10.4319/lo.2001.46.4.0892

Adams, H., Ye, J., Persaud, B.D., Slowinski, S., Kheyrollah Pour, H., Van Cappellen, P., 2022. Rates and timing of chlorophyll *a* increases and related environmental variables in global temperate and cold-temperate lakes. *Earth Syst. Sci. Data* **14**, 5139–5156. https://doi.org/10.5194/essd-14-5139-2022

Adrian, R., O'Reilly, C.M., Zagarese, H., Baines, S.B., Hessen, D.O., Keller, W., Livingstone, D.M., Sommaruga, R., Straile, D., Van Donk, E., Weyhenmeyer, G.A., Winder, M., 2009. Lakes as sentinels of climate change. *Limnol. Oceanogr.* **54**, 2283–2297. https://doi.org/10.4319/lo.2009.54.6_part_2.2283

Ahonen, S.A., Vuorio, K.M., Jones, R.I., Hämäläinen, H., Rantamo, K., Tiirola, M., Vähätalo, A.V., 2024. Assessing and predicting the influence of chromophoric dissolved organic matter on light absorption by phytoplankton in boreal lakes. *Limnol. Oceanogr.* **69**, 422–433. https://doi.org/10.1002/lno.12495

Alam, M.S., Barthod, B., Li, J., Liu, H., Zastepa, A., Liu, X., Dittrich, M., 2020. Geochemical controls on internal phosphorus loading in Lake of the Woods. *Chem. Geol.* **558**, 119873. https://doi.org/10.1016/j.chemgeo.2020.119873

Alambo, K.I., 2016. *Cyanobacteria North of 60°: Environmental DNA Approaches* (Master's). University of Ottawa, Department of Biology, Ottawa.

Albright, E.A., 2022. *Mechanisms of Internal Phosphorus Loading in Lentic Ecosystems* (PhD). The University of Wisconsin – Madison, Madison, WI.

Albright, E.A., Rachel, F.K., Shingai, Q.K., Wilkinson, G.M., 2022. High inter- and intra-lake variation in sediment phosphorus pools in shallow lakes. *J. Geophys. Res. Biogeosci.* **127**. https://doi.org/10.1029/2022JG006817

Amirbahman, A., Fitzgibbon, K.N., Norton, S.A., Bacon, L.C., Birkel, S.D., 2022. Controls on the epilimnetic phosphorus concentration in small temperate lakes. *Environ. Sci. Process. Impacts* **24**, 89–101. https://doi.org/10.1039/D1EM00353D

Amirbahman, A., Lake, B.A., Norton, S.A., 2013. Seasonal phosphorus dynamics in the surficial sediment of two shallow temperate lakes: A solid-phase and pore-water study. *Hydrobiologia* **701**, 65–77. https://doi.org/10.1007/s10750-012-1257-z

Amorim, C.A., Moura, A. do N., 2021. Ecological impacts of freshwater algal blooms on water quality, plankton biodiversity, structure, and ecosystem functioning. *Sci. Total Environ.* **758**, 143605. https://doi.org/10.1016/j.scitotenv.2020.143605

Andersen, I.M., Williamson, T.J., González, M.J., Vanni, M.J., 2020. Nitrate, ammonium, and phosphorus drive seasonal nutrient limitation of chlorophytes, cyanobacteria, and diatoms in a hyper-eutrophic reservoir. *Limnol. Oceanogr.* **65**, 962–978. https://doi.org/10.1002/lno.11363

Andersen, J.M., 1976. An ignition method for determination of total phosphorus in lake sediments. *Water Res.* **10**, 329–331. https://doi.org/10.1016/0043-1354(76)90175-5

Anderson, D.M., Cembella, A.D., Hallegraeff, G.M., 2012. Progress in understanding harmful algal blooms: Paradigm shifts and new technologies for research, monitoring, and management. *Annu. Rev. Mar. Sci.* **4**, 143–176. https://doi.org/10.1146/annurev-marine-120308-081121

Anderson, D.M., Glibert, P.M., Burkholder, J.M., 2002. Harmful algal blooms and eutrophication: Nutrient sources, composition, and consequences. *Estuaries* **25**, 704–726. https://doi.org/10.1007/BF02804901

Andersson, G., Graneli, W., Stenson, J., 1988. The influence of animals on phosphorus cycling in lake ecosystems. *Hydrobiologia.* **170**, 267–284. https://doi.org/10.1007/BF00024909

Antunes, J.T., Leão, P.N., Vasconcelos, V.M., 2015. Cylindrospermopsis raciborskii: Review of the distribution, phylogeography, and ecophysiology of a global invasive species. *Front. Microbiol.* **6**. https://doi.org/10.3389/fmicb.2015.00473

Arhonditsis, G.B., Neumann, A., Shimoda, Y., Kim, D.-K., Dong, F., Onandia, G., Yang, C., Javed, A., Brady, M., Visha, A., Ni, F., Cheng, V., 2019. Castles built on sand or predictive limnology in action? Part A: Evaluation of an integrated modelling framework to guide adaptive management implementation in Lake Erie. *Ecol. Inform.* **53**, 100968. https://doi.org/10.1016/j.ecoinf.2019.05.014

Ashley, K.I., 1983. Hypolimnetic aeration of a naturally eutrophic lake: Physical and chemical effects. *Can. J. Fish. Aquat. Sci.* **40**, 1343–1359. https://doi.org/10.1139/f83-157

Bachmann, R.W., Hoyer, M.V., Canfield, Jr., D.E., 2000. The potential for wave disturbance in shallow Florida lakes. *Lake Reserv. Manage.* **16**, 281–291. https://doi.org/10.1080/07438140009354236

Bachmann, R.W., Jones, J.R., 1974. Phosphorus inputs and algal blooms in lakes. *Iowas State J. Res.* **49**, 155–160.

Badger, M.R., Price, G.D., 2003. CO2 concentrating mechanisms in cyanobacteria: Molecular components, their diversity and evolution. *J. Exp. Bot.* **54**, 609–622. https://doi.org/10.1093/jxb/erg076

Baer, M.M., Godwin, C.M., Johengen, T.H., 2023. The effect of single versus dual nutrient decreases on phytoplankton growth rates, community composition, and Microcystin concentration in the western basin of Lake Erie. *Harmful Algae* **123**, 102382. https://doi.org/10.1016/j.hal.2023.102382

Bahlai, C.A., Hart, C., Kavanaugh, M.T., White, J.D., Ruess, R.W., Brinkman, T.J., Ducklow, H.W., Foster, D.R., Fraser, W.R., Genet, H., Groffman, P.M., Hamilton, S.K., Johnstone, J.F., Kielland, K., Landis, D.A., Mack, M.C., Sarnelle, O., Thompson, J.R., 2021. Cascading effects: Insights from the U.S. long term ecological research network. *Ecosphere* **12**. https://doi.org/10.1002/ecs2.3430

Bakker, E.S., Hilt, S., 2016. Impact of water-level fluctuations on cyanobacterial blooms: Options for management. *Aquat. Ecol.* **50**, 485–498. https://doi.org/10.1007/s10452-015-9556-x

Bakker, E.S., Van Donk, E., Immers, A.K., 2015. Lake restoration by in-lake iron addition: A synopsis of iron impact on aquatic organisms and shallow lake ecosystems. *Aquat. Ecol.* 1–15. https://doi.org/10.1007/s10452-015-9552-1

Baldwin, D.S., Gigney, H., Wilson, J.S., Watson, G., Boulding, A.N., 2008. Drivers of water quality in a large water storage reservoir during a period of extreme drawdown. *Water Res.* **42**, 4711–4724. https://doi.org/10.1016/j.watres.2008.08.020

Barbiero, R.P., Kann, J., 1994. The importance of benthic recruitment to the population development of Aphanizomenon flos-aquae and internal loading in a shallow lake. *J. Plankton Res.* **16**, 1581–1588. https://doi.org/10.1093/plankt/16.11.1581

Barnard, M.A., Chaffin, J.D., Plaas, H.E., Boyer, G.L., Wei, B., Wilhelm, S.W., Rossignol, K.L., Braddy, J.S., Bullerjahn, G.S., Bridgeman, T.B., Davis, T.W., Wei, J., Bu, M., Paerl, H.W., 2021. Roles of nutrient limitation on Western Lake Erie cyanohab toxin production. *Toxins* **13**, 47. https://doi.org/10.3390/toxins13010047

Bartosiewicz, M., Przytulska, A., Lapierre, J., Laurion, I., Lehmann, M.F., Maranger, R., 2019. Hot tops, cold bottoms: Synergistic climate warming and shielding effects increase carbon burial in lakes. *Limnol. Oceanogr. Lett.* https://doi.org/10.1002/lol2.10117

Batt, R.D., Carpenter, S.R., Ives, A.R., 2017. Extreme events in lake ecosystem time series. *Limnol. Oceanogr. Lett.* **2**, 63–69. https://doi.org/10.1002/lol2.10037

Beaulieu, J.J., DelSontro, T., Downing, J.A., 2019. Eutrophication will increase methane emissions from lakes and impoundments during the 21st century. *Nat. Commun.* **10**, 1375. https://doi.org/10.1038/s41467-019-09100-5

Beaulieu, M., Pick, F., Gregory-Eaves, I., 2013. Nutrients and water temperature are significant predictors of cyanobacterial biomass in a 1147 lakes data set. *Limnol. Oceanogr*. **58**, 1736–1746. https://doi.org/10.4319/lo.2013.58.5.1736

Beaulieu, M., Pick, F., Palmer, M., Watson, S., Winter, J., Zurawell, R., Gregory-Eaves, I., Prairie, Y., 2014. Comparing predictive cyanobacterial models from temperate regions. *Can. J. Fish. Aquat. Sci.* **71**, 1830–1839. https://doi.org/10.1139/cjfas-2014-0168

Beaver, J.R., Tausz, C.E., Scotese, K.C., Pollard, A.I., Mitchell, R.M., 2018. Environmental factors influencing the quantitative distribution of microcystin and common potentially toxigenic cyanobacteria in U.S. lakes and reservoirs. *Harmful Algae*. **78**, 118–128. https://doi.org/10.1016/j.hal.2018.08.004

Beduhn, R.J., 1994. The effects of destratification aeration on five minnesota lakes. *Lake Reserv. Manag.* **9**, 105–110. https://doi.org/10.1080/07438149409354737

Beklioğlu, M., Bucak, T., Levi, E.E., Erdoğan, Ş., Özen, A., Filiz, N., Bezirci, G., Çakıroğlu, A.İ., Tavşanoğlu, Ü.N., Gökçe, D., Demir, N., Özuluğ, M., Duran, M., Özkan, K., Brucet, S., Jeppesen, E., 2020. Influences of climate and nutrient enrichment on the multiple trophic levels of Turkish shallow lakes. *Inland Waters*. 1–13. https://doi.org/10.1080/20442041.2020.1746599

Bellinger, B.J., Richter, A., Porras, A., Davis, S.L., 2018. Drought and management effects on biophysicochemistry in a rapidly-flushed reservoir. *Lake Reserv. Manag.* **34**, 182–198. https://doi.org/10.1080/10402381.2017.1384770

Berkowitz, J., Anderson, M.A., Amrhein, C., 2006. Influence of aging on phosphorus sorption to alum floc in lake water. *Water Res.* **40**, 911–916. https://doi.org/10.1016/j.watres.2005.12.018

Bernard, C., Ballot, A., Thomazeau, S., Maloufi, S., Furey, A., Mankiewicz-Boczek, J., Pawlik-Skowrońska, B., Capelli, C., Salmaso, N., 2016. Appendix 2: Cyanobacteria associated with the production of cyanotoxins. In *Handbook of Cyanobacterial Monitoring and Cyanotoxin Analysis*. Wiley Online Library, Chichester, West Sussex, pp. 501–525.

Bernhardt, J., Kirillin, G., Hupfer, M., 2014. Periodic convection within littoral lake sediments on the background of seiche-driven oxygen fluctuations. *Limnol. Oceanogr. Fluids Environ.* **4**, 17–33. https://doi.org/10.1215/21573689-2683238

Berthold, M., Campbell, D.A., 2021. Restoration, conservation and phytoplankton hysteresis. *Conserv. Physiol.* **9**. https://doi.org/10.1093/conphys/coab062

Berthold, M., Karstens, S., Buczko, U., Schumann, R., 2018. Potential export of soluble reactive phosphorus from a coastal wetland in a cold-temperate lagoon system: Buffer capacities of macrophytes and impact on phytoplankton. *Sci. Total Environ.* **616–617**, 46–54. https://doi.org/10.1016/j.scitotenv.2017.10.244

Beutel, M.W., Duvil, R., Cubas, F.J., Matthews, D.A., Wilhelm, F.M., Grizzard, T.J., Austin, D., Horne, A.J., Gebremariam, S., 2016. A review of managed nitrate addition to enhance surface water quality. *Crit. Rev. Environ. Sci. Technol.* 1–28. https://doi.org/10.1080/10643389.2016.1151243

Beutel, M.W., Horne, A.J., Taylor, W.D., Losee, R.F., Whitney, R.D., 2008. Effects of oxygen and nitrate on nutrient release from profundal sediments of a large, oligo-mesotrophic reservoir, Lake Mathews, California. *Lake Reserv. Manag.* **24**, 18–29. https://doi.org/10.1080/07438140809354047

Bever, A.J., Friedrichs, M.A., Friedrichs, C.T., Scully, M.E., 2018. Estimating hypoxic volume in the Chesapeake Bay using two continuously sampled oxygen profiles. *J. Geophys. Res. Oceans*. **123**, 6392–6407. https://doi.org/10.1029/2018JC014129

Binding, C.E., Zeng, C., Pizzolato, L., Booth, C., Valipour, R., Fong, P., Zastepa, A., Pascoe, T., 2023. Reporting on the status, trends, and drivers of algal blooms on Lake of the Woods using satellite-derived bloom indices (2002–2021). *J. Gt. Lakes Res.* **49**, 32–43. https://doi.org/10.1016/j.jglr.2022.12.007

Bingham, M., Sinha, S.K., Lupi, F., 2015. *Economic Benefits of Reducing Harmful Algal Blooms in Lake Erie*. Environmental Consulting & Technology, Inc., Gainesville, FL.

Bishop, W.M., McNabb, T., Cormican, I., Willis, B.E., Hyde, S., 2014. Operational evaluation of Phoslock phosphorus locking technology in Laguna Niguel Lake, California. *Water. Air. Soil Pollut.* **225**, 2018–2029. https://doi.org/10.1007/s11270-014-2018-6

Björnerås, C., Škerlep, M., Floudas, D., Persson, P., Kritzberg, E.S., 2019. High sulfate concentration enhances iron mobilization from organic soil to water. *Biogeochemistry*. https://doi.org/10.1007/s10533-019-00581-6

Björnerås, C., Weyhenmeyer, G.A., Evans, C.D., Gessner, M.O., Grossart, H.-P., Kangur, K., Kokorite, I., Kortelainen, P., Laudon, H., Lehtoranta, J., Lottig, N., Monteith, D.T., Nõges, P., Nõges, T., Oulehle, F., Riise, G., Rusak, J.A., Räike, A., Sire, J., Sterling, S., Kritzberg, E.S., 2017. Widespread increases in iron concentration in European and North American freshwaters: Increasing iron concentrations. *Glob. Biogeochem. Cycles*. **31**, 1488–1500. https://doi.org/10.1002/2017GB005749

Blais, J.M., Kalff, J., 1995. The influence of lake morphometry on sediment focusing. *Limnol. Oceanogr.* **40**, 582–588. https://doi.org/10.4319/lo.1995.40.3.0582

Blanchet, C.C., Arzel, C., Davranche, A., Kahilainen, K.K., Secondi, J., Taipale, S., Lindberg, H., Loehr, J., Manninen-Johansen, S., Sundell, J., Maanan, M., Nummi, P., 2022. Ecology and extent of freshwater browning – What we know and what should be studied next in the context of global change. *Sci. Total Environ.* **812**, 152420. https://doi.org/10.1016/j.scitotenv.2021.152420

Bocaniov, S.A., Scavia, D., Van Cappellen, P., 2023. Long-term phosphorus mass-balance of Lake Erie (Canada-USA) reveals a major contribution of in-lake phosphorus loading. *Ecol. Inform.* **77**, 102131. https://doi.org/10.1016/j.ecoinf.2023.102131

Bocaniov, S.A., Ullmann, C., Rinke, K., Lamb, K.G., Boehrer, B., 2014. Internal waves and mixing in a stratified reservoir: Insights from three-dimensional modeling. *Limnologica* **49**, 52–67. https://doi.org/10.1016/j.limno.2014.08.004

Bogard, M.J., Vogt, R.J., Hayes, N.M., Leavitt, P.R., 2020. Unabated nitrogen pollution favors growth of toxic cyanobacteria over chlorophytes in most hypereutrophic lakes. *Environ. Sci. Technol.* **54**, 3219–3227. https://doi.org/10.1021/acs.est.9b06299

Bonilla, I., Garcia-González, M., Mateo, P., 1990. Boron requirement in cyanobacteria: Its possible role in the early evolution of photosynthetic organisms. *Plant Physiol.* **94**, 1554–1560. https://doi.org/10.1104/pp.94.4.1554

Bonilla, S., Aguilera, A., Aubriot, L., Huszar, V., Almanza, V., Haakonsson, S., Izaguirre, I., O'Farrell, I., Salazar, A., Becker, V., 2023. Nutrients and not temperature are the key drivers for cyanobacterial biomass in the Americas. *Harmful Algae*. **121**, 102367. https://doi.org/10.1016/j.hal.2022.102367

Bonilla, S., Pick, F.R., 2017. *Freshwater Bloom-Forming Cyanobacteria and Anthropogenic Change – Bonilla – 2017 – Limnology and Oceanography e-Lectures – Wiley Online Library* [WWW Document]. http://onlinelibrary.wiley.com/doi/10.1002/loe2.10006/full (accessed 2.28.18).

Bormans, M., Maršálek, B., Jančula, D., 2016. Controlling internal phosphorus loading in lakes by physical methods to reduce cyanobacterial blooms: A review. *Aquat. Ecol.* **50**, 407–422. https://doi.org/10.1007/s10452-015-9564-x

Bormans, M., Sherman, B.S., Webster, I.T., 1999. Is buoyancy regulation in cyanobacteria an adaptation to exploit separation of light and nutrients? *Mar. Freshw. Res.* https://doi.org/10.1071/MF99105

Bouma-Gregson, K., Kudela, R.M., Power, M.E., 2018. Widespread anatoxin-a detection in benthic cyanobacterial mats throughout a river network. *PLoS ONE* **13**, e0197669. https://doi.org/10.1371/journal.pone.0197669

Brandenburg, F., Schoffman, H., Kurz, S., Krämer, U., Keren, N., Weber, A.P., Eisenhut, M., 2017. The Synechocystis manganese exporter Mnx is essential for manganese homeostasis in cyanobacteria. *Plant Physiol.* **173**, 1798–1810. https://doi.org/10.1104/pp.16.01895

Brandenburg, K., Siebers, L., Keuskamp, J., Jephcott, T.G., Van de Waal, D.B., 2020. Effects of nutrient limitation on the synthesis of N-rich phytoplankton toxins: A meta-analysis. *Toxins*. **12**, 221. https://doi.org/10.3390/toxins12040221

Brasil, J., Attayde, J.L., Vasconcelos, F.R., Dantas, D.D.F., Huszar, V.L.M., 2016. Drought-induced water-level reduction favors cyanobacteria blooms in tropical shallow lakes. *Hydrobiologia*. **770**, 145–164. https://doi.org/10.1007/s10750-015-2578-5

Brett, M.T., Benjamin, M.M., 2008. A review and reassessment of lake phosphorus retention and the nutrient loading concept. *Freshw. Biol.* **53**, 194–211. https://doi.org/10.1111/j.1365-2427.2007.01862.x

Bridgeman, T.B., Chaffin, J.D., Filbrun, J.E., 2013. A novel method for tracking western Lake Erie Microcystis blooms, 2002–2011. *J. Gt. Lakes Res.* **39**, 83–89. https://doi.org/10.1016/j.jglr.2012.11

Bristow, C.E., Morin, A., Hesslein, R.H., Podemski, C.L., 2008. Phosphorus budget and productivity of an experimental lake during the initial three years of cage aquaculture. *Can. J. Fish. Aquat. Sci.* **65**, 2485–2495. https://doi.org/10.1139/F08-155

Brothers, S., Köhler, J., Attermeyer, K., Grossart, H.-P., Mehner, T., Meyer, N., Scharnweber, K., Hilt, S., 2014. A feedback loop links brownification and anoxia in a temperate, shallow lake. *Limnol. Oceanogr.* **59**, 1388–1398. https://doi.org/10.4319/lo.2014.59.4.1388

Brunberg, A.-K., Blomqvist, P., 2002. Benthic overwintering of Microcystis colonies under different environmental conditions. *J. Plankton. Res.* **24**, 1247–1252. https://doi.org/10.1093/plankt/24.11.1247

Brunberg, A.K., Bostrom, B., 1992. Coupling between benthic biomass of Microcystis and phosphorus release from the sediments of a highly eutrophic lake. *Hydrobiologia* **235/236**, 375–385. https://doi.org/10.1007/978-94-011-2783-7_32

Bryant, L.D., Lorrai, C., McGinnis, D.F., Brand, A., Est, A.W., Little, J.C., 2010. Variable sediment oxygen uptake in response to dynamic forcing. *Limnol. Oceanogr.* **55**, 950–964. https://doi.org/10.4319/lo.2010.55.2.0950

Bryhn, A.C., Håkanson, L., 2009. Eutrophication: Model before acting. *Science*. **324**, 723–723. https://doi.org/10.1126/science.324_723a

Bukaveckas, P.A., Buikema, L., Stewart, C., 2024. Effects of climate change and variability on thermal regime and dissolved oxygen resources of oligotrophic lakes in the Adirondack Mountain region. *Aquat. Sci.* **86**, 9. https://doi.org/10.1007/s00027-023-01021-2

Bullerjahn, G.S., McKay, R.M., Davis, T.W., Baker, D.B., Boyer, G.L., D'Anglada, L.V., Doucette, G.J., Ho, J.C., Irwin, E.G., Kling, C.L., Kudela, R.M., Kurmayer, R., Michalak, A.M., Ortiz, J.D., Otten, T.G., Paerl, H.W., Qin, B., Sohngen, B.L., Stumpf, R.P., Visser, P.M., Wilhelm, S.W., 2016. Global solutions to regional problems: Collecting global expertise to address the problem of harmful cyanobacterial blooms. A Lake Erie case study. *Harmful Algae*. **54**, 223–238. https://doi.org/10.1016/j.hal.2016.01.003

Burford, M.A., Carey, C.C., Hamilton, D.P., Huisman, J., Paerl, H.W., Wood, S.A., Wulff, A., 2020. Perspective: Advancing the research agenda for improving understanding of cyanobacteria in a future of global change. *Harmful Algae*. **91**, 101601. https://doi.org/10.1016/j.hal.2019.04.004

Burger, D.F., Hamilton, D.P., Pilditch, C.A., 2008. Modelling the relative importance of internal and external nutrient loads on water column nutrient concentrations and phytoplankton biomass in a shallow polymictic lake. *Ecol. Model.* **211**, 411–423. https://doi.org/10.1016/j.ecolmodel.2007.09.028

Burns, N., McIntosh, J., Scholes, P., 2009. Managing the lakes of the Rotorua District, New Zealand. *Lake Reserv. Manage.* **25**, 284–296. https://doi.org/10.1080/07438140903083815

Bush, E., Lemmen, D.S., 2019. *Canada's Changing Climate Report* (Government of Canada). Ottawa, ON, Canada.

Bykova, O., Laursen, A., Bostan, V., Bautista, J., McCarthy, L., 2006. Do zebra mussels (*Dreissena polymorpha*) alter lake water chemistry in a way that favours Microcystis growth? *Sci. Total Environ.* **371**, 362–372. https://doi.org/10.1016/j.scitotenv.2006.08.022

Calomeni, A.J., McQueen, A.D., Kinley-Baird, C.M., Clyde, G.A., 2023. Identification and prioritization of sites with overwintering cyanobacteria to inform preventative management of harmful algal blooms. *J. Aquat. Plant Manage.* **61**, 30–41.

Camarero, L., Catalan, J., 2012. Atmospheric phosphorus deposition may cause lakes to revert from phosphorus limitation back to nitrogen limitation. *Nat. Commun.* **3**, 1118. https://doi.org/10.1038/ncomms2125

Carey, C.C., Hanson, P.C., Thomas, R.Q., Gerling, A.B., Hounshell, A.G., Lewis, A.S.L., Lofton, M.E., McClure, R.P., Wander, H.L., Woelmer, W.M., Niederlehner, B.R., Schreiber, M.E., 2022. Anoxia decreases the magnitude of the carbon, nitrogen, and phosphorus sink in freshwaters. *Glob. Change Biol.* **28**, 4861–4881. https://doi.org/10.1111/gcb.16228

Carey, C.C., Ibelings, B.W., Hoffmann, E.P., Hamilton, D.P., Brookes, J.D., 2012. Eco-physiological adaptations that favour freshwater cyanobacteria in a changing climate. *Water Res.* **46**, 1394–1407. https://doi.org/10.1016/j.hal.2016.01.003

Carey, C.C., Weathers, K.C., Cottingham, K.L., 2008. Gloeotrichia echinulata blooms in an oligotrophic lake: Helpful insights from eutrophic lakes. *J. Plankton Res.* **30**, 893–904. https://doi.org/10.1093/plankt/fbn055

Carleton, J.N., Lee, S.S., 2023. Modeling lake recovery lag times following influent phosphorus loading reduction. *Environ. Model. Softw.* **162**, 105642. https://doi.org/10.1016/j.envsoft.2023.105642

Carmack, E.C., Gray, C.B.J., Pharo, C.H., Daley, R.J., 1979. Importance of lake-river interaction on seasonal patterns in the general circulation of Kamloops Lake, British Columbia: Lake-river interaction. *Limnol. Oceanogr.* **24**, 634–644. https://doi.org/10.4319/lo.1979.24.4.0634

Carneiro, B.L.D.S., de Jesus D. Rocha, M., Barros, M.U.G., Paulino, W.D., Lima Neto, I.E., 2023. Predicting anoxia in the wet and dry periods of tropical semiarid reservoirs. *J. Environ. Manage.* **326**, 116720. https://doi.org/10.1016/j.jenvman.2022.116720

Carpenter, S.R., 1980. Enrichment of Lake Wingra, Wisconsin, by submersed macrophyte decay. *Ecol.* **61**, 1145–1155. https://doi.org/10.2307/1936834

Carpenter, S.R., Brock, W.A., 2006. Rising variance: A leading indicator of ecological transition. *Ecol. Lett.* **9**, 311–318. https://doi.org/10.1111/j.1461-0248.2005.00877.x

Carvalho, L., Poikane, S., Solheim, A.L., Phillips, G., Borics, G., Catalan, J., De Hoyos, C., Drakare, S., Dudley, B.J., Järvinen, M., et al., 2013. Strength and uncertainty of phytoplankton metrics for assessing eutrophication impacts in lakes. *Hydrobiologia* **704**, 127–140. https://doi.org/10.1007/s10750-012-1344-1

Cha, Y., Stow, C.A., Bernhardt, E.S., 2013. Impacts of dreissenid mussel invasions on chlorophyll and total phosphorus in 25 lakes in the USA. *Freshw. Biol.* **58**, 192–206. https://doi.org/10.1111/fwb.12050

Cha, Y., Stow, C.A., Nalepa, T.F., Reckhow, K.H., 2011. Do invasive mussels restrict offshore phosphorus transport in Lake Huron? *Env. Sci. Technol.* **45**, 7226–7231. https://doi.org/10.1021/es2014715

Chaffin, J.D., Bridgeman, T.B., Bade, D.L., 2013. Nitrogen constrains the growth of late summer cyanobacterial blooms in Lake Erie. *Adv. Microbiol.* **2013**. https://doi.org/10.4236/aim.2013.36A003

Chaffin, J.D., Kane, D.D., 2010. Burrowing mayfly (Ephemeroptera: Ephemeridae: *Hexagenia spp.*) bioturbation and bioirrigation: A source of internal phosphorus loading in Lake Erie. *J. Gt. Lakes Res.* **36**, 57–63. https://doi.org/10.1016/j.jglr.2009.09.003

Chaffin, J.D., Stanislawczyk, K., Kane, D.D., Lambrix, M.M., 2020. Nutrient addition effects on chlorophyll a, phytoplankton biomass, and heterocyte formation in Lake Erie's Central Basin during 2014–2017: Insights into diazotrophic blooms in high nitrogen water. *Freshw. Biol.* **65**, 2154–2168. https://doi.org/10.1111/fwb.13610

Chaffin, J.D., Westrick, J.A., Reitz, L.A., Bridgeman, T.B., 2023. Microcystin congeners in Lake Erie follow the seasonal pattern of nitrogen availability. *Harmful Algae.* **127**, 102466. https://doi.org/10.1016/j.hal.2023.102466

Chapra, S.C., 1997. *Surface Water-Quality Modeling*. McGraw-Hill Companies, Inc., Boston.

Charlton, M.N., Milne, J.E., Booth, W.G., Chiocchio, F., 1993. Lake Erie offshore in 1990: Restoration and resilience in the central basin. *J. Gt. Lakes Res.* **19**, 291–309. https://doi.org/10.1016/S0380-1330(93)71218-6

Chen, M., Ding, S., Chen, X., Sun, Q., Fan, X., Lin, J., Ren, M., Yang, L., Zhang, C., 2018. Mechanisms driving phosphorus release during algal blooms based on hourly changes in iron and phosphorus concentrations in sediments. *Water Res.* **133**, 153–164. https://doi.org/10.1016/j.watres.2018.01.040

Chen, M., Ding, S., Wu, Y., Fan, X., Jin, Z., Tsang, D.C.W., Wang, Y., Zhang, C., 2019a. Phosphorus mobilization in lake sediments: Experimental evidence of strong control by iron and negligible influences of manganese redox reactions. *Environ. Pollut.* **246**, 472–481. https://doi.org/10.1016/j.envpol.2018.12.031

Chen, M., Huang, Y., Wang, Y., Liu, C., He, Y., Li, N., 2023. Inhibitory effects and mechanisms of insoluble humic acids on internal phosphorus release from the sediments. *Water Res.* 121074. https://doi.org/10.1016/j.watres.2023.121074

Chen, Q., Wang, M., Zhang, J., Shi, W., Mynett, A.E., Yan, H., Hu, L., 2019b. Physiological effects of nitrate, ammonium, and urea on the growth and microcystins contamination of Microcystis aeruginosa: Implication for nitrogen mitigation. *Water Res.* **163**, 114890. https://doi.org/10.1016/j.watres.2019.114890

Cheng, D., He, Q., 2020. *Microbial Photosynthesis, Chapter 3: Iron Deficiency in Cyanobacteria.* Springer, Singapore. https://doi.org/10.1007/978-981-15-3110-1

Chorus, I., Fastner, J., Welker, M., 2021. Cyanobacteria and cyanotoxins in a changing environment: Concepts, controversies, challenges. *Water.* **13**, 2463. https://doi.org/10.3390/w13182463

Chorus, I., Köhler, A., Beulker, C., Fastner, J., van de Weyer, K., Hegewald, T., Hupfer, M., 2020. Decades needed for ecosystem components to respond to a sharp and drastic phosphorus load reduction. *Hydrobiologia.* **847**, 4621–4651. https://doi.org/10.1007/s10750-020-04450-4

Chorus, I., Spijkerman, E., 2020. What Colin Reynolds could tell us about nutrient limitation, N:P ratios and eutrophication control. *Hydrobiologia.* https://doi.org/10.1007/s10750-020-04377-w

Chorus, I., Welker, M., 2021. *Toxic Cyanobacteria in Water; A Guide to Their Public Health Consequences, Monitoring and Management*, 2nd ed. World Health Organization, London. https://doi.org/10.1201/9781003081449.

Christensen, V.G., Khan, E., 2020. Freshwater neurotoxins and concerns for human, animal, and ecosystem health: A review of anatoxin-a and saxitoxin. *Sci. Total Environ.* **736**, 139515. https://doi.org/10.1016/j.scitotenv.2020.139515

Christianson, K.R., Loria, K.A., Blanken, P.D., Caine, N., Johnson, P.T.J., 2021. On thin ice: Linking elevation and long-term losses of lake ice cover. *Limnol. Oceanogr. Lett.* **6**, 77–84. https://doi.org/10.1002/lol2.10181

Cianci-Gaskill, J.A., Klug, J.L., Merrell, K.C., Millar, E.E., Wain, D.J., Kramer, L., van Wijk, D., Paule-Mercado, M.C.A., Finlay, K., Glines, M.R., Munthali, E.M., Teurlincx, S., Borre, L., Yan, N.D., 2024. A lake management framework for global application: Monitoring, restoring, and protecting lakes through community engagement. *Lake Reserv. Manag.* **1**, 1–27. https://doi.org/10.1080/10402381.2023.2299868

Cirés, S., Wörmer, L., Agha, R., Quesada, A., 2013. Overwintering populations of Anabaena, Aphanizomenon and Microcystis as potential inocula for summer blooms. *J. Plankton Res.* **35**, 1254–1266. https://doi.org/10.1093/plankt/fbt081

Conley, D.J., Björck, S., Bonsdorff, E., Carstensen, J., Destouni, G., Gustafsson, B.G., Hietanen, S., Kortekaas, M., Kuosa, H., Markus Meier, H.E., Müller-Karulis, B., Nordberg, K., Norkko, A., Nürnberg, G., Pitkänen, H., Rabalais, N.N., Rosenberg, R., Savchuk, O.P., Slomp, C.P., Voss, M., Wulff, F., Zillén, L., 2009b. Critical review: Hypoxia-related processes in the Baltic Sea. *Environ. Sci. Technol.* **43**, 3412–3420. https://doi.org/10.1021/es802762a

Conley, D.J., Humborg, C., Rahm, L., Savchuk, O., Wulff, F., 2002. Hypoxia in the Baltic Sea and basin-scale changes in phosphorus biochemistry. *Env. Sci. Technol.* **36**, 5315–5320. https://doi.org/10.1021/es025763w

Conley, D.J., Paerl, H.W., Howarth, R.R., Boesch, D.F., Seitzinger, S.P., Havens, K.E., Lancelot, C., Likens, G.E., 2009a. Controlling eutrophication: Nitrogen and phosphorus. *Science.* **323**, 1014–1015. https://doi.org/10.1126/science.1167755

Conley, D.J., Paerl, H.W., Howarth, R.W., Boesch, D.F., Seitzinger, S.P., Havens, K.E., Lancelot, C., Likens, G.E., 2009c. Eutrophication: Time to adjust expectations – response. *Science.* **324**, 724–725. https://doi.org/10.1126/science.324_7

Cooke, G.D., Welch, E.B., Martin, A.B., Fulmer, D.G., Hyde, J.B., Schrieve, G.D., 1993. Effectiveness of Al, Ca, and Fe salts for control of internal phosphorus loading in shallow and deep lakes. *Hydrobiologia.* **253**, 323–335. https://doi.org/10.1007/BF00050758

Cooke, G.D., Welch, E.B., Peterson, S.A., Nichols, S.A., 2005. *Restoration and Management of Lakes and Reservoirs*, 3nd ed. CRC, Boca Raton, FL.

Cooke, S.J., Rytwinski, T., Taylor, J.J., Nyboer, E.A., Nguyen, V.M., Bennett, J.R., Young, N., Aitken, S., Auld, G., Lane, J.-F., Prior, K.A., Smokorowski, K.E., Smith, P.A., Jacob, A.L., Browne, D.R., Blais, J.M., Kerr, J.T., Ormeci, B., Alexander, S.M., Burn, C.R., Buxton, R.T., Orihel, D.M., Vermaire, J.C., Murray, D.L., Simon, P., Edwards, K.A., Clarke, J., Xenopoulos, M.A., Gregory-Eaves, I., Bennett, E.M., Smol, J.P., 2020. On "success" in applied environmental research – What is it, how can it be achieved, and how does one know when it has been achieved? *Environ. Rev.* **28**, 357–372. https://doi.org/10.1139/er-2020-0045

Copetti, D., Finsterle, K., Marziali, L., Stefani, F., Tartari, G., Douglas, G., Reitzel, K., Spears, B.M., Winfield, I.J., Crosa, G., D'Haese, P., Yasseri, S., Lürling, M., 2016. Eutrophication management in surface waters using lanthanum modified bentonite: A review. *Water Res.* **97**, 162–174. https://doi.org/10.1016/j.watres.2015.11.056

Coppens, J., Özen, A., Tavşanoğlu, Ü.N., Erdoğan, Ş., Levi, E.E., Yozgatlıgil, C., Jeppesen, E., Beklioğlu, M., 2016. Impact of alternating wet and dry periods on long-term seasonal phosphorus and nitrogen budgets of two shallow Mediterranean lakes. *Sci. Total Environ.* **563–564**, 456–467. https://doi.org/10.1016/j.scitotenv.2016.04.028

Cordell, D., Drangert, J.-O., White, S., 2009. The story of phosphorus: Global food security and food for thought. *Glob. Environ. Change.* **19**, 292–305. www.sciencedirect.com/science/article/pii/S095937800800099X

Cottingham, K.L., Weathers, K.C., Ewing, H.A., Greer, M.L., Carey, C.C., 2021. Predicting the effects of climate change on freshwater cyanobacterial blooms requires consideration of the complete cyanobacterial life cycle. *J. Plankton Res.* **43**, 10–19. https://doi.org/10.1093/plankt/fbaa059

Couture, R.-M., Moe, S.J., Lin, Y., Kaste, Ø., Haande, S., Lyche Solheim, A., 2018. Simulating water quality and ecological status of Lake Vansjø, Norway, under land-use and climate change by linking process-oriented models with a Bayesian network. *Sci. Total Environ.* **621**, 713–724. https://doi.org/10.1016/j.scitotenv.2017.11.303

Creed, I.F., Bergström, A.-K., Trick, C.G., Grimm, N.B., Hessen, D.O., Karlsson, J., Kidd, K.A., Kritzberg, E., McKnight, D.M., Freeman, E.C., 2018. Global change-driven effects on dissolved organic matter composition: Implications for food webs of northern lakes. *Glob. Change Biol.* **24**, 3692–3714. https://doi.org/10.1111/gcb.14129

Cyr, H., McCabe, S.K., Nürnberg, G.K., 2009. Phosphorus sorption experiments and the potential for internal phosphorus loading in littoral areas of a stratified lake. *Water Res.* **43**, 1654–1666. https://doi.org/10.1016/j.watres.2008.12.050

Dadi, T., Schultze, M., Kong, X., Seewald, M., Rinke, K., Friese, K., 2023. Sudden eutrophication of an aluminum sulphate treated lake due to abrupt increase of internal phosphorus loading after three decades of mesotrophy. *Water Res.* **235**, 119824. https://doi.org/10.1016/j.watres.2023.119824

Dahlke, F.T., Wohlrab, S., Butzin, M., Pörtner, H.-O., 2020. Thermal bottlenecks in the life cycle define climate vulnerability of fish. *Science.* **369**, 65–70. https://doi.org/10.1126/science.aaz3658

Dai, Y., Yang, S., Zhao, D., Hu, C., Xu, W., Anderson, D.M., Li, Y., Song, X.-P., Boyce, D.G., Gibson, L., Zheng, C., Feng, L., 2023. Coastal phytoplankton blooms expand and intensify in the 21st century. *Nature.* 1–5. https://doi.org/10.1038/s41586-023-05760-y

Davidson, T.A., Sayer, C.D., Jeppesen, E., Søndergaard, M., Lauridsen, T.L., Johansson, L.S., Baker, A., Graeber, D., 2023. Bimodality and alternative equilibria do not help explain long-term patterns in shallow lake chlorophyll-a. *Nat. Commun.* **14**, 398. https://doi.org/10.1038/s41467-023-36043-9

Davis, T.W., Berry, D.L., Boyer, G.L., Gobler, C.J., 2009. The effects of temperature and nutrients on the growth and dynamics of toxic and non-toxic strains of Microcystis during cyanobacteria blooms. *Harmful Algae.* **8**, 715–725. https://doi.org/10.1016/j.hal.2009.02.004

Davis, T.W., Harke, M.J., Marcoval, M.A., Goleski, J., Orano-Dawson, C., Berry, D.L., Gobler, C.J., 2010. Effects of nitrogenous compounds and phosphorus on the growth of toxic and non-toxic strains of Microcystis during cyanobacterial blooms. *Aquat. Microb. Ecol.* **61**, 149–162. https://doi.org/10.3354/ame01445

Deemer, B.R., Harrison, J.A., 2019. Summer redox dynamics in a eutrophic reservoir and sensitivity to a summer's end drawdown event. *Ecosystems*. **22**, 1618–1632. https://doi.org/10.1007/s10021-019-00362-0

Deng, J., Paerl, H.W., Qin, B., Zhang, Y., Zhu, G., Jeppesen, E., Cai, Y., Xu, H., 2018. Climatically-modulated decline in wind speed may strongly affect eutrophication in shallow lakes. *Sci. Total Environ.* **645**, 1361–1370. https://doi.org/10.1016/j.scitotenv.2018.07.208

Dengg, M., Stirling, C.H., Safi, K., Lehto, N.J., Wood, S.A., Seyitmuhhamedov, K., Reid, M.R., Verburg, P., 2023. Bioavailable iron concentrations regulate phytoplankton growth and bloom formation in low-nutrient lakes. *Sci. Total Environ.* **902**, 166399. https://doi.org/10.1016/j.scitotenv.2023.166399

de Vicente, I., Huang, P., Andersen, F.O., Jensen, H.S., 2008. Phosphate adsorption by fresh and aged aluminum hydroxide. Consequences for lake restoration. *Env. Sci. Technol.* **42**, 6650–6655. https://doi.org/10.1021/es800503s

Devito, K.J., Dillon, P.J., 1993. Importance of runoff and winter anoxia to the P and N dynamics of a beaver pond. *Can. J. Fish. Aquat. Sci.* **50**, 2222–2234. https://doi.org/10.1139/f93-248

Devito, K.J., Dillon, P.J., LaZerte, B.D., 1989. Phosphorus and nitrogen retention in five Precambrian shield wetlands. *Biogeochem.* **8**, 185–204. https://doi.org/10.1007/BF00002888

Diaz, R.J., 2001. Overview of hypoxia around the world. *J. Environ. Qual.* **30**, 275–281. https://doi.org/10.2134/jeq2001.302275x

Diaz, R.J., Rosenberg, R., 2008. Spreading dead zones and consequences for marine ecosystems. *Science*. **1156401**, 321. www.gfishasia.com/doc/Hypoxia%20in%20oceans.pdf

Dibi-Anoh, P.A., Koné, M., Gerdener, H., Kusche, J., N'Da, C.K., 2023. Hydrometeorological extreme events in West Africa: Droughts. *Surv. Geophys.* **44**, 173–195. https://doi.org/10.1007/s10712-022-09748-7

Dillon, P.J., Rigler, F.H., 1974. The phosphorus-chlorophyll relationship in lakes. *Limnol. Oceanogr.* **19**, 767–773. https://doi.org/10.4319/lo.1974.19.5.0767

Ding, S., Chen, M., Gong, M., Fan, X., Qin, B., Xu, H., Gao, S., Jin, Z., Tsang, D., Zhang, C., 2018. Internal phosphorus loading from sediments causes seasonal nitrogen limitation for harmful algal blooms. *Sci. Total Environ.* **625**, 872–884. https://doi.org/10.1016/j.scitotenv.2017.12.348

Dobberfuhl, D.R., 2003. Cylindrospermopsis racihorskii in three central Florida lakes: Population dynamics, controls, and management implications. *Lake Reserv. Manag.* **19**, 341–348.

Dodds, W.K., Bouska, W.W., Eitzmann, J.L., Pilger, T.J., Pitts, K.L., Riley, A.J., Schloesser, J.T., Thornbrugh, D.J., 2009. Eutrophication of U.S. freshwaters: Analysis of potential economic damages. *Env. Sci. Technol.* **43**, 12–19. https://doi.org/10.1021/es801217q

Dodds, W.K., Jones, H.R., Welch, E.B., 1998. Suggested classification of stream trophic state: Distributions of temperate stream types by chlorophyll, total nitrogen, and phosphorus. *Water Res.* **32**, 1455–1462. https://doi.org/10.1016/S0043-1354(97)00370-9

Dokulil, M.T., Teubner, K., 2000. Cyanobacterial dominance in lakes. *Hydrobiologia*. **438**, 1–12. https://doi.org/10.1023/A:1004155810302

Dolman, A.M., Rücker, J., Pick, F.R., Fastner, J., Rohrlack, T., Mischke, U., Wiedner, C., 2012. Cyanobacteria and cyanotoxins: The influence of nitrogen versus phosphorus. *PLoS ONE*. **7**, e38757. https://doi.org/10.1371/journal.pone.0038757

Dondajewska, R., Kowalczewska-Madura, K., Gołdyn, R., Kozak, A., Messyasz, B., Cerbin, S., 2019. Long-term water quality changes as a result of a sustainable restoration – a case study of dimictic Lake Durowskie. *Water*. **11**, 616. https://doi.org/10.3390/w11030616

Donis, D., Mantzouki, E., McGinnis, D.F., Vachon, D., Gallego, I., Grossart, H., Senerpont Domis, L.N., Teurlincx, S., Seelen, L., Lürling, M., Verstijnen, Y., Maliaka, V., Fonvielle, J., Visser, P.M., Verspagen, J., Herk, M., Antoniou, M.G., Tsiarta, N., McCarthy, V., Perello, V.C., Machado-Vieira, D., Oliveira, A.G., Maronić, D.Š., Stević, F., Pfeiffer, T.Ž., Vucelić, I.B., Žutinić, P., Udovič, M.G., Plenković-Moraj, A., Bláha, L., Geriš, R., Fránková, M., Christoffersen, K.S., Warming, T.P., Feldmann, T., Laas, A., Panksep, K., Tuvikene, L., Kangro, K., Koreivienė, J., Karosienė, J., Kasperovičienė, J., Savadova-Ratkus, K., Vitonytė, I., Häggqvist, K., Salmi, P., Arvola, L., Rothhaupt, K., Avagianos, C., Kaloudis, T., Gkelis, S., Panou, M., Triantis, T., Zervou, S., Hiskia, A., Obertegger, U., Boscaini, A., Flaim, G., Salmaso, N., Cerasino, L., Haande, S., Skjelbred, B., Grabowska, M., Karpowicz, M., Chmura, D., Nawrocka, L., Kobos, J.,

Mazur-Marzec, H., Alcaraz-Párraga, P., Wilk-Woźniak, E., Krztoń, W., Walusiak, E., Gagala-Borowska, I., Mankiewicz-Boczek, J., Toporowska, M., Pawlik-Skowronska, B., Niedźwiecki, M., Pęczuła, W., Napiórkowska-Krzebietke, A., Dunalska, J., Sieńska, J., Szymański, D., Kruk, M., Budzyńska, A., Goldyn, R., Kozak, A., Rosińska, J., Szeląg-Wasielewska, E., Domek, P., Jakubowska-Krepska, N., Kwasizur, K., Messyasz, B., Pełechata, A., Pełechaty, M., Kokocinski, M., Madrecka-Witkowska, B., Kostrzewska-Szlakowska, I., Frąk, M., Bańkowska-Sobczak, A., Wasilewicz, M., Ochocka, A., Pasztaleniec, A., Jasser, I., Antão-Geraldes, A.M., Leira, M., Vasconcelos, V., Morais, J., Vale, M., Raposeiro, P.M., Gonçalves, V., Aleksovski, B., Krstić, S., Nemova, H., Drastichova, I., Chomova, L., Remec-Rekar, S., Elersek, T., Hansson, L., Urrutia-Cordero, P., Bravo, A.G., Buck, M., Colom-Montero, W., Mustonen, K., Pierson, D., Yang, Y., Richardson, J., Edwards, C., Cromie, H., Delgado-Martín, J., García, D., Cereijo, J.L., Gomà, J., Trapote, M.C., Vegas-Vilarrúbia, T., Obrador, B., García-Murcia, A., Real, M., Romans, E., Noguero-Ribes, J., Duque, D.P., Fernández-Morán, E., Úbeda, B., Gálvez, J.Á., Catalán, N., Pérez-Martínez, C., Ramos-Rodríguez, E., Cillero-Castro, C., Moreno-Ostos, E., Blanco, J.M., Rodríguez, V., Montes-Pérez, J.J., Palomino, R.L., Rodríguez-Pérez, E., Hernández, A., Carballeira, R., Camacho, A., Picazo, A., Rochera, C., Santamans, A.C., Ferriol, C., Romo, S., Soria, J.M., Özen, A., Karan, T., Demir, N., Beklioğlu, M., Filiz, N., Levi, E., Iskin, U., Bezirci, G., Tavşanoğlu, Ü.N., Çelik, K., Ozhan, K., Karakaya, N., Koçer, M.A.T., Yilmaz, M., Maraşlıoğlu, F., Fakioglu, Ö., Soylu, E.N., Yağcı, M.A., Çınar, Ş., Çapkın, K., Yağcı, A., Cesur, M., Bilgin, F., Bulut, C., Uysal, R., Latife, K., Akçaalan, R., Albay, M., Alp, M.T., Özkan, K., Sevindik, T.O., Tunca, H., Önem, B., Paerl, H., Carey, C.C., Ibelings, B.W., 2021. Stratification strength and light climate explain variation in chlorophyll a at the continental scale in a European multilake survey in a heatwave summer. *Limnol. Oceanogr.* **66**, 4314–4333. https://doi.org/10.1002/lno.11963

Douglas, G.B., 2002. *US Patent 6350383: Remediation Material and Remediation Process for Sediments*. US6350383 B1.

Douville, H., Allan, R.P., Arias, P.A., Betts, R.A., Caretta, M.A., Cherchi, A., Mukherji, A., Raghavan, K., Renwick, J., 2022. Water remains a blind spot in climate change policies. *PLoS Water*. **1**, e0000058. https://doi.org/10.1371/journal.pwat.0000058

Douville, H., Raghavan, K., Renwick, J., Allan, R.P., Arias, P.A., Barlow, M., Cerezo-Mota, R., Cherchi, A., Gan, T.Y., Gergis, J., Jiang, D., Khan, A., Pokam Mba, W., Rosenfeld, D., Tierney, J., Zolina, O., 2023. Water cycle changes. In *Climate Change 2023: The Physical Science Basis. Contribution of Working Group I to the Sixth Assessment Report of the Intergovernmental Panel on Climate Change*. Cambridge University Press. https://www.cambridge.org/core/product/identifier/9781009157896/type/book

Downing, J.A., Watson, S.B., McCauley, E., 2001. Predicting cyanobacteria dominance in lakes. *Can. J. Fish. Aquat. Sci.* **58**, 1905–1908. https://doi.org/10.1139/f01-143

Du, X.L., Creed, I.F., Sorichetti, R.J., Trick, C.G., 2019. Cyanobacteria biomass in shallow eutrophic lakes is linked to the presence of iron-binding ligands. *Can. J. Fish. Aquat. Sci.* 1–12. https://doi.org/10.1139/cjfas-2018-0261

Duan, Z., Tan, X., Paerl, H.W., Waal, D.B.V. de, 2021. Ecological stoichiometry of functional traits in a colonial harmful cyanobacterium. *Limnol. Oceanogr.* **66**, 2051–2062. https://doi.org/10.1002/lno.11744

Dugener, N.M., Weinke, A.D., Stone, I.P., Biddanda, B.A., 2023. Recurringly hypoxic: Bottom water oxygen depletion is linked to temperature and precipitation in a Great Lakes Estuary. *Hydrobiology*. **2**, 410–430. https://doi.org/10.3390/hydrobiology2020027

Dupas, R., Tittel, J., Jordan, P., Musolff, A., Rode, M., 2018. Non-domestic phosphorus release in rivers during low-flow: Mechanisms and implications for sources identification. *J. Hydrol.* **560**, 141–149. https://doi.org/10.1016/j.jhydrol.2018.03.023

Dyble, J., Fahnenstiel, G.L., Litaker, R.W., Millie, D.F., Tester, P.A., 2008. Microcystin concentrations and genetic diversity of *Microcystis* in the lower Great Lakes. *Environ. Toxicol.* **23**, 507–516. https://doi.org/10.1002/tox.20370

Edmondson, W.T., 1970. Phosphorus, nitrogen, and algae in lake Washington after diversion of sewage. *Science*. **169**, 690–691. https://doi.org/10.1126/science.169.3946.690

Edmondson, W.T., Lehman, J.T., 1981. The effect of changes in the nutrient income on the condition of Lake Washington1. *Limnol. Oceanogr.* **26**, 1–29. https://doi.org/10.4319/lo.1981.26.1.0001

Einsele, W., 1936. Über die Beziehungen des Eisenkreislaufs zum Phosphatkreislauf im eutrophen See. *Arch. Hydrobiol.* **29**, 664–686.

Eisenhut, M., 2020. Manganese homeostasis in cyanobacteria. *Plants.* **9**, 18. https://doi.org/10.3390/plants9010018

Ekström, S.M., Regnell, O., Reader, H.E., Nilsson, P.A., Löfgren, S., Kritzberg, E.S., 2016. Increasing concentrations of iron in surface waters as a consequence of reducing conditions in the catchment area. *J. Geophys. Res. Biogeosci.* **121**, 479–493. https://doi.org/10.1002/2015JG003141

Ekvall, M.K., de la Calle Martin, J., Faassen, E.J., Gustafsson, S., Lürling, M., Hansson, L.-A., 2013. Synergistic and species-specific effects of climate change and water colour on cyanobacterial toxicity and bloom formation. *Freshw. Biol.* **58**, 2414–2422. https://doi.org/10.1111/fwb.12220

Elliott, J.A., 2010. The seasonal sensitivity of cyanobacteria and other phytoplankton to changes in flushing rate and water temperature. *Glob. Change Biol.* **16**, 864–876. https://doi.org/10.1111/j.1365-2486.2009.01998.x

Elliott, J.A., 2012. Is the future blue-green? A review of the current model predictions of how climate change could affect pelagic freshwater cyanobacteria. *Water Res.* **46**, 1364–1371. https://doi.org/10.1016/j.watres.2011.12.018

Epe, T.S., Finsterle, K., Yasseri, S., 2017. Nine years of phosphorus management with lanthanum modified bentonite (Phoslock) in a eutrophic, shallow swimming lake in Germany. *Lake Reserv. Manag.* **33**, 119–129. https://doi.org/10.1080/10402381.2016.1263693

Erratt, K.J., Creed, I.F., Favot, E.J., Todoran, I., Tai, V., Smol, J.P., Trick, C.G., 2022a. Paleolimnological evidence reveals climate-related preeminence of cyanobacteria in a temperate meromictic lake. *Can. J. Fish. Aquat. Sci.* **79**, 558–565. https://doi.org/10.1139/cjfas-2021-0095

Erratt, K.J., Creed, I.F., Freeman, E.C., Trick, C.G., Westrick, J., Birbeck, J.A., Watson, L.C., Zastepa, A., 2022b. Deep cyanobacteria layers: An overlooked aspect of managing risks of cyanobacteria. *Environ. Sci. Technol.* acs.est.2c06928. https://doi.org/10.1021/acs.est.2c06928

Erratt, K.J., Creed, I.F., Trick, C.G., 2018. Comparative effects of ammonium, nitrate and urea on growth and photosynthetic efficiency of three bloom-forming cyanobacteria. *Freshw. Biol.* **63**, 626–638. https://doi.org/10.1111/fwb.13099

Ettahiri, Y., Bouargane, B., Fritah, K., Akhsassi, B., Pérez-Villarejo, L., Aziz, A., Bouna, L., Benlhachemi, A., Novais, R.M., 2023. A state-of-the-art review of recent advances in porous geopolymer: Applications in adsorption of inorganic and organic contaminants in water. *Constr. Build. Mater.* **395**, 132269. https://doi.org/10.1016/j.conbuildmat.2023.132269

Evans, S., Saleh, M., 2015. Cyanobacteria diversity in blooms from the Greater Sudbury area. *J. Water Resour. Prot.* **7**, 871–882. https://doi.org/10.4236/jwarp.2015.711071

Ewing, H.A., Weathers, K.C., Cottingham, K.L., Leavitt, P.R., Greer, M.L., Carey, C.C., Steele, B.G., Fiorillo, A.U., Sowles, J.P., 2020. "New" cyanobacterial blooms are not new: Two centuries of lake production are related to ice cover and land use. *Ecosphere.* **11**. https://doi.org/10.1002/ecs2.3170

Fadel, A., Sharaf, N., Siblini, M., Slim, K., Kobaissi, A., 2019. A simple modelling approach to simulate the effect of different climate scenarios on toxic cyanobacterial bloom in a eutrophic reservoir. *Ecohydrol. Hydrobiol.* https://doi.org/10.1016/j.ecohyd.2019.02.005

Farrell, K.J., Ward, N.K., Krinos, A.I., Hanson, P.C., Daneshmand, V., Figueiredo, R.J., Carey, C.C., 2020. Ecosystem-scale nutrient cycling responses to increasing air temperatures vary with lake trophic state. *Ecol. Model.* **430**, 109134. https://doi.org/10.1016/j.ecolmodel.2020.109134

Fastner, J., Abella, S., Litt, A., Morabito, G., Vörös, L., Pálffy, K., Straile, D., Kümmerlin, R., Matthews, D., Phillips, M.G., Chorus, I., 2016. Combating cyanobacterial proliferation by avoiding or treating inflows with high P load – experiences from eight case studies. *Aquat. Ecol.* **50**, 367–383. https://doi.org/10.1007/s10452-015-9558-8

Favot, E.J., Holeton, C., DeSellas, A.M., Paterson, A.M., 2023. Cyanobacterial blooms in Ontario, Canada: Continued increase in reports through the 21st century. *Lake Reserv. Manag.* **39**, 1–20. https://doi.org/10.1080/10402381.2022.2157781

Favot, E.J., Rühland, K.M., DeSellas, A.M., Ingram, R., Paterson, A.M., Smol, J.P., 2019. Climate variability promotes unprecedented cyanobacterial blooms in a remote, oligotrophic Ontario lake: Evidence from paleolimnology. *J. Paleolimnol.* https://doi.org/10.1007/s10933-019-00074-4

Favot, E.J., Rühland, K.M., Paterson, A.M., Smol, J.P., 2024. Sediment records from Lake Nipissing register a lake-wide multi-trophic response to climate change and reveal its possible role for increased cyanobacterial blooms. *J. Gt. Lakes Res.* **50**, 102268. https://doi.org/10.1016/j.jglr.2023.102268

Ficker, H., Luger, M., Gassner, H., 2017. From dimictic to monomictic: Empirical evidence of thermal regime transitions in three deep alpine lakes in Austria induced by climate change. *Freshw. Biol.* **62**, 1335–1345.

Filazzola, A., Blagrave, K., Imrit, M.A., Sharma, S., 2020. Climate change drives increases in extreme events for lake ice in the northern hemisphere. *Geophys. Res. Lett.* **47**, e2020GL089608. https://doi.org/10.1029/2020GL089608

Fleischer, S., 1978. Evidence for the anaerobic release of phosphorus from lake sediments as a biological process. *Naturwissenschaften.* **65**, 109–110. https://doi.org/10.1007/BF00440553

Flood, B., Wells, M., Midwood, J.D., Brooks, J., Kuai, Y., Li, J., 2021. Intense variability of dissolved oxygen and temperature in the internal swash zone of Hamilton Harbour, Lake Ontario. *Inland Waters.* **11**, 162–179. https://doi.org/10.1080/20442041.2020.1843930

Foley, B., Jones, I.D., Maberly, S.C., Rippey, B., 2012. Long-term changes in oxygen depletion in a small temperate lake: Effects of climate change and eutrophication. *Freshw. Biol.* **57**, 278–289.

Fortin, N., Aranda-Rodriguez, R., Jing, H., Pick, F., Bird, D., Greer, C.W., 2010. Detection of microcystin-producing cyanobacteria in Missisquoi Bay, Quebec, Canada, using quantitative PCR. *Appl. Environ. Microbiol.* **76**, 5105–5112. https://doi.org/10.1128/AEM.00183-10

Frost, P.C., Pearce, N.J.T., Berger, S.A., Gessner, M.O., Makower, A.K., Marzetz, V., Nejstgaard, J.C., Pralle, A., Schälicke, S., Wacker, A., Wagner, N.D., Xenopoulos, M.A., 2023. Interactive effects of nitrogen and phosphorus on growth and stoichiometry of lake phytoplankton. *Limnol. Oceanogr.* **68**, 1172–1184. https://doi.org/10.1002/lno.12337

Funkey, C.P., Conley, D.J., Reuss, N.S., Humborg, C., Jilbert, T., Slomp, C.P., 2014. Hypoxia sustains cyanobacteria blooms in the Baltic Sea. *Environ. Sci. Technol.* **48**, 2598–2602.

Gächter, R., Mares, A., 1985. Does settling seston release soluble reactive phosphorus in the hypolimnion of lakes? *Limnol. Ocean.* **30**, 364–371. https://doi.org/10.4319/lo.1985.30.2.0364

Gächter, R., Wehrli, B., 1998. Ten years of artificial mixing and oxygenation: No effect on the internal phosphorus loading of two eutrophic lakes. *Environ. Sci. Technol.* **32**, 3659–3665. https://doi.org/10.1021/es980418l

Gantzer, P.A., Bryant, L.D., Little, J.C., 2009. Controlling soluble iron and manganese in a water-supply reservoir using hypolimnetic oxygenation. *Water Res.* **43**, 1285–1294. https://doi.org/10.1016/j.watres.2008.12.019

Gao, J., Zhao, J., Zhang, J., Li, Q., Gao, J., Cai, M., Zhang, J., 2020. Preparation of a new low-cost substrate prepared from drinking water treatment sludge (DWTS)/bentonite/zeolite/fly ash for rapid phosphorus removal in constructed wetlands. *J. Clean. Prod.* **261**, 121110. https://doi.org/10.1016/j.jclepro.2020.121110

Gao, Y., Cornwell, J.C., Stoecker, D.K., Owens, M.S., 2014. Influence of cyanobacteria blooms on sediment biogeochemistry and nutrient fluxes. *Limnol. Ocean.* **59**, 959–971. https://doi.org/10.4319/lo.2014.59.3.0959

Gardner, W.S., Newell, S.E., McCarthy, M.J., Hoffman, D.K., Lu, K., Lavrentyev, P.J., Hellweger, F.L., Wilhelm, S.W., Liu, Z., Bruesewitz, D.A., Paerl, H.W., 2017. Community biological ammonium demand: A conceptual model for cyanobacteria blooms in eutrophic lakes. *Environ. Sci. Technol.* **51**, 7785–7793. https://doi.org/10.1021/acs.est.6b06296

Gebremariam, S.Y., McCormick, P., Rochelle, P., 2021. Evidence of a rapid phosphorus-induced regime shift in a large deep reservoir. *Sci. Total Environ.* **782**, 146755. https://doi.org/10.1016/j.scitotenv.2021.146755

Gensemer, R.W., Playle, R.C., 1999. The bioavailability and toxicity of aluminum in aquatic environments. *CRC Crit. Rev. Environ. Sci. Technol.* **29**, 315–450. https://doi.org/10.1080/10643389991259245

Georges des Aulnois, M., Roux, P., Caruana, A., Réveillon, D., Briand, E., Hervé, F., Savar, V., Bormans, M., Amzil, Z., 2019. Physiological and metabolic responses of freshwater and brackish-water strains of microcystis aeruginosa acclimated to a salinity gradient: Insight into salt tolerance. *Appl. Environ. Microbiol.* **85**, e01614–e01619. https://doi.org/10.1128/AEM.01614-19

Gerling, A.B., Browne, R.G., Gantzer, P.A., Mobley, M.H., Little, J.C., Carey, C.C., 2014. First report of the successful operation of a side stream supersaturation hypolimnetic oxygenation system in a eutrophic, shallow reservoir. *Water Res.* **67**, 129–143. https://doi.org/10.1016/j.watres.2014.09.002

Gerten, D., Adrian, R., 2000. Climate-driven changes in spring plankton dynamics and the sensitivity of shallow polymictic lakes to the North Atlantic Oscillation. *Limnol. Oceanogr.* **45**, 1058–1066. https://doi.org/10.4319/lo.2000.45.5.1058

Gervais, F., Siedel, U., Heilmann, B., Weithoff, G., Heisig-Gunkel, G., Nicklisch, A., 2003. Small-scale vertical distribution of phytoplankton, nutrients and sulphide below the oxycline of a mesotrophic lake. *J. Plankton Res.* **25**, 273–278. https://doi.org/10.1093/plankt/25.3.273

Giani, A., Bird, D.F., Prairie, Y.T., Lawrence, J.F., 2005. Empirical study of cyanobacterial toxicity along a trophic gradient of lakes. *Can. J. Fish. Aquat. Sci.* **62**, 2100–2109. https://doi.org/10.1139/f05-124

Gilbert, D., 2017. Environmental science: Oceans lose oxygen. *Nature.* **542**, 303–304. https://doi.org/10.1038/542303a

Gilbert, N., 2009. Environment: The disappearing nutrient. *Nature.* **461**, 716–718. https://doi.org/10.1038/461716a

Giles, C.D., Isles, P.D.F., Manley, T., Xu, Y., Druschel, G.K., Schroth, A.W., 2016. The mobility of phosphorus, iron, and manganese through the sediment – water continuum of a shallow eutrophic freshwater lake under stratified and mixed water-column conditions. *Biogeochemistry.* **127**, 15–34. https://doi.org/10.1007/s10533-015-0144-x

Glass, J.B., Axler, R.P., Chandra, S., Goldman, C.R., 2012. Molybdenum limitation of microbial nitrogen assimilation in aquatic ecosystems and pure cultures. *Front. Microbiol.* **3**. https://doi.org/10.3389/fmicb.2012.00331

Glibert, P.M., 2017. Eutrophication, harmful algae and biodiversity — Challenging paradigms in a world of complex nutrient changes. *Mar. Pollut. Bull.* **124**, 591–606. https://doi.org/10.1016/j.marpolbul.2017.04.027

Glibert, P.M., 2020. Harmful algae at the complex nexus of eutrophication and climate change. *Harmful Algae.* **91**, 101583. https://doi.org/10.1016/j.hal.2019.03.001

Glibert, P.M., Icarus Allen, J., Artioli, Y., Beusen, A., Bouwman, L., Harle, J., Holmes, R., Holt, J., 2014a. Vulnerability of coastal ecosystems to changes in harmful algal bloom distribution in response to climate change: Projections based on model analysis. *Glob. Change Biol.* **20**, 3845–3858. https://doi.org/10.1111/gcb.12662

Glibert, P.M., Maranger, R., Sobota, D.J., Bouwman, L., 2014b. The Haber Bosch – harmful algal bloom (HB – HAB) link. *Environ. Res. Lett.* **9**, 105001. https://doi.org/10.1088/1748-9326/9/10/105001

Gobler, C.J., Burkholder, J.M., Davis, T.W., Harke, M.J., Johengen, T., Stow, C.A., Van de Waal, D.B., 2016. The dual role of nitrogen supply in controlling the growth and toxicity of cyanobacterial blooms. *Harmful Algae.* **54**, 87–97. https://doi.org/10.1016/j.hal.2016.01.010

Godwin, W., Coveney, M., Lowe, E., Battoe, L., 2011. Improvements in water quality following biomanipulation of gizzard shad (Dorosoma cepedianum) in Lake Denham, Florida. *Lake Reserv. Manage.* **27**, 287–297. https://doi.org/10.1080/07438141.2011.609633

Golterman, H.L., Oude, N.T., 1991. Eutrophication of lakes, rivers and coastal seas. In *Water Pollution.* Springer Berlin Heidelberg, Berlin, Heidelberg, pp. 79–124. https://doi.org/10.1007/978-3-540-46685-7_3

Graham, J.L., Dubrovsky, N.M., Foster, G.M., King, L.R., Loftin, K.A., Rosen, B.H., Stelzer, E.A., 2020. Cyanotoxin occurrence in large rivers of the United States. *Inland Waters.* **10**, 109–117. https://doi.org/10.1080/20442041.2019.1700749

Graham, J.L., Jones, J.R., Jones, S.B., Downing, J.A., Clevenger, T.E., 2004. Environmental factors influencing microcystin distribution and concentration in the Midwestern United States. *Water Res.* **38**, 4395–4404. https://doi.org/10.1016/j.watres.2004.08.00

Graneli, W., Solander, D., 1988. Influence of aquatic macrophytes on phosphorus cycling in lakes. *Hydrobiologia.* **170**, 245–266. https://doi.org/10.1007/978-94-009-3109-1_15

Griffith, A.W., Gobler, C.J., 2020. Harmful algal blooms: A climate change co-stressor in marine and freshwater ecosystems. *Harmful Algae.* **91**, 101590. https://doi.org/10.1016/j.hal.2019.03.008

Haaland, S., Eikebrokk, B., Riise, G., Vogt, R.D., 2023. Browning of Scottish surface water sources exposed to climate change. *PLoS Water*. **2**, e0000172. https://doi.org/10.1371/journal.pwat.0000172

Hadley, K.R., Paterson, A.M., Stainsby, E.A., Michelutti, N., Yao, H., Rusak, J.A., Ingram, R., McConnell, C., Smol, J.P., 2014. Climate warming alters thermal stability but not stratification phenology in a small north-temperate lake. *Hydrol. Process*. **28**, 6309–6319. https://doi.org/10.1002/hyp.10120

Haggard, B.E., Grantz, E., Austin, B.J., Lasater, A.L., Haddock, L., Ferri, A., Wagner, N., Scott, J.T., 2023. Microcystin shows thresholds and hierarchical structure with physicochemical properties at Lake Fayetteville, Arkansas, May through September 2020. *J. Asabe*. **66**, 307–317. https://doi.org/10.13031/ja.15273

Håkanson, L., 1982. Lake bottom dynamics and morphometry: The dynamic ratio. *Water Resour. Res*. **18**, 1444–1450. https://doi.org/10.1029/WR018i005p01444

Haldna, M., Möls, T., Buhvestova, O., Kangur, K., 2013. Predictive model for phosphorus in large shallow Lake Peipsi: Approach based on covariance structures. *Aquat. Ecosyst. Health Manag*. **16**, 222–226. https://doi.org/10.1080/14634988.2013.789766

Hall, K.J., Murphy, T.P.D., Mawhinney, M., Ashley, K.I., 1994. Iron treatment for eutrophication control in Black Lake, British Columbia. *NALMS Lake Reserv. Manag*. **9**, 114–117. https://doi.org/10.1080/07438149409354739

Hallegraeff, G.M., Anderson, D.M., Belin, C., Bottein, M.-Y.D., Bresnan, E., Chinain, M., Enevoldsen, H., Iwataki, M., Karlson, B., McKenzie, C.H., Sunesen, I., Pitcher, G.C., Provoost, P., Richardson, A., Schweibold, L., Tester, P.A., Trainer, V.L., Yñiguez, A.T., Zingone, A., 2021. Perceived global increase in algal blooms is attributable to intensified monitoring and emerging bloom impacts. *Commun. Earth Environ*. **2**, 117. https://doi.org/10.1038/s43247-021-00178-8

Hamilton, D.S., Perron, M.M.G., Bond, T.C., Bowie, A.R., Buchholz, R.R., Guieu, C., Ito, A., Maenhaut, W., Myriokefalitakis, S., Olgun, N., Rathod, S.D., Schepanski, K., Tagliabue, A., Wagner, R., Mahowald, N.M., 2022. Earth, wind, fire, and pollution: Aerosol nutrient sources and impacts on ocean biogeochemistry. *Annu. Rev. Mar. Sci*. **14**, 303–330. https://doi.org/10.1146/annurev-marine-031921-013612

Hamilton, G.S., Fielding, F., Chiffings, A.W., Hart, B.T., Johnstone, R.W., Mengersen, K., 2007. Investigating the use of a Bayesian network to model the risk of Lyngbya majuscula bloom initiation in deception bay, Queensland, Australia. *Hum. Ecol. Risk Assess. Int. J*. **13**, 1271–1287. https://doi.org/10.1080/10807030701655616

Hamilton, P.B., Ley, L.M., Dean, S., Pick, F.R., 2005. The occurrence of the cyanobacterium *Cylindrospermopsis raciborskii* in Constance Lake: An exotic cyanoprokaryote new to Canada. *Phycologia*. **44**, 17–25. https://doi.org/10.2216/0031-8884(2005)44[17:TOOTCC]2.0.CO;2

Hampel, J.J., McCarthy, M.J., Neudeck, M., Bullerjahn, G.S., McKay, R.M.L., Newell, S.E., 2019. Ammonium recycling supports toxic Planktothrix blooms in Sandusky Bay, Lake Erie: Evidence from stable isotope and metatranscriptome data. *Harmful Algae*. **81**, 42–52. https://doi.org/10.1016/j.hal.2018.11.011

Hampton, S.E., Baron, J.S., Ladwig, R., McClure, R.P., Meyer, M.F., Oleksy, I.A., Shampain, A., 2023. Warming-induced changes in benthic redox as a potential driver of increasing benthic algal blooms in high-elevation lakes. *Limnol. Oceanogr. Lett*. **9**, 1–6. https://doi.org/10.1002/lol2.10357

Hampton, S.E., Scheuerell, M.D., Schindler, D.E., 2006. Coalescence in the Lake Washington story: Interaction strengths in a planktonic food web. *Limnol. Oceanogr*. **51**, 2042–2051. https://doi.org/10.4319/lo.2006.51.5.2042

Han, Y., Jeppesen, E., Lürling, M., Zhang, Y., Ma, T., Li, W., Chen, K., Li, K., 2022. Combining lanthanum-modified bentonite (LMB) and submerged macrophytes alleviates water quality deterioration in the presence of omni-benthivorous fish. *J. Environ. Manage*. **314**, 115036. https://doi.org/10.1016/j.jenvman.2022.115036

Han, Y., Zhang, Y., Li, Q., Lürling, M., Li, W., He, H., Gu, J., Li, K., 2021. Submerged macrophytes benefit from lanthanum modified bentonite treatment under juvenile omni-benthivorous fish disturbance: Implications for shallow lake restoration. *Freshw. Biol*. fwb.13871. https://doi.org/10.1111/fwb.13871

Hansen, C.H., Burian, S.J., Dennison, P.E., Williams, G.P., 2020. Evaluating historical trends and influences of meteorological and seasonal climate conditions on lake chlorophyll a using remote sensing. *Lake Reserv. Manag*. **36**, 45–63. https://doi.org/10.1080/10402381.2019.1632397

Hanson, P.C., Ladwig, R., Buelo, C., Albright, E.A., Delany, A.D., Carey, C.C., 2023. Legacy phosphorus and ecosystem memory control future water quality in a eutrophic lake. *J. Geophys. Res. Biogeosci.* **128**, e2023JG007620. https://doi.org/10.1029/2023JG007620

Hanson, P.C., Stillman, A.B., Jia, X., Karpatne, A., Dugan, H.A., Carey, C.C., Stachelek, J., Ward, N.K., Zhang, Y., Read, J.S., Kumar, V., 2020. Predicting lake surface water phosphorus dynamics using process-guided machine learning. *Ecol. Model.* **430**, 109136. https://doi.org/10.1016/j.ecolmodel.2020.109136

Harke, M.J., Davis, T.W., Watson, S.B., Gobler, C.J., 2016. Nutrient-controlled niche differentiation of western Lake Erie cyanobacterial populations revealed via metatranscriptomic surveys. *Environ. Sci. Technol.* **50**, 604–615. https://doi.org/10.1021/acs.est.5b03931

Harris, G.P., 1994. Pattern, process and prediction in aquatic ecology. A limnological view of some general ecological problems. *Freshw. Biol.* **32**, 143–160. https://doi.org/10.1111/j.1365-2427.1994.tb00874.x

Harris, G.P., 1999. This is not the end of limnology (or of science): The world may well be a lot simpler than we think: This is not the end of limnology. *Freshw. Biol.* **42**, 689–706. https://doi.org/10.1046/j.1365-2427.1999.00486.x

Havens, K.E., Ji, G., Beaver, J.R., Fulton, R.S., Teacher, C.E., 2019. Dynamics of cyanobacteria blooms are linked to the hydrology of shallow Florida lakes and provide insight into possible impacts of climate change. *Hydrobiologia* **829**, 43–59. https://doi.org/10.1007/s10750-017-3425-7

Havens, K.E., Nürnberg, G.K., 2004. The phosphorus-chlorophyll relationship in lakes: Potential influences of color and mixing regime. *Lake Reserv. Manag.* **20**, 188–196. https://doi.org/10.1080/07438140409354243

Hayes, N.M., Deemer, B.R., Corman, J.R., Razavi, N.R., Strock, K.E., 2017. Key differences between lakes and reservoirs modify climate signals: A case for a new conceptual model. *Limnol. Oceanogr. Lett.* **2**, 47–62. https://doi.org/10.1002/lol2.10036

Hayes, N.M., Haig, H.A., Simpson, G.L., Leavitt, P.R., 2020. Effects of lake warming on the seasonal risk of toxic cyanobacteria exposure. *Limnol. Oceanogr. Lett.* **5**, 393–402. https://doi.org/10.1002/lol2.10164

Haygarth, P.M., Condron, L.M., Heathwaite, A.L., Turner, B.L., Harris, G.P., 2005. The phosphorus transfer continuum: Linking source to impact with an interdisciplinary and multi-scaled approach. *Sci. Total Environ.* **344**, 5–14. https://doi.org/10.1016/j.scitotenv.2005.02.001

Head, R.M., Jones, R.I., Bailey-Watts, A.E., 1999. An assessment of the influence of recruitment form the sediment on the development of planktonic populations of cyanobacteria in a temperate mesotrophic lake. *Freshw. Biol.* **41**, 759–769. https://doi.org/10.1046/j.1365-2427.1999.00421.x

Hecky, R.E., Smith, R.E., Barton, D.R., Guildford, S.J., Taylor, W.D., Charlton, M.N., Howell, T., 2004. The nearshore phosphorus shunt: A consequence of ecosystem engineering by dreissenids in the Laurentian Great Lakes. *Can. J. Fish. Aquat. Sci.* **61**, 1285–1293. https://doi.org/10.1139/f04-065

Hejzlar, J., Šamalova, K., Boers, P., Kronvang, B., 2006. Modelling phosphorus retention in lakes and reservoirs. *Wat. Air. Soil. Poll.* **6**, 487–494. https://doi.org/10.1007/978-1-4020-5478-5_13

Helliwell, K.E., 2017. The roles of B vitamins in phytoplankton nutrition: New perspectives and prospects. *New Phytol.* **216**, 62–68. https://doi.org/10.1111/nph.14669

Hellweger, F.L., Kravchuk, E.S., Novotny, V., Gladyshev, M.I., 2008. Agent-based modeling of the complex life cycle of a cyanobacterium (Anabaena) in a shallow reservoir. *Limnol. Oceanogr.* **53**, 1227–1241. https://doi.org/10.4319/lo.2008.53.4.1227

Helmond, N.A.G.M., Lougheed, B.C., Vollebregt, A., Peterse, F., Fontorbe, G., Conley, D.J., Slomp, C.P., 2020. Recovery from multi-millennial natural coastal hypoxia in the Stockholm Archipelago, Baltic Sea, terminated by modern human activity. *Limnol. Oceanogr.* **65**, 3085–3097. https://doi.org/10.1002/lno.11575

Hickey, C.W., Gibbs, M.M., 2009. Lake sediment phosphorus release management – Decision support and risk assessment framework. *NZ J. Mar. Freshw. Res.* **43**, 819–856. https://doi.org/10.1080/00288330909510043

Hillebrand, H., Steinert, G., Boersma, M., Malzahn, A., Léo Meunier, C., Plum, C., Ptacnik, R., 2013. Goldman revisited: Faster growing phytoplankton has lower N:P and lower stoichiometric flexibility. *Limnol. Oceanogr.* **58**, 2076–2088. https://doi.org/10.4319/lo.2013.58.6.2076

Hintz, W.D., Arnott, S.E., Symons, C.C., Greco, D.A., McClymont, A., Brentrup, J.A., Cañedo-Argüelles, M., Derry, A.M., Downing, A.L., Gray, D.K., Melles, S.J., Relyea, R.A., Rusak, J.A., Searle, C.L., Astorg, L., Baker, H.K., Beisner, B.E., Cottingham, K.L., Ersoy, Z., Espinosa, C., Franceschini, J., Giorgio, A.T., Göbeler, N., Hassal, E., Hébert, M.-P., Huynh, M., Hylander, S., Jonasen, K.L., Kirkwood, A.E., Langenheder, S., Langvall, O., Laudon, H., Lind, L., Lundgren, M., Proia, L., Schuler, M.S., Shurin, J.B., Steiner, C.F., Striebel, M., Thibodeau, S., Urrutia-Cordero, P., Vendrell-Puigmitja, L., Weyhenmeyer, G.A., 2022. Current water quality guidelines across North America and Europe do not protect lakes from salinization. *Proc. Natl. Acad. Sci.* **119**. https://doi.org/10.1073/pnas.2115033119

Ho, J.C., Michalak, A.M., 2020. Exploring temperature and precipitation impacts on harmful algal blooms across continental U.S. lakes. *Limnol. Oceanogr.* **65**, 992–1009. https://doi.org/10.1002/lno.11365

Ho, J.C., Michalak, A.M., Pahlevan, N., 2019. Widespread global increase in intense lake phytoplankton blooms since the 1980s. *Nature*. **574**, 667–670. https://doi.org/10.1038/s41586-019-1648-7

Hoffman, D.K., McCarthy, M.J., Boedecker, A.R., Myers, J.A., Newell, S.E., 2022. The role of internal nitrogen loading in supporting non-N-fixing harmful cyanobacterial blooms in the water column of a large eutrophic lake. *Limnol. Oceanogr.* **67**, 2028–2041. https://doi.org/10.1002/lno.12185

Holgerson, M.A., Richardson, D.C., Roith, J., Bortolotti, L.E., Finlay, K., Hornbach, D.J., Gurung, K., Ness, A., Andersen, M.R., Bansal, S., Finlay, J.C., Cianci-Gaskill, J.A., Hahn, S., Janke, B.D., McDonald, C., Mesman, J.P., North, R.L., Roberts, C.O., Sweetman, J.N., Webb, J.R., 2022. Classifying mixing regimes in ponds and shallow lakes. *Water Resour. Res.* **58**, e2022WR032522. https://doi.org/10.1029/2022WR032522

Hölker, F., Vanni, M.J., Kuiper, J.J., Meile, C., Grossart, H.-P., Stief, P., Adrian, R., Lorke, A., Dellwig, O., Brand, A., 2015. Tube-dwelling invertebrates: Tiny ecosystem engineers have large effects in lake ecosystems. *Ecol. Monogr.* **85**, 333–351. https://doi.org/10.1890/14-1160.1

Holmroos, H., Niemistö, J., Weckström, K., Horppila, J., 2009. Seasonal variation of resuspension-mediated aerobic release of phosphorus. *Boreal. Environ. Res.* **14**.

Horne, A.J., Beutel, M., 2019. Hypolimnetic oxygenation 3: An engineered switch from eutrophic to a meso-/oligotrophic state in a California reservoir. *Lake Reserv. Manag.* **35**, 338–353. https://doi.org/10.1080/10402381.2019.1648613

Horppila, J., 2019. Sediment nutrients, ecological status and restoration of lakes. *Water Res.* **160**, 206–208. https://doi.org/10.1016/j.watres.2019.05.074

Horppila, J., Holmroos, H., Niemistö, J., Massa, I., Nygrén, N., Schönach, P., Tapio, P., Tammeorg, O., 2017. Variations of internal phosphorus loading and water quality in a hypertrophic lake during 40 years of different management efforts. *Ecol. Eng.* **103**, 264–274. https://doi.org/10.1016/j.ecoleng.2017.04.018

Horppila, J., Keskinen, S., Nurmesniemi, M., Nurminen, L., Pippingsköld, E., Rajala, S., Sainio, K., Estlander, S., 2023. Factors behind the threshold-like changes in lake ecosystems along a water colour gradient: The effects of dissolved organic carbon and iron on euphotic depth, mixing depth and phytoplankton biomass. *Freshw. Biol.* **68**, 1031–1040. https://doi.org/10.1111/fwb.14083

Horppila, J., Köngäs, P., Niemistö, J., Hietanen, S., 2015. Oxygen flux and penetration depth in the sediments of aerated and non-aerated lake basins. *Int. Rev. Hydrobiol.* **100**, 106–115.

Hosper, H., 1997. *Clearing Lakes, an Ecosystem Approach to the Restoration and Management of Shallow Lakes in the Netherlands* (PhD Thesis). Wageningen University, The Netherlands.

Hou, X., Feng, L., Dai, Y., Hu, C., Gibson, L., Tang, J., Lee, Z., Wang, Y., Cai, X., Liu, J., Zheng, Y., Zheng, C., 2022. Global mapping reveals increase in lacustrine algal blooms over the past decade. *Nat. Geosci.* **15**, 130–134. https://doi.org/10.1038/s41561-021-00887-x

Hrycik, A.R., Isles, P.D.F., Adrian, R., Albright, M., Bacon, L.C., Berger, S.A., Bhattacharya, R., Grossart, H., Hejzlar, J., Hetherington, A.L., Knoll, L.B., Laas, A., McDonald, C.P., Merrell, K., Nejstgaard, J.C., Nelson, K., Nõges, P., Paterson, A.M., Pilla, R.M., Robertson, D.M., Rudstam, L.G., Rusak, J.A., Sadro, S., Silow, E.A., Stockwell, J.D., Yao, H., Yokota, K., Pierson, D.C., 2021. Earlier winter/spring runoff and snowmelt during warmer winters lead to lower summer chlorophyll- *a* in north temperate lakes. *Glob. Change Biol.* **27**, 4615–4629. https://doi.org/10.1111/gcb.15797

Huang, J., Xu, Q., Wang, X., Ji, H., Quigley, E.J., Sharbatmaleki, M., Li, S., Xi, B., Sun, B., Li, C., 2021. Effects of hydrological and climatic variables on cyanobacterial blooms in four large shallow lakes fed by the Yangtze River. *Environ. Sci. Ecotechnol.* **5**, 100069. https://doi.org/10.1016/j.ese.2020.100069

Huber, V., Wagner, C., Gerten, D., Adrian, R., 2012. To bloom or not to bloom: Contrasting responses of cyanobacteria to recent heat waves explained by critical thresholds of abiotic drivers. *Oecologia.* **169**, 245–256. https://doi.org/10.1007/s00442-011-2186-7

Hudon, C., De Sève, M., Cattaneo, A., 2014. Increasing occurrence of the benthic filamentous cyanobacterium Lyngbya wollei: A symptom of freshwater ecosystem degradation. *Freshw. Sci.* **33**, 606–618. https://doi.org/10.1086/675932

Huisman, J., Codd, G.A., Paerl, H.W., Ibelings, B.W., Verspagen, J.M.H., Visser, P.M., 2018. Cyanobacterial blooms. *Nat. Rev. Microbiol.* **16**, 471–483. https://doi.org/10.1038/s41579-018-0040-1

Huot, Y., Brown, C.A., Potvin, G., Antoniades, D., Baulch, H.M., Beisner, B.E., Bélanger, S., Brazeau, S., Cabana, H., Cardille, J.A., del Giorgio, P.A., Gregory-Eaves, I., Fortin, M.-J., Lang, A.S., Laurion, I., Maranger, R., Prairie, Y.T., Rusak, J.A., Segura, P.A., Siron, R., Smol, J.P., Vinebrooke, R.D., Walsh, D.A., 2019. The NSERC Canadian lake pulse network: A national assessment of lake health providing science for water management in a changing climate. *Sci. Total Environ.* **695**, 133668. https://doi.org/10.1016/j.scitotenv.2019.133668

Hupfer, M., Lewandowski, J., 2005. Retention and early diagenetic transformation of phosphorus in Lake Arendsee (Germany) – consequences for management strategies. *Arch. Hydrobiol.* **164**, 143–167. https://doi.org/10.1127/0003-9136/2005/0164-0143

Hupfer, M., Lewandowski, J., 2008. Oxygen controls the phosphorus release from lake sediments – a long-lasting paradigm in limnology. *Int. Rev. Hydrobiol.* **93**, 415–432. https://doi.org/10.1002/iroh.200711054

Hupfer, M., Zak, D., Roßberg, R., Herzog, C., Pöthig, R., 2009. Evaluation of a well-established sequential phosphorus fractionation technique for use in calcite-rich lake sediments: Identification and prevention of artifacts due to apatite formation. *Limnol. Oceanogr. Methods.* **7**, 399–410. https://doi.org/10.4319/lom.2009.7.399

Hurst, N.R., VanZomeren, C.M., Berkowitz, J.F., 2022. Temperature, redox, and amendments alter wetland soil inorganic phosphorus retention dynamics in a Laurentian Great Lakes priority watershed. *J. Gt. Lakes Res.* **48**, 935–943. https://doi.org/10.1016/j.jglr.2022.05.010

Huser, B.J., Bajer, P.G., Chizinski, C.J., Sorensen, P.W., 2016a. Effects of common carp (*Cyprinus carpio*) on sediment mixing depth and mobile phosphorus mass in the active sediment layer of a shallow lake. *Hydrobiologia.* **763**, 23–33. https://doi.org/10.1007/s10750-015-2356-4

Huser, B.J., Bajer, P.G., Kittelson, S., Christenson, S., Menken, K., 2022. Changes to water quality and sediment phosphorus forms in a shallow, eutrophic lake after removal of common carp (*Cyprinus carpio*). *Inland Waters.* **12**, 33–46. https://doi.org/10.1080/20442041.2020.1850096

Huser, B.J., Egemose, S., Harper, H., Hupfer, M., Jensen, H., Pilgrim, K.M., Reitzel, K., Rydin, E., Futter, M., 2016b. Longevity and effectiveness of aluminum addition to reduce sediment phosphorus release and restore lake water quality. *Water Res.* **97**, 122–132. https://doi.org/10.1016/j.watres.2015.06.051

Huser, B.J., Malmaeus, M., Karlsson, M., Almstrand, R., Witter, E., 2023. *Handbook – A Decision Support Tool for Measures Against Internal Phosphorus Loading in Lakes (LIFE IP Rich Waters).* Swedish Agency for Marine and Water Management, Sweden. https://www.researchgate.net/publication/376851897_Handbook_-a_decision_support_tool_for_measures_against_internal_phosphorus_loading_in_lakes_Title_Handbook_

Huser, B.J., Rydin, E., 2005. Phosphorus inactivation by aluminum in Lakes Gårdsjön and Härsvatten sediment during the industrial acidification period in Sweden. *Can. J. Fish. Aquat. Sci.* **62**, 1702–1709. https://doi.org/10.1139/f05-083

Huttunen, I., Huttunen, M., Piirainen, V., Korppoo, M., Lepistö, A., Räike, A., Tattari, S., Vehviläinen, B., 2016. A national-scale nutrient loading model for finnish watersheds – vemala. *Environ. Model. Assess.* **21**, 83–109. https://doi.org/10.1007/s10666-015-9470-6

Hyenstrand, P., Rydin, E., Gunnerhed, M., Linder, J., Blomqvist, P., 2001. Response of the cyanobacterium *Gloeotrichia echinulata* to iron and boron additions – an experiment from Lake Erken. *Freshw. Biol.* **46**, 735–741. https://doi.org/10.1046/j.1365-2427.2001.00710.x

Ibelings, B.W., Bormans, M., Fastner, J., Visser, P.M., 2016a. CYANOCOST special issue on cyanobacterial blooms: Synopsis – a critical review of the management options for their prevention, control and mitigation. *Aquat. Ecol.* **50**, 595–605. https://doi.org/10.1007/s10452-016-9596-x

Ibelings, B.W., Fastner, J., Bormans, M., Visser, P.M., 2016b. Cyanobacterial blooms. Ecology, prevention, mitigation and control: Editorial to a CYANOCOST special issue. *Aquat. Ecol.* **50**, 327–331. https://doi.org/10.1007/s10452-016-9595-y

Ibelings, B.W., Havens, K.E., 2008. Cyanobacterial toxins: A qualitative meta – analysis of concentrations, dosage and effects in freshwater, estuarine and marine biota. In *'Cyanobacterial Harmful Algal Blooms: State of the Science and Research Needs' 'Advances in Experimental Medicine and Biology'* (Ed. Hudnell, H.K.). Springer, New York, pp. 675–732.

Intergovernmental Panel on Climate Change, 2022. *Climate Change 2022 – Mitigation of Climate Change (No. AR6)*. https://doi.org/10.1017/9781009157926

Intergovernmental Panel on Climate Change, 2023. *Climate Change 2021 – The Physical Science Basis: Working Group I Contribution to the Sixth Assessment Report of the Intergovernmental Panel on Climate Change*, 1st ed. Cambridge University Press, Cambridge. https://doi.org/10.1017/9781009157896

Isles, P.D.F., 2020. The misuse of ratios in ecological stoichiometry. *Ecology*. **101**. https://doi.org/10.1002/ecy.3153

Isles, P.D.F., Creed, I.F., Jonsson, A., Bergström, A.-K., 2021. Trade-offs between light and nutrient availability across gradients of dissolved organic carbon lead to spatially and temporally variable responses of lake phytoplankton biomass to browning. *Ecosystems*. **24**, 1837–1852. https://doi.org/10.1007/s10021-021-00619-7

Isles, P.D.F., Giles, C.D., Gearhart, T.A., Xu, Y., Druschel, G.K., Schroth, A.W., 2015. Dynamic internal drivers of a historically severe cyanobacteria bloom in Lake Champlain revealed through comprehensive monitoring. *J. Gt. Lakes Res.* **41**, 818–829. https://doi.org/10.1016/j.jglr.2015.06.006

Isles, P.D.F., Pomati, F., 2021. An operational framework for defining and forecasting phytoplankton blooms. *Front. Ecol. Environ.* fee.2376. https://doi.org/10.1002/fee.2376

Isles, P.D.F., Rizzo, D.M., Xu, Y., Schroth, A.W., 2017a. Modeling the drivers of interannual variability in cyanobacterial bloom severity using self-organizing maps and high-frequency data. *Inland Waters*. **7**, 333–347. https://doi.org/10.1080/20442041.2017.1318640

Isles, P.D.F., Xu, Y., Stockwell, J.D., Schroth, A.W., 2017b. Climate-driven changes in energy and mass inputs systematically alter nutrient concentration and stoichiometry in deep and shallow regions of Lake Champlain. *Biogeochemistry*. **133**, 201–217. https://doi.org/10.1007/s10533-017-0327-8

Istvánovics, V., Honti, M., Torma, P., Kousal, J., 2022. Record-setting algal bloom in polymictic Lake Balaton (Hungary): A synergistic impact of climate change and (mis)management. *Freshw. Biol.* fwb.13903. https://doi.org/10.1111/fwb.13903

Istvánovics, V., Osztoics, A., Honti, M., 2004. Dynamics and ecological significance of daily internal load of phosphorus in shallow Lake Balaton, Hungary. *Freshw. Biol.* **49**, 232–252. https://doi.org/10.1111/j.1365-2427.2004.01180.x

Istvánovics, V., Petterson, K., 1998. Phosphorus release in relation to composition and isotopic exchangeability of sediment phosphorus. *Arch. Hydrobiol. Spec. Issues Adv. Limnol.* **51**, 91–104.

Istvánovics, V., Petterson, K., Rodrigo, M.A., Pierson, D., Padisák, J., Colom, W., 1993. *Gloeotrichia echinulata*, a colonial cyanobacterium with a unique phosphorus uptake and life strategy. *J. Plankton. Res.* **15**, 531–552. https://doi.org/10.1007/978-94-011-1598-8_15

Jabbari, A., Ackerman, J.D., Boegman, L., Zhao, Y., 2021. Increases in Great Lake winds and extreme events facilitate interbasin coupling and reduce water quality in Lake Erie. *Sci. Rep.* **11**, 5733. https://doi.org/10.1038/s41598-021-84961-9

Jacoby, C.A., Frazer, T.K., 2009. Eutrophication: Time to adjust expectations. *Science*. **324**, 723–724. https://doi.org/10.1126/science.324_723b

Jacoby, J.M., Gibbons, H.L., Stoops, K.B., Bouchard, D.D., 1994. Response of a shallow, polymictic lake to buffered alum treatment. *Lake Reserv. Manage.* **10**, 103–112. https://doi.org/10.1080/07438149409354181

James, W.F., 2011. Variations in the aluminum:phosphorus binding ratio and alum dosage considerations for Half Moon Lake, Wisconsin. *Lake Reserv. Manage.* **27**, 128–137. https://doi.org/10.1080/07438141.2011.572232

James, W.F., 2017a. Internal phosphorus loading contributions from deposited and resuspended sediment to the Lake of the Woods. *Lake Reserv. Manag.* **33**, 347–359. https://doi.org/10.1080/10402381.2017.1312647

James, W.F., 2017b. Diffusive phosphorus fluxes in relation to the sediment phosphorus profile in Big Traverse Bay, Lake of the Woods. *Lake Reserv. Manag.* **33**, 360–368. https://doi.org/10.1080/10402381.2017.1346010

James, W.F., Barko, J.W., 1990. Macrophyte influences on the zonation of sediment accretion and composition in a north-temperate reservoir. *Arch. Hydrobiol.* **120**, 129–142.

James, W.F., Barko, J.W., 1991. Littoral-pelagic phosphorus dynamics during nighttime convective circulation. *Limnol. Oceanogr.* **36**, 949–960. https://doi.org/10.4319/lo.1991.36.5.0949

James, W.F., Barko, J.W., 2005. Biologically labile and refractory phosphorus loads from the agriculturally-managed upper Eau Galle River watershed, Wisconsin. *Lake Reserv. Manage.* **21**, 165–173. https://doi.org/10.1080/07438140509354426

James, W.F., Barko, J.W., Field, S.J., 1996. Phosphorus mobilization from littoral sediments of an inlet region in Lake Delavan, Wisconsin. *Arch. Hydrobiol.* **138**, 245–257.

James, W.F., Bischoff, J.M., 2020. Sediment aluminum:phosphorus binding ratios and internal phosphorus loading characteristics 12 years after aluminum sulfate application to Lake McCarrons, Minnesota. *Lake Reserv. Manag.* **36**, 1–13. https://doi.org/10.1080/10402381.2019.1661554

James, W.F., Kennedy, R.H., Montgomery, R.H., Nix, J., 1987. Seasonal and longitudinal variations in apparent deposition rates within an Arkansas reservoir1. *Limnol. Oceanogr.* **32**, 1169–1176. https://doi.org/10.4319/lo.1987.32.5.1169

James, W.F., Larson, C.E., 2008. Phosphorus dynamics and loading in the turbid Minnesota River (USA): Controls and recycling potential. *Biogeochemistry.* **90**, 75–92. https://doi.org/10.1007/s10533-008-9232-5

James, W.F., Sorge, P.W., Garrison, P.J., 2015. Managing internal phosphorus loading and vertical entrainment in a weakly stratified eutrophic lake. *Lake Reserv. Manag.* **31**, 292–305. https://doi.org/10.1080/10402381.2015.1079755

James, W.F., Taylor, W.D., Barko, J.W., 1992. Production and vertical migration of *Ceratium hirundinella* in relation to phosphorus availability in Eau Galle Reservoir, Wisconsin. *Can. J. Fish. Aquat. Sci.* **49**, 694–700. https://doi.org/10.1139/f92-078

Janatian, N., Olli, K., Cremona, F., Laas, A., Nõges, P., 2020. Atmospheric stilling offsets the benefits from reduced nutrient loading in a large shallow lake. *Limnol. Oceanogr.* **65**, 717–731. https://doi.org/10.1002/lno.11342

Janatian, N., Olli, K., Nõges, P., 2021. Phytoplankton responses to meteorological and hydrological forcing at decadal to seasonal time scales. *Hydrobiologia.* **848**, 2745–2759. https://doi.org/10.1007/s10750-021-04594-x

Jane, S.F., Hansen, G.J.A., Kraemer, B.M., Leavitt, P.R., Mincer, J.L., North, R.L., Pilla, R.M., Stetler, J.T., Williamson, C.E., Woolway, R.I., Arvola, L., Chandra, S., DeGasperi, C.L., Diemer, L., Dunalska, J., Erina, O., Flaim, G., Grossart, H.-P., Hambright, K.D., Hein, C., Hejzlar, J., Janus, L.L., Jenny, J.-P., Jones, J.R., Knoll, L.B., Leoni, B., Mackay, E., Matsuzaki, S.-I.S., McBride, C., Müller-Navarra, D.C., Paterson, A.M., Pierson, D., Rogora, M., Rusak, J.A., Sadro, S., Saulnier-Talbot, E., Schmid, M., Sommaruga, R., Thiery, W., Verburg, P., Weathers, K.C., Weyhenmeyer, G.A., Yokota, K., Rose, K.C., 2021. Widespread deoxygenation of temperate lakes. *Nature.* **594**, 66–70. https://doi.org/10.1038/s41586-021-03550-y

Jane, S.F., Mincer, J.L., Lau, M.P., Lewis, A.S.L., Stetler, J.T., Rose, K.C., 2023. Longer duration of seasonal stratification contributes to widespread increases in lake hypoxia and anoxia. *Glob. Change Biol.* **29**, 1009–1023. https://doi.org/10.1111/gcb.16525

Jankowiak, J., Hattenrath-Lehmann, T., Kramer, B.J., Ladds, M., Gobler, C.J., 2019. Deciphering the effects of nitrogen, phosphorus, and temperature on cyanobacterial bloom intensification, diversity, and toxicity in western Lake Erie. *Limnol. Oceanogr.* **64**, 1347–1370.

Jansen, L.S., Sobota, D., Pan, Y., Strecker, A.L., 2024. Watershed, lake, and food web factors influence diazotrophic cyanobacteria in mountain lakes. *Limnol. Oceanogr.* **n/a**. https://doi.org/10.1002/lno.12523

Janssen, A.B., Janse, J.H., Beusen, A.H., Chang, M., Harrison, J.A., Huttunen, I., Kong, X., Rost, J., Teurlincx, S., Troost, T.A., van Wijk, D., Mooij, W.M., 2019. How to model algal blooms in any lake on earth. *Curr. Opin. Environ. Sustain.* **36**, 1–10. https://doi.org/10.1016/j.cosust.2018.09.001

Janssen, E.M.-L., 2019. Cyanobacterial peptides beyond microcystins – A review on co-occurrence, toxicity, and challenges for risk assessment. *Water Res.* **151**, 488–499. https://doi.org/10.1016/j.watres.2018.12.048

Jargal, N., Lee, E.-H., An, K.-G., 2023. Monsoon-induced response of algal chlorophyll to trophic state, light availability, and morphometry in 293 temperate reservoirs. *J. Environ. Manage.* **337**, 117737. https://doi.org/10.1016/j.jenvman.2023.117737

Jarvie, H.P., Johnson, L.T., Sharpley, A.N., Smith, D.R., Baker, D.B., Bruulsema, T.W., Confesor, R., 2017. Increased soluble phosphorus loads to Lake Erie: Unintended consequences of conservation practices? *J. Environ. Qual.* **46**, 123–132. https://doi.org/10.2134/jeq2016.07.0248

Jarvie, H.P., Sharpley, A.N., Spears, B., Buda, A.R., May, L., Kleinman, P.J.A., 2013. Water quality remediation faces unprecedented challenges from "legacy phosphorus". *Environ. Sci. Technol.* **47**, 8997–8998. https://doi.org/10.1021/es403160a

Jensen, H.S., Mortensen, P.B., Andersen, F.O., Rasmussen, E., Jensen, A., 1995. Phosphorus cycling in a coastal marine sediment, Aarhus Bay, Denmark. *Limnol. Ocean.* **40**, 908–917. https://doi.org/10.4319/lo.1995.40.5.0908

Jeong, I., Nakashita, S., Hibino, T., Kim, K., 2022. Effect of sediment deposition on phosphate and hydrogen sulfide removal by granulated coal ash in coastal sediments. *Mar. Pollut. Bull.* **179**, 113679. https://doi.org/10.1016/j.marpolbul.2022.113679

Jeppesen, E., Brucet, S., Naselli-Flores, L., Papastergiadou, E., Stefanidis, K., Nõges, T., Nõges, P., Attayde, J.L., Zohary, T., Coppens, J., Bucak, T., Menezes, R.F., Freitas, F.R.S., Kernan, M., Søndergaard, M., Beklioğlu, M., 2015. Ecological impacts of global warming and water abstraction on lakes and reservoirs due to changes in water level and related changes in salinity. *Hydrobiologia.* **750**, 201–227. https://doi.org/10.1007/s10750-014-2169-x

Jeppesen, E., Canfield, D.E., Bachmann, R.W., Søndergaard, M., Havens, K.E., Johansson, L.S., Lauridsen, T.L., Sh, T., Rutter, R.P., Warren, G., Ji, G., Hoyer, M.V., 2020. Toward predicting climate change effects on lakes: A comparison of 1656 shallow lakes from Florida and Denmark reveals substantial differences in nutrient dynamics, metabolism, trophic structure, and top-down control. *Inland Waters.* **10**, 197–211. https://doi.org/10.1080/20442041.2020.1711681

Jeppesen, E., Meerhoff, M., Davidson, T.A., Trolle, D., Søndergaard, M., Lauridsen, T.L., Beklioglu, M., Brucet, S., Volta, P., González-Bergonzoni, I., Nielsen, A., 2014. Climate change impacts on lakes: An integrated ecological perspective based on a multi-faceted approach, with special focus on shallow lakes. *J. Limnol.* **73**. https://doi.org/10.4081/jlimnol.2014.844

Jeppesen, E., Sørensen, P.B., Johansson, L.S., Søndergaard, M., Lauridsen, T.L., Nielsen, A., Mejlhede, P., 2024a. Recovery of lakes from eutrophication: Changes in nitrogen retention capacity and the role of nitrogen legacy in 10 Danish lakes studied over 30 years. *Hydrobiologia.* https://doi.org/10.1007/s10750-024-05478-6

Jeppesen, E., Volta, P., Mao, Z., 2024b. Chapter 22: Fish. In *Wetzel's Limnology*, 4th ed. Academic Press, San Diego, pp. 359–425. https://doi.org/10.1016/B978-0-12-822701-5.00015-X

Ji, G., Havens, K., 2019. Periods of extreme shallow depth hinder but do not stop long-term improvements of water quality in Lake Apopka, Florida (USA). *Water.* **11**, 538. https://doi.org/10.3390/w11030538

Jöhnk, K.D., Huisman, J., Sharples, J., Sommeijer, B., Visser, P.M., Stroom, J.M., 2008. Summer heatwaves promote blooms of harmful cyanobacteria. *Glob. Change Biol.* **14**, 495–512. https://doi.org/10.1111/j.1365-2486.2007.01510.x

Jones, H.P., Jones, P.C., Barbier, E.B., Blackburn, R.C., Rey Benayas, J.M., Holl, K.D., McCrackin, M., Meli, P., Montoya, D., Mateos, D.M., 2018. Restoration and repair of Earth's damaged ecosystems. *Proc. R. Soc. B Biol. Sci.* **285**, 20172577. https://doi.org/10.1098/rspb.2017.2577

Jones, J.R., Bachmann, R.W., 1976. Prediction of phosphorus and chlorophyll levels in lakes. *J. Water Pollut. Control. Fed.* **48**, 2176–2182.

Jones, L., Gorst, A., Elliott, J., Fitch, A., Illman, H., Evans, C., Thackeray, S., Spears, B., Gunn, I., Carvalho, L., May, L., Schonrogge, K., Clilverd, H., Mitchell, Z., Garbutt, A., Taylor, P., Fletcher, D., Giam, G., Aron, J., Ray, D., Berenice-Wilmes, S., King, N., Malham, S., Fung, F., Tinker, J., Wright, P., Smale, R., 2020. Climate driven threshold effects in the natural environment. *Report to the Climate Change Committee (UK).* https://www.ukclimaterisk.org/wp-content/uploads/2020/07/Thresholds-in-the-natural-environment_CEH.pdf

Kakouei, K., Kraemer, B.M., Adrian, R., 2022. Variation in the predictability of lake plankton metric types. *Limnol. Oceanogr*. **67**, 608–620. https://doi.org/10.1002/lno.12021

Kaltenberg, E.M., Matisoff, G., 2020. Chapter 19: Internal loading of phosphorus to Lake Erie: Significance, measurement methods, and available data. In *Internal Phosphorus Loading: Causes, Case Studies, and Management* (Eds. Steinman, A.D., Spears, B.M.). J. Ross Publishing, Plantation, FL, pp. 359–376.

Kane, D.D., Conroy, J.D., Peter Richards, R., Baker, D.B., Culver, D.A., 2014. Re-eutrophication of Lake Erie: Correlations between tributary nutrient loads and phytoplankton biomass. *J. Gt. Lakes Res*. **40**, 496–501. https://doi.org/10.1016/j.jglr.2014.04.004

Kang, L., Haasler, S., Mucci, M., Korving, L., Dugulan, A.I., Prot, T., Waajen, G., Lürling, M., 2023. Comparison of dredging, lanthanum-modified bentonite, aluminium-modified zeolite, and FeCl2 in controlling internal nutrient loading. *Water Res*. **244**, 120391. https://doi.org/10.1016/j.watres.2023.120391

Kang, L., Mucci, M., Lürling, M., 2022a. Compounds to mitigate cyanobacterial blooms affect growth and toxicity of *Microcystis aeruginosa. Harmful Algae*. **118**, 102311. https://doi.org/10.1016/j.hal.2022.102311

Kang, L., Mucci, M., Lürling, M., 2022b. Influence of temperature and pH on phosphate removal efficiency of different sorbents used in lake restoration. *Sci. Total Environ.* **812**, 151489. https://doi.org/10.1016/j.scitotenv.2021.151489

Karatayev, A.Y., Burlakova, L.E., Padilla, D.K., 2015. Zebra versus quagga mussels: A review of their spread, population dynamics, and ecosystem impacts. *Hydrobiologia*. **746**, 97–112. https://doi.org/10.1007/s10750-014-1901-x

Karlsson-Elfgren, I., Brunberg, A.-K., 2004. The importance of shallow sediments in the recruitment of Anabaena and Aphanizomenon (cyanophyceae)1. *J. Phycol.* **40**, 831–836. https://doi.org/10.1111/j.1529-8817.2004.04070.x

Karstens, S., Buczko, U., Glatzel, S., 2015. Phosphorus storage and mobilization in coastal Phragmites wetlands: Influence of local-scale hydrodynamics. *Estuar. Coast. Shelf Sci*. **164**, 124–133. https://doi.org/10.1016/j.ecss.2015.07.014

Kasprzak, P., Gonsiorczyk, T., Grossart, H.-P., Hupfer, M., Koschel, R., Petzoldt, T., Wauer, G., 2018. Restoration of a eutrophic hard-water lake by applying an optimised dosage of poly-aluminium chloride (PAC). *Limnologica*. **70**, 33–48. https://doi.org/10.1016/j.limno.2018.04.002

Kasprzak, P., Shatwell, T., Gessner, M.O., Gonsiorczyk, T., Kirillin, G., Selmeczy, G., Padisák, J., Engelhardt, C., 2017. Extreme weather event triggers cascade towards extreme turbidity in a clear-water lake. *Ecosystems*. **20**, 1407–1420. https://doi.org/10.1007/s10021-017-0121-4

Katsev, S., Dittrich, M., 2013. Modeling of decadal scale phosphorus retention in lake sediment under varying redox conditions. *Ecol. Model.* **251**, 246–259. https://doi.org/10.1016/j.ecolmodel.2012.12.008

Kaushal, S.S., Likens, G.E., Pace, M.L., Utz, R.M., Haq, S., Gorman, J., Grese, M., 2018. Freshwater salinization syndrome on a continental scale. *Proc. Natl. Acad. Sci.* **115**, E574 – E583. https://doi.org/10.1073/pnas.1711234115

Kaushal, S.S., Mayer, P.M., Likens, G.E., Reimer, J.E., Maas, C.M., Rippy, M.A., Grant, S.B., Hart, I., Utz, R.M., Shatkay, R.R., Wessel, B.M., Maietta, C.E., Pace, M.L., Duan, S., Boger, W.L., Yaculak, A.M., Galella, J.G., Wood, K.L., Morel, C.J., Nguyen, W., Querubin, S.E.C., Sukert, R.A., Lowien, A., Houde, A.W., Roussel, A., Houston, A.J., Cacopardo, A., Ho, C., Talbot-Wendlandt, H., Widmer, J.M., Slagle, J., Bader, J.A., Chong, J.H., Wollney, J., Kim, J., Shepherd, L., Wilfong, M.T., Houlihan, M., Sedghi, N., Butcher, R., Chaudhary, S., Becker, W.D., 2023. Five state factors control progressive stages of freshwater salinization syndrome. *Limnol. Oceanogr. Lett.* **8**, 190–211. https://doi.org/10.1002/lol2.10248

Kelly, N.E., Javed, A., Shimoda, Y., Zastepa, A., Watson, S., Mugalingam, S., Arhonditsis, G.B., 2019. A Bayesian risk assessment framework for microcystin violations of drinking water and recreational standards in the Bay of Quinte, Lake Ontario, Canada. *Water Res.* **162**, 288–301. https://doi.org/10.1016/j.watres.2019.06.005

Kiani, M., Tammeorg, P., Niemistö, J., Simojoki, A., Tammeorg, O., 2020. Internal phosphorus loading in a small shallow lake: Response after sediment removal. *Sci. Total Environ.* **725**, 138279. https://doi.org/10.1016/j.scitotenv.2020.138279

Kiani, M., Zrim, J., Simojoki, A., Tammeorg, O., Penttinen, P., Markkanen, T., Tammeorg, P., 2023. Recycling eutrophic lake sediments into grass production: A four-year field experiment on agronomical and environmental implications. *Sci. Total Environ.* **870**, 161881. https://doi.org/10.1016/j.scitotenv.2023.161881

Kinsman-Costello, L., Bean, E., Goeckner, A., Matthews, J.W., O'Driscoll, M., Palta, M.M., Peralta, A.L., Reisinger, A.J., Reyes, G.J., Smyth, A.R., Stofan, M., 2022. Mud in the city: Effects of freshwater salinization on inland urban wetland nitrogen and phosphorus availability and export. *Limnol. Oceanogr. Lett.* **n/a**. https://doi.org/10.1002/lol2.10273

Kirol, A.P., Morales-Williams, A.M., Braun, D.C., Marti, C.L., Pierson, O.E., Wagner, K.J., Schroth, A.W., 2024. Linking sediment and water column phosphorus dynamics to oxygen, temperature, and aeration in shallow eutrophic lakes. *Water Resour. Res.* **60**, e2023WR034813. https://doi.org/10.1029/2023WR034813

Kisand, A., Nõges, P., 2003. Sediment phosphorus release in phytoplankton dominated versus macrophyte dominated shallow lakes: Importance of oxygen conditions. *Hydrobiologia.* **506**, 129–133. https://doi.org/10.1023/B:HYDR.0000008620.87704.3b

Kitchens, C.M., Johengen, T.H., Davis, T.W., 2018. Establishing spatial and temporal patterns in Microcystis sediment seed stock viability and their relationship to subsequent bloom development in Western Lake Erie. *PLoS ONE.* **13**, e0206821. https://doi.org/10.1371/journal.pone.0206821

Klaus, M., Karlsson, J., Seekell, D., 2021. Tree line advance reduces mixing and oxygen concentrations in arctic-alpine lakes through wind sheltering and organic carbon supply. *Glob. Change Biol.* gcb.15660. https://doi.org/10.1111/gcb.15660

Klausmeier, C.A., Litchman, E., Daufresne, T., Levin, S.A., 2004. Optimal nitrogen-to-phosphorus stoichiometry of phytoplankton. *Nature.* **429**, 171–174. https://doi.org/10.1038/nature02454

Kleeberg, A., Nixdorf, B., Mathes, J., 2000. Lake Jabel restoration project: Phosphorus status and possibilities and limitations of diversion of its nutrient-rich main inflow. *Lake Reserv. Res. Manag.* **5**, 23–33. https://doi.org/10.1046/j.1440-1770.2000.00092.x

Kleinman, P., Sharpley, A., Buda, A., McDowell, R., Allen, A., 2011. Soil controls of phosphorus in runoff: Management barriers and opportunities. *Can. J. Soil. Sci.* **91**, 329–338. https://doi.org/10.4141/cjss09106

Knapp, D., Fernández Castro, B., Marty, D., Loher, E., Köster, O., Wüest, A., Posch, T., 2021. The red harmful plague in times of climate change: Blooms of the cyanobacterium Planktothrix rubescens triggered by stratification dynamics and irradiance. *Front. Microbiol.* **12**, 705914. https://doi.org/10.3389/fmicb.2021.705914

Knoll, L.B., Fry, B., Hayes, N.M., Sauer, H.M., 2024. Reduced snow and increased nutrients show enhanced ice-associated photoautotrophic growth using a modified experimental under-ice design. *Limnol. Oceanogr.* **69**, 203–216. https://doi.org/10.1002/lno.12469

Knoll, L.B., Sarnelle, O., Hamilton, S.K., Kissman, C.E.H., Wilson, A.E., Rose, J.B., Morgan, M.R., 2008. Invasive zebra mussels (*Dreissena polymorpha*) increase cyanobacterial toxin concentrations in low-nutrient lakes. *Can. J. Fish. Aquat. Sci.* **65**, 448–455. https://doi.org/10.1139/f07-181

Knoll, L.B., Williamson, C.E., Pilla, R.M., Leach, T.H., Brentrup, J.A., Fisher, T.J., 2018. Browning-related oxygen depletion in an oligotrophic lake. *Inland Waters.* **8**, 255–263. https://doi.org/10.1080/20442041.2018.1452355

Knorr, K.-H., 2012. DOC-dynamics in a small headwater catchment as driven by redox fluctuations and hydrological flow paths – are DOC exports mediated by iron reduction/oxidation cycles? *Biogeosciences Discuss.* **9**. https://doi.org/10.5194/bgd-9-12951-2012

Köhler, J., Wang, L., Guislain, A., Shatwell, T., 2018. Influence of vertical mixing on light-dependency of phytoplankton growth. *Limnol. Oceanogr.* **63**, 1156–1167. https://doi.org/10.1002/lno.10761

Kopáček, J., Grill, S., Hejzlar, J., Porcal, P., Turek, J., 2024. Tree dieback and subsequent changes in water quality accelerated the climate-related warming of a central European forest lake. *J. Water Clim. Change.* 127–138. https://doi.org/10.2166/wcc.2023.581

Kopáček, J., Hejzlar, J., Kaňa, J., Norton, S.A., Stuchlík, E., 2015. Effects of acidic deposition on in-lake phosphorus availability: A lesson from lakes recovering from acidification. *Environ. Sci. Technol.* **49**, 2895–2903. https://doi.org/10.1021/es5058743

Koschel, R.H., Dittrich, M., Casper, P., Heiser, A., Roßberg, R., 2001. Induced hypolimnetic calcite precipitation – ecotechnology for restoration of stratified eutrophic hardwater lakes. *SIL Proc.* **27**, 3644–3649. https://doi.org/10.1080/03680770.1998.11902507

Koski-Vähälä, J., Hartikainen, H., 2001. Assessment of the risk of phosphorus loading due to resuspended sediment. *J. Environ. Qual.* **30**, 960–966. https://doi.org/10.2134/jeq2001.303960x

Kosten, S., Huszar, V.L., Bécares, E., Costa, L.S., Donk, E., Hansson, L.-A., Jeppesen, E., Kruk, C., Lacerot, G., Mazzeo, N., et al., 2012. Warmer climates boost cyanobacterial dominance in shallow lakes. *Glob. Change Biol.* **18**, 118–126. https://doi.org/10.1111/j.1365-2486.2011.02488.x/full

Kowalczewska-Madura, K., Dondajewska-Pielka, R., Gołdyn, R., 2022. The assessment of external and internal nutrient loading as a basis for lake management. *Water.* **14**, 2844. https://doi.org/10.3390/w14182844

Kowalczewska-Madura, K., Rosińska, J., Dondajewska-Pielka, R., Gołdyn, R., Kaczmarek, L., 2020. The effects of limiting restoration treatments in a shallow urban lake. *Water.* **12**, 1383. https://doi.org/10.3390/w12051383

Kraemer, B.M., Anneville, O., Chandra, S., Dix, M., Kuusisto, E., Livingstone, D.M., Rimmer, A., Schladow, S.G., Silow, E., Sitoki, L.M., Tamatamah, R., Vadeboncoeur, Y., McIntyre, P.B., 2015. Morphometry and average temperature affect lake stratification responses to climate change. *Geophys. Res. Lett.* **42**, 4981–4988. https://doi.org/10.1002/2015GL064097

Kraemer, B.M., Kakouei, K., Munteanu, C., Thayne, M.W., Adrian, R., 2022. Worldwide moderate-resolution mapping of lake surface chl-a reveals variable responses to global change (1997–2020). *PLoS Water.* **1**, e0000051. https://doi.org/10.1371/journal.pwat.0000051

Kraemer, B.M., Mehner, T., Adrian, R., 2017. Reconciling the opposing effects of warming on phytoplankton biomass in 188 large lakes. *Sci. Rep.* **7**, 10762. https://doi.org/10.1038/s41598-017-11167-3

Kraemer, B.M., Pilla, R.M., Woolway, R.I., Anneville, O., Ban, S., Colom-Montero, W., Devlin, S.P., Dokulil, M.T., Gaiser, E.E., Hambright, K.D., Hessen, D.O., Higgins, S.N., Jöhnk, K.D., Keller, W., Knoll, L.B., Leavitt, P.R., Lepori, F., Luger, M.S., Maberly, S.C., Müller-Navarra, D.C., Paterson, A.M., Pierson, D.C., Richardson, D.C., Rogora, M., Rusak, J.A., Sadro, S., Salmaso, N., Schmid, M., Silow, E.A., Sommaruga, R., Stelzer, J.A.A., Straile, D., Thiery, W., Timofeyev, M.A., Verburg, P., Weyhenmeyer, G.A., Adrian, R., 2021. Climate change drives widespread shifts in lake thermal habitat. *Nat. Clim. Change.* **11**, 521–529. https://doi.org/10.1038/s41558-021-01060-3

Kraemer, B.M., Seimon, A., Adrian, R., McIntyre, P.B., 2020. Worldwide lake level trends and responses to background climate variation. *Hydrol. Earth Syst. Sci.* **24**, 2593–2608. https://doi.org/10.5194/hess-24-2593-2020

Kritzberg, E.S., 2017. Centennial-long trends of lake browning show major effect of afforestation. *Limnol. Oceanogr. Lett.* **2**, 105–112. https://doi.org/10.1002/lol2.10041

Kröger, B., Selmeczy, G.B., Casper, P., Soininen, J., Padisák, J., 2023. Long-term phytoplankton community dynamics in Lake Stechlin (north-east Germany) under sudden and heavily accelerating eutrophication. *Freshw. Biol.* **68**, 737–751. https://doi.org/10.1111/fwb.14060

Kruk, C., Devercelli, M., Huszar, V.L., 2020. Reynolds functional groups: A trait-based pathway from patterns to predictions. *Hydrobiologia.* https://doi.org/10.1007/s10750-020-04340-9

Kumagai, M., Nakano, S., Jiao, C., Hayakawa, K., Tsujimura, S., Nakajima, T., Frenette, J.-J., Quesada, A., 2000. Effect of cyanobacterial blooms on thermal stratification. *Limnology*. **1**, 191–195. https://doi.org/10.1007/s102010070006

Kurmayer, R., Deng, L., Entfellner, E., 2016. Role of toxic and bioactive secondary metabolites in colonization and bloom formation by filamentous cyanobacteria Planktothrix. *Harmful Algae*. **54**, 69–86. https://doi.org/10.1016/j.hal.2016.01.004

Labrecque, V., Nürnberg, G.K., Tremblay, R., Pienitz, R., 2012. Caractérisation de la charge interne de phosphore du lac Nairne, Charlevoix (Québec), (Internal phosphorus load assessment of Lake Nairne, Charlevoix, Quebec). *Rev. Sci. L'Eau*. **25**, 77–93. https://doi.org/10.7202/1008537ar

Ladwig, R., Rock, L.A., Dugan, H.A., 2021. Impact of salinization on lake stratification and spring mixing. *Limnol. Oceanogr. Lett*. **n/a**. https://doi.org/10.1002/lol2.10215

Laiveling, A., Nauman, C., Stanislawczyk, K., Bair, H.B., Kane, D.D., Chaffin, J.D., 2022. Potamoplankton of the Maumee River during 2018 and 2019: The relationship between cyanobacterial toxins and environmental factors. *J. Gt. Lakes Res*. **48**, 1587–1598. https://doi.org/10.1016/j.jglr.2022.08.015

Lang, P., Meis, S., Procházková, L., Carvalho, L., Mackay, E.B., Woods, H.J., Pottie, J., Milne, I., Taylor, C., Maberly, S.C., et al., 2016. Phytoplankton community responses in a shallow lake following lanthanum-bentonite application. *Water Res*. **97**, 55–68. https://doi.org/10.1016/j.watres.2016.03.018

Larsen, D.P., Schults, D.W., Malueg, K.W., 1981. Summer internal phosphorus supplies in Shagawa Lake, Minnesota. *Limnol. Ocean*. **26**, 740–753. https://doi.org/10.4319/lo.1981.26.4.0740

Larson, J.H., James, W.F., Fitzpatrick, F.A., Frost, P.C., Evans, M.A., Reneau, P.C., Xenopoulos, M.A., 2020. Phosphorus, nitrogen and dissolved organic carbon fluxes from sediments in freshwater rivermouths entering Green Bay (Lake Michigan; USA). *Biogeochemistry*. **147**, 179–197. https://doi.org/10.1007/s10533-020-00635-0

Larson, J.H., Trebitz, A.S., Steinman, A.D., Wiley, M.J., Mazur, M.C., Pebbles, V., Braun, H.A., Seelbach, P.W., 2013. Great Lakes rivermouth ecosystems: Scientific synthesis and management implications. *J. Gt. Lakes Res*. **39**, 513–524. https://doi.org/10.1016/j.jglr.2013.06.002

Latour, D., Salençon, M.-J., Reyss, J.-L., Giraudet, H., 2007. Sedimentary imprint of Microcystis Aeruginosa (cyanobacteria) Bbooms in Grangent Reservoir (Loire, France). *J. Phycol*. **43**, 417–425. https://doi.org/10.1111/j.1529-8817.2007.00343.x

Lau, M.P., del Giorgio, P., 2020. Reactivity, fate and functional roles of dissolved organic matter in anoxic inland waters. *Biol. Lett*. **16**, 20190694. https://doi.org/10.1098/rsbl.2019.0694

Layden, T.J., Kremer, C.T., Brubaker, D.L., Kolk, M.A., Trout-Haney, J.V., Vasseur, D.A., Fey, S.B., 2022. Thermal acclimation influences the growth and toxin production of freshwater cyanobacteria. *Limnol. Oceanogr. Lett*. **7**, 34–42. https://doi.org/10.1002/lol2.10197

Leão, P.N., Vasconcelos, M.T.S.D., Vasconcelos, V.M., 2009. Allelopathy in freshwater cyanobacteria. *Crit. Rev. Microbiol*. **35**, 271–282. https://doi.org/10.3109/10408410902823705

LeBlanc, S., Pick, F.R., Hamilton, P.B., 2008. Fall cyanobacterial blooms in oligotrophic-to-mesotrophic temperate lakes and the role of climate change. *SIL Proc. 1922–2010*. **30**, 90–94. https://doi.org/10.1080/03680770.2008.11902091

Lee, C., Fletcher, T.D., Sun, G., 2009. Nitrogen removal in constructed wetland systems. *Eng. Life Sci*. **9**, 11–22. https://doi.org/10.1002/elsc.200800049

Lee, S.-K., Mapes, B.E., Wang, C., Enfield, D.B., Weaver, S.J., 2014. Springtime ENSO phase evolution and its relation to rainfall in the continental U.S.: Springtime ENSO Phase and U.S. Rainfall. *Geophys. Res. Lett*. **41**, 1673–1680. https://doi.org/10.1002/2013GL059137

Lehman, J.T., 2023. Opinion: Inappropriate applications of phosphorus loading models in lake management regulatory policy. *Lake Reserv. Manag*. **39**, 273–277. https://doi.org/10.1080/10402381.2023.2279042

Lehner, B., 2024. Chapter 4: Rivers and lakes – their distribution, origins, and forms. In *Wetzel's Limnology*, 4th ed. (Eds. Jones, I.D., Smol, J.P.). Academic Press, San Diego, pp. 25–56. https://doi.org/10.1016/B978-0-12-822701-5.00004-5

Le Moal, M., Pannard, A., Brient, L., Richard, B., Chorin, M., Mineaud, E., Wiegand, C., 2021. Is the cyanobacterial bloom composition shifting due to climate forcing or nutrient changes? Example of a shallow eutrophic reservoir. *Toxins*. **13**, 351. https://doi.org/10.3390/toxins13050351

Lepori, F., Bartosiewicz, M., Simona, M., Veronesi, M., 2018. Effects of winter weather and mixing regime on the restoration of a deep perialpine lake (Lake Lugano, Switzerland and Italy). *Hydrobiologia*. **824**, 229–242. https://doi.org/10.1007/s10750-018-3575-2

Lepori, F., Capelli, C., 2021. Effects of phosphorus control on primary productivity and deep-water oxygenation: Insights from Lake Lugano (Switzerland and Italy). *Hydrobiologia*. **848**, 613–629.

Leung, T., Wilkinson, G.M., Swanner, E.D., 2021. Iron availability allows sustained cyanobacterial blooms: A dual-lake case study. *Inland Waters*. **11**, 417–429. https://doi.org/10.1080/20442041.2021.1904762

Lévesque, D., Cattaneo, A., Hudon, C., Gagnon, P., Weyhenmeyer, G., 2012. Predicting the risk of proliferation of the benthic cyanobacterium Lyngbya wollei in the St. Lawrence River. *Can. J. Fish. Aquat. Sci*. **69**, 1585–1595. https://doi.org/10.1139/f2012-087

Lewandowski, J., Hupfer, M., 2005. Effect of macrozoobenthos on two-dimensional small-scale heterogeneity of pore water phosphorus concentrations in lake sediments: A laboratory study. *Limnol. Ocean*. **50**, 1106–1118. https://doi.org/10.4319/lo.2005.50.4.1106

Lewandowski, J., Schauser, I., Hupfer, M., 2003. Long term effects of phosphorus precipitation with alum in hypereutrophic Lake Süsser See (Germany). *Water Res*. **37**, 3194–3204. https://doi.org/10.1016/S0043-1354(03)00137-4

Lewis, A.S.L., Breef-Pilz, A., Howard, D.W., Lofton, M.E., Olsson, F., Wander, H.L., Wood, C.E., Schreiber, M.E., Carey, C.C., 2024a. Reservoir drawdown highlights the emergent effects of water level change on reservoir physics, chemistry, and biology. *J. Geophys. Res. Biogeosci*. **129**, e2023JG007780. https://doi.org/10.1029/2023JG007780

Lewis, A.S.L., Lau, M.P., Jane, S.F., Rose, K.C., Be'eri-Shlevin, Y., Burnet, S.H., Clayer, F., Feuchtmayr, H., Grossart, H., Howard, D.W., Mariash, H., Delgado Martin, J., North, R.L., Oleksy, I., Pilla, R.M., Smagula, A.P., Sommaruga, R., Steiner, S.E., Verburg, P., Wain, D., Weyhenmeyer, G.A., Carey, C.C., 2024b. Anoxia begets anoxia: A positive feedback to the deoxygenation of temperate lakes. *Glob. Change Biol*. **30**, e17046. https://doi.org/10.1111/gcb.17046

Li, H., Yang, G., Ma, J., Wei, Y., Kang, L., He, Y., He, Q., 2019. The role of turbulence in internal phosphorus release: Turbulence intensity matters. *Environ. Pollut*. **252**, 84–93. https://doi.org/10.1016/j.envpol.2019.05.068

Li, J., Bai, Y., Bear, K., Joshi, S., Jaisi, D., 2017. Phosphorus availability and turnover in the Chesapeake Bay: Insights from nutrient stoichiometry and phosphate oxygen isotope ratios. *J. Geophys. Res. Biogeosci*. **122**, 811–824. https://doi.org/10.1002/2016JG003589

Li, S., Arnscheidt, J., Cassidy, R., Douglas, R.W., McGrogan, H.J., Jordan, P., 2023. The spatial and temporal dynamics of sediment phosphorus attenuation and release in impacted stream catchments. *Water Res*. **245**, 120663. https://doi.org/10.1016/j.watres.2023.120663

Li, X., Peng, S., Xi, Y., Woolway, R.I., Liu, G., 2022. Earlier ice loss accelerates lake warming in the Northern Hemisphere. *Nat. Commun*. **13**, 5156. https://doi.org/10.1038/s41467-022-32830-y

Li, Y., Wang, L., Yan, Z., Chao, C., Yu, H., Yu, D., Liu, C., 2020. Effectiveness of dredging on internal phosphorus loading in a typical aquacultural lake. *Sci. Total Environ*. **744**, 140883. https://doi.org/10.1016/j.scitotenv.2020.140883

Liboriussen, L., Søndergaard, M., Jeppesen, E., Thorsgaard, I., Grünfeld, S., Jakobsen, T., Hansen, K., 2009. Effects of hypolimnetic oxygenation on water quality: Results from five Danish lakes. *Hydrobiologia*. **625**, 157–172. https://doi.org/10.1007/s10750-009-9705-0

Lindim, C., Becker, A., Grüneberg, B., Fischer, H., 2015. Modelling the effects of nutrient loads reduction and testing the N and P control paradigm in a German shallow lake. *Ecol. Eng*. **82**, 415–427. https://doi.org/10.1016/j.ecoleng.2015.05.009

Linz, D.M., Sienkiewicz, N., Struewing, I., Stelzer, E.A., Graham, J.L., Lu, J., 2023. Metagenomic mapping of cyanobacteria and potential cyanotoxin producing taxa in large rivers of the United States. *Sci. Rep*. **13**, 2806. https://doi.org/10.1038/s41598-023-29037-6

Litchman, E., 2023. Understanding and predicting harmful algal blooms in a changing climate: A trait-based framework. *Limnol. Oceanogr. Lett.* **8**, 229–246. https://doi.org/10.1002/lol2.10294

Liu, B., Liu, X.G., Yang, J., Garman, D., Zhang, K., Zhang, H.G., 2012. Research and application of in-situ control technology for sediment rehabilitation in eutrophic water bodies. *Wat. Sci. Technol.* **65**, 1190–1199. https://doi.org/10.2166/wst.2012.927

Liu, C., Du, Y., Zhong, J., Zhang, L., Huang, W., Han, C., Chen, K., Gu, X., 2022. From macrophyte to algae: Differentiated dominant processes for internal phosphorus release induced by suspended particulate matter deposition. *Water Res.* **224**, 119067. https://doi.org/10.1016/j.watres.2022.119067

Liu, Z., Hu, J., Zhong, P., Zhang, X., Ning, J., Larsen, S.E., Chen, D., Gao, Y., He, H., Jeppesen, E., 2018. Successful restoration of a tropical shallow eutrophic lake: Strong bottom-up but weak top-down effects recorded. *Water Res.* **146**, 88–97. https://doi.org/10.1016/j.watres.2018.09.007

Loewen, M.R., Ackerman, J.D., Hamblin, P.F., 2007. Environmental implications of stratification and turbulent mixing in a shallow lake basin. *Can. J. Fish. Aquat. Sci.* **64**, 43–57. https://doi.org/10.1139/f06-165

Lofton, M.E., Brentrup, J.A., Beck, W.S., Zwart, J.A., Bhattacharya, R., Brighenti, L.S., Burnet, S.H., McCullough, I.M., Steele, B.G., Carey, C.C., Cottingham, K.L., Dietze, M.C., Ewing, H.A., Weathers, K.C., LaDeau, S.L., 2022. Using near-term forecasts and uncertainty partitioning to inform prediction of oligotrophic lake cyanobacterial density. *Ecol. Appl.* https://doi.org/10.1002/eap.2590

Lofton, M.E., Leach, T.H., Beisner, B.E., Carey, C.C., 2020. Relative importance of top-down vs. bottom-up control of lake phytoplankton vertical distributions varies among fluorescence-based spectral groups. *Limnol. Oceanogr.* **65**, 2485–2501. https://doi.org/10.1002/lno.11465

Lowe, E.F., Steward, J.S., 2011. Cutting through complexity with Vollenweider's razor. *Aquat. Ecosyst. Health Manag.* **14**, 209–213. https://doi.org/10.1080/14634988.2011.577704

Lukkari, K., Hartikainen, H., Leivuori, M., 2007. Fractionation of sediment phosphorus revisited: I Fractionation steps and their biogeochemical basis. *Limnol. Oceanogr. Methods.* **5**, 433–444. https://doi.org/10.4319/lom.2007.5.433

Luo, A., Chen, H., Gao, X., Carvalho, L., Xue, Y., Jin, L., Yang, J., 2022. Short-term rainfall limits cyanobacterial bloom formation in a shallow eutrophic subtropical urban reservoir in warm season. *Sci. Total Environ.* **827**, 154172. https://doi.org/10.1016/j.scitotenv.2022.154172

Lürling, M., Beekman, W., 2006. Palmelloids formation in *Chlamydomonas reinhardtii*: Defence against rotifer predators? *Ann. Limnol. – Int. J. Limnol.* **42**, 65–72. https://doi.org/10.1051/limn/2006010

Lürling, M., Eshetu, F., Faassen, E.J., Kosten, S., Huszar, V.L.M., 2013. Comparison of cyanobacterial and green algal growth rates at different temperatures. *Freshw. Biol.* **58**, 552–559. https://doi.org/10.1111/j.1365-2427.2012.02866.x

Lürling, M., Mackay, E., Reitzel, K., Spears, B.M., 2016. Editorial – A critical perspective on geo-engineering for eutrophication management in lakes. *Water Res.* **97**, 1–10. https://doi.org/10.1016/j.watres.2016.03.035

Lürling, M., Mello, M.M. e, van Oosterhout, F., de Senerpont Domis, L., Marinho, M.M., 2018. Response of natural cyanobacteria and algae assemblages to a nutrient pulse and elevated temperature. *Front. Microbiol.* **9**. https://doi.org/10.3389/fmicb.2018.01851

Lürling, M., Mucci, M., 2020. Mitigating eutrophication nuisance: In-lake measures are becoming inevitable in eutrophic waters in the Netherlands. *Hydrobiologia*. https://doi.org/10.1007/s10750-020-04297-9

Lürling, M., Smolders, A.J.P., Douglas, G., 2020. Chapter 5: Methods for the management of internal phosphorus loading in lakes. In *Internal Phosphorus Loading: Causes, Case Studies, and Management* (Eds. Steinman, A.D., Spears, B.M.). J. Ross Publishing, Plantation, FL, pp. 77–107.

Lürling, M., Tolman, Y., 2010. Effects of lanthanum and lanthanum-modified clay on growth, survival and reproduction of *Daphnia magna*. *Water Res.* **44**, 309. https://doi.org/10.1016/j.watres.2009.09.034

Lürling, M., van Oosterhout, F., 2013. Case study on the efficacy of a lanthanum-enriched clay (Phoslock®) in controlling eutrophication in Lake Het Groene Eiland (The Netherlands). *Hydrobiologia.* **710**, 253–263. https://doi.org/10.1007/s10750-012-1141-x

Lürling, M., van Oosterhout, F., Faassen, E., 2017. Eutrophication and warming boost cyanobacterial biomass and microcystins. *Toxins.* **9**, 64. https://doi.org/10.3390/toxins9020064

Lürling, M., van Oosterhout, F., Mucci, M., Waajen, G., 2023. Combination of measures to restore eutrophic urban ponds in The Netherlands. *Water*. **15**, 3599. https://doi.org/10.3390/w15203599

Lyche Solheim, A., Gundersen, H., Mischke, U., Skjelbred, B., Nejstgaard, J.C., Guislain, A.L.N., Sperfeld, E., Giling, D.P., Haande, S., Ballot, A., Moe, S.J., Stephan, S., Walles, T.J.W., Jechow, A., Minguez, L., Ganzert, L., Hornick, T., Hansson, T.H., Stratmann, C.N., Järvinen, M., Drakare, S., Carvalho, L., Grossart, H.-P., Gessner, M.O., Berger, S.A., 2024. Lake browning counteracts cyanobacteria responses to nutrients: Evidence from phytoplankton dynamics in large enclosure experiments and comprehensive observational data. *Glob. Change Biol.* **30**, e17013. https://doi.org/10.1111/gcb.17013

Ma, S.-N., Dong, X.-M., Jeppesen, E., Søndergaard, M., Cao, J.-Y., Li, Y.-Y., Wang, H.-J., Xu, J.-L., 2022. Responses of coastal sediment phosphorus release to elevated urea loading. *Mar. Pollut. Bull.* **174**, 113203. https://doi.org/10.1016/j.marpolbul.2021.113203

Maberly, S.C., O'Donnell, R.A., Woolway, R.I., Cutler, M.E.J., Gong, M., Jones, I.D., Merchant, C.J., Miller, C.A., Politi, E., Scott, E.M., Thackeray, S.J., Tyler, A.N., 2020. Global lake thermal regions shift under climate change. *Nat. Commun*. **11**, 1–9. https://doi.org/10.1038/s41467-020-15108-z

MacKeigan, P.W., Taranu, Z.E., Pick, F.R., Beisner, B.E., Gregory-Eaves, I., 2023. Both biotic and abiotic predictors explain significant variation in cyanobacteria biomass across lakes from temperate to subarctic zones. *Limnol. Oceanogr*. **68**, 1360–1375. https://doi.org/10.1002/lno.12352

Madsen, J.D., Chambers, P.A., James, W.F., Koch, E.W., Westlake, D.F., 2001. The interaction between water movement, sediment dynamics and submersed macrophytes. *Hydrobiologia*. **444**, 71–84. https://doi.org/10.1023/A:1017520800568

Magee, M.R., Wu, C.H., 2017. Response of water temperatures and stratification to changing climate in three lakes with different morphometry. *Hydrol. Earth Syst. Sci*. **21**, 6253–6274. https://doi.org/10.5194/hess-21-6253-2017

Magnuson, J.J., 2007. Perspectives on the long-term dynamics of lakes in the landscape. *Lake Reserv. Manag*. **23**, 452–456. https://doi.org/10.1080/07438140709354030

Magnuson, J.J., Robertson, D.M., Benson, B.J., Wynne, R.H., Livingstone, D.M., Arai, T., Assel, R.A., Barry, R.G., Card, V., Kuusisto, E., Granin, N.G., Prowse, T.D., Stewart, K.M., Vuglinski, V.S., 2000. Historical trends in lake and river ice cover in the northern hemisphere. *Science*. **289**, 1743–1746. https://doi.org/10.1126/science.289.5485.174

Malmaeus, J.M., Rydin, E., 2006. A time-dynamic phosphorus model for the profundal sediments of Lake Erken, Sweden. *Aquat. Sci*. **68**, 16–27. https://doi.org/10.1007/s00027-005-0801-6

Mammarella, I., Gavrylenko, G., Zdorovennova, G., Ojala, A., Erkkilä, K.-M., Zdorovennov, R., Stepanyuk, O., Palshin, N., Terzhevik, A., Vesala, T., Heiskanen, J., 2018. Effects of similar weather patterns on the thermal stratification, mixing regimes and hypolimnetic oxygen depletion in two boreal lakes with different water transparency. *Boreal. Environ. Res*. **23**, 237–247.

Mantzouki, E., Lürling, M., Fastner, J., de Senerpont Domis, L., Wilk-Wózniak, E., Seelen, L., Teurlincx, S., Verstijnen, Y., Krzton, W., Walusiak, E., Karosienė, J., Savadova, K., Vitonytė, I., Cillero-Castro, C., Budzyńska, A., Goldyn, R., Kozak, A., Rosińska, J., 2018. Temperature effects explain continental scale distribution of cyanobacterial toxins. *Toxins*. **10**, 156. https://doi.org/10.3390/toxins10040156

Mantzouki, E., Visser, P.M., Bormans, M., Ibelings, B.W., 2016. Understanding the key ecological traits of cyanobacteria as a basis for their management and control in changing lakes. *Aquat. Ecol*. **50**, 333–350. https://doi.org/10.1007/s10452-015-9526-3

Marcé, R., Armengol, J., 2010. Water quality in reservoirs under a changing climate. In *'Water Scarcity in the Mediterranean' 'The Handbook of Environmental Chemistry'* (Eds. Sabater, S., Barceló, D.). Springer, Berlin, Heidelberg, pp. 73–94. https://doi.org/10.1007/698_2009_38

Marcé, R., Rodríguez-Arias, M.À., García, J.C., Armengol, J., 2010. El Niño Southern Oscillation and climate trends impact reservoir water quality. *Glob. Change Biol*. **16**, 2857–2865. https://doi.org/10.1111/j.1365-2486.2010.02163.x

Marino, J.A., Denef, V.J., Dick, G.J., Duhaime, M.B., James, T.Y., 2022. Fungal community dynamics associated with harmful cyanobacterial blooms in two Great Lakes. *J. Gt. Lakes Res*. **48**, 1021–1031. https://doi.org/10.1016/j.jglr.2022.05.007

Markelov, I., Couture, R.-M., Fischer, R., Haande, S., Cappellen, P.V., 2019. Coupling water column and sediment biogeochemical dynamics: Modeling internal phosphorus loading, climate change responses, and mitigation measures in Lake Vansjø, Norway. *J. Geophys. Res. Biogeosci.* **124**, 3847–3866. https://doi.org/10.1029/2019JG005254

Matthews, D., Matt, M., Mueller, N., June, S., 2021. Citizen science advances the understanding of: Cyanohabs in New York State lakes. *LakeLine*. **41**, 34–38.

Mays, C., McLoughlin, S., Frank, T.D., Fielding, C.R., Slater, S.M., Vajda, V., 2021. Lethal microbial blooms delayed freshwater ecosystem recovery following the end-Permian extinction. *Nat. Commun.* **12**, 5511. https://doi.org/10.1038/s41467-021-25711-3

McCarthy, M.J., Gardner, W.S., Lehmann, M.F., Guindon, A., Bird, D.F., 2016. Benthic nitrogen regeneration, fixation, and denitrification in a temperate, eutrophic lake: Effects on the nitrogen budget and cyanobacteria blooms: Sediment N cycling in Missisquoi Bay. *Limnol. Oceanogr.* **61**, 1406–1423. https://doi.org/10.1002/lno.10306

McCarty, J.A., 2019. Algal demand drives sediment phosphorus release in a shallow eutrophic cove. *Trans. ASABE*. **62**, 1315–1324. https://doi.org/10.13031/trans.13300

McCullough, I.M., Cheruvelil, K.S., Collins, S.M., Soranno, P.A., 2019. Geographic patterns of the climate sensitivity of lakes. *Ecol. Appl.* **29**, e01836. https://doi.org/10.1002/eap.1836

McElarney, Y., Rippey, B., Miller, C., Allen, M., Unwin, A., 2021. The long-term response of lake nutrient and chlorophyll concentrations to changes in nutrient loading in Ireland's largest lake, Lough Neagh. *Biol. Environ. Proc. R. Ir. Acad.* **121**, 47–60. https://doi.org/10.1353/bae.2021.0002

Me, W., Hamilton, D.P., McBride, C.G., Abell, J.M., Hicks, B.J., 2018. Modelling hydrology and water quality in a mixed land use catchment and eutrophic lake: Effects of nutrient load reductions and climate change. *Environ. Model. Softw.* **109**, 114–133. https://doi.org/10.1016/j.envsoft.2018.08.001

Medrano, E.A., Uittenbogaard, R.E., van de Wiel, B.J.H., Dionisio Pires, L.M., Clercx, H.J.H., 2016. An alternative explanation for cyanobacterial scum formation and persistence by oxygenic photosynthesis. *Harmful Algae*. **60**, 27–35. https://doi.org/10.1016/j.hal.2016.10.002

Mehner, T., Diekmann, M., Gonsiorczyk, T., Kasprzak, P., Koschel, R., Krienitz, L., Rumpf, M., Schulz, M., Wauer, G., 2008. Rapid recovery from eutrophication of a stratified lake by disruption of internal nutrient load. *Ecosystems*. **11**, 1142–1156. https://doi.org/10.1007/s10021-008-9185-5

Meis, S., Spears, B.M., Maberly, S.C., O'Malley, M.B., Perkins, R.G., 2012. Sediment amendment with Phoslock® in Clatto Reservoir (Dundee, UK): Investigating changes in sediment elemental composition and phosphorus fractionation. *J. Env. Manag.* **93**, 185–193. https://doi.org/10.1016/j.jenvman.2011.09.015

Mesman, J.P., Ayala, A.I., Goyette, S., Kasparian, J., Marcé, R., Markensten, H., Stelzer, J.A.A., Thayne, M.W., Thomas, M.K., Pierson, D.C., Ibelings, B.W., 2022. Drivers of phytoplankton responses to summer wind events in a stratified lake: A modeling study. *Limnol. Oceanogr.* **67**, 856–873. https://doi.org/10.1002/lno.12040

Messager, M.L., Lehner, B., Cockburn, C., Lamouroux, N., Pella, H., Snelder, T., Tockner, K., Trautmann, T., Watt, C., Datry, T., 2021. Global prevalence of non-perennial rivers and streams. *Nature*. **594**, 391–397. https://doi.org/10.1038/s41586-021-03565-5

Messina, N.J., Couture, R.-M., Norton, S.A., Birkel, S.D., Amirbahman, A., 2020. Modeling response of water quality parameters to land-use and climate change in a temperate, mesotrophic lake. *Sci. Total Environ.* **713**, 136549. https://doi.org/10.1016/j.scitotenv.2020.136549

Meyer, K.A., Davis, T.W., Watson, S.B., Denef, V.J., Berry, M.A., Dick, G.J., 2017. Genome sequences of lower Great Lakes Microcystis sp. reveal strain-specific genes that are present and expressed in western Lake Erie blooms. *PLoS ONE*. **12**, e0183859. https://doi.org/10.1371/journal.pone.0183859

Meyer, M.F., Topp, S.N., King, T.V., Ladwig, R., Pilla, R.M., Dugan, H.A., Eggleston, J.R., Hampton, S.E., Leech, D.M., Oleksy, I.A., Ross, J.C., Ross, M.R.V., Woolway, R.I., Yang, X., Brousil, M.R., Fickas, K.C., Padowski, J.C., Pollard, A.I., Ren, J., Zwart, J.A., 2024. National-scale remotely sensed lake trophic state from 1984 through 2020. *Sci. Data*. **11**, 77. https://doi.org/10.1038/s41597-024-02921-0

Meyer-Jacob, C., Labaj, A.L., Paterson, A.M., Edwards, B.A., Keller, W. (Bill), Cumming, B.F., Smol, J.P., 2020. Re-browning of Sudbury (Ontario, Canada) lakes now approaches pre-acid deposition lake-water dissolved organic carbon levels. *Sci. Total Environ.* **725**, 138347. https://doi.org/10.1016/j.scitotenv.2020.138347

Meyer-Jacob, C., Michelutti, N., Paterson, A.M., Cumming, B.F., Keller, W. (Bill), Smol, J.P., 2019. The browning and re-browning of lakes: Divergent lake-water organic carbon trends linked to acid deposition and climate change. *Sci. Rep.* **9**, 16676. https://doi.org/10.1038/s41598-019-52912-0

Michalak, A.M., 2016. Study role of climate change in extreme threats to water quality. *Nat. News.* **535**, 349–350. https://doi.org/10.1038/535349a

Mishra, A.K., Singh, V.P., 2010. A review of drought concepts. *J. Hydrol.* **391**, 202–216. https://doi.org/10.1016/j.jhydrol.2010.07.012

Mobley, M., Gantzer, P., Benskin, P., Hannoun, I., McMahon, S., Austin, D., Scharf, R., 2019. Hypolimnetic oxygenation of water supply reservoirs using bubble plume diffusers. *Lake Reserv. Manag.* **35**, 247–265. https://doi.org/10.1080/10402381.2019.1628134

Molot, L.A., Depew, D.C., Zastepa, A., Arhonditsis, G.B., Watson, S.B., Verschoor, M.J., 2022. Long-term and seasonal nitrate trends illustrate potential prevention of large cyanobacterial biomass by sediment oxidation in Hamilton Harbour, Lake Ontario. *J. Gt. Lakes Res.* **48**, 971–984. https://doi.org/10.1016/j.jglr.2022.05.014

Molot, L.A., Higgins, S.N., Schiff, S.L., Venkiteswaran, J.J., Paterson, M.J., Baulch, H.M., 2021a. Phosphorus-only fertilization rapidly initiates large nitrogen-fixing cyanobacteria blooms in two oligotrophic lakes. *Environ. Res. Lett.* **16**, 064078. https://doi.org/10.1088/1748-9326/ac0564

Molot, L.A., Schiff, S.L., Venkiteswaran, J.J., Baulch, H.M., Higgins, S.N., Zastepa, A., Verschoor, M.J., Walters, D., 2021b. Low sediment redox promotes cyanobacteria blooms across a trophic range: Implications for management. *Lake Reserv. Manag.* **37**, 120–142. https://doi.org/10.1080/10402381.2020.1854400

Molot, L.A., Watson, S.B., Creed, I.F., Trick, C.G., McCabe, S.K., Verschoor, M.J., Sorichetti, R.J., Powe, C., Venkiteswaran, J.J., Schiff, S.L., 2014. A novel model for cyanobacteria bloom formation: The critical role of anoxia and ferrous iron. *Freshw. Biol.* **59**, 1323–1340. https://doi.org/10.1111/fwb.12334

Monaco, C.J., McQuaid, C.D., 2019. Climate warming reduces the reproductive advantage of a globally invasive intertidal mussel. *Biol. Invasions.* **21**, 2503–2516. https://doi.org/10.1007/s10530-019-01990-2

Monchamp, M.-E., Spaak, P., Domaizon, I., Dubois, N., Bouffard, D., Pomati, F., 2018. Homogenization of lake cyanobacterial communities over a century of climate change and eutrophication. *Nat. Ecol. Evol.* **2**, 317–324. https://doi.org/10.1038/s41559-017-0407-0

Monk, W.A., Baird, D.J., 2014. Biodiversity in Canadian lakes and rivers: Canadian biodiversity: ecosystem status and trends 2010 (Technical Thematic Report No. 19. Canadian Councils of Resource Ministers). *Technical Thematic Report No. 20*, Canadian Councils of Resource Ministers, Ottawa, ON. https://doi.org/10.5203/LWBIN.EC.2010.4

Mooij, W.M., Trolle, D., Jeppesen, E., Arhonditsis, G., Belolipetsky, P.V., Chitamwebwa, D.B.R., Degermendzhy, A.G., DeAngelis, D.L., Domis, L.N.D.S., Downing, A.S., Elliott, J.A., Fragoso, C.R., Gaedke, U., Genova, S.N., Gulati, R.D., Håkanson, L., Hamilton, D.P., Hipsey, M.R., Hoen, J. 't, Hülsmann, S., Los, F.H., Makler-Pick, V., Petzoldt, T., Prokopkin, I.G., Rinke, K., Schep, S.A., Tominaga, K., Dam, A.A.V., Nes, E.H.V., Wells, S.A., Janse, J.H., 2010. Challenges and opportunities for integrating lake ecosystem modelling approaches. *Aquat. Ecol.* **44**, 633–667. https://doi.org/10.1007/s10452-010-9339-3

Moore, B.C., Cross, B.K., Beutel, M., Dent, S., Preece, E., Swanson, M., 2012. Newman Lake restoration: A case study Part III. Hypolimnetic oxygenation. *Lake Reserv. Manage.* **28**, 311–327. https://doi.org/10.1080/07438141.2012.738463

Moore, B.C., Funk, W.H., Lafer, J., 1988. Long-term effects of dredging on phosphorus availability from liberty lake sediments. *Lake Reserv. Manag.* **4**, 293–301. https://doi.org/10.1080/07438148809354839

Moore, T.N., 2020. *Predicting In-Lake Responses to Short and Long-Term Changes Using Lake Physical Models* (PhD Thesis). Dundalk Institute of Technology, Ireland.

Mortimer, C.H., 1941. The exchange of dissolved substances between mud and water in lakes. *J. Ecol.* **29**, 280–329. https://doi.org/10.2307/2256395

Mortimer, C.H., 1987. Fifty years of physical investigations and related limnological studies on Lake Erie, 1928–1977. *J. Gt. Lakes Res.* **13**, 407–435. https://doi.org/10.1016/S0380-1330(87)71664-5

Mosley, L.M., 2015. Drought impacts on the water quality of freshwater systems; review and integration. *Earth-Sci. Rev.* **140**, 203–214. https://doi.org/10.1016/j.earscirev.2014.11.010

Moy, N.J., Dodson, J., Tassone, S.J., Bukaveckas, P.A., Bulluck, L.P., 2016. Biotransport of algal toxins to riparian food webs. *Environ. Sci. Technol.* **50**, 10007–10014. https://doi.org/10.1021/acs.est.6b02760

Mucci, M., Maliaka, V., Noyma, N.P., Marinho, M.M., Lürling, M., 2018. Assessment of possible solid-phase phosphate sorbents to mitigate eutrophication: Influence of pH and anoxia. *Sci. Total Environ.* **619**, 1431–1440. https://doi.org/0.1016/j.scitotenv.2017.11.198

Mucci, M., Noyma, N.P., de Magalhães, L., Miranda, M., van Oosterhout, F., Guedes, I.A., Huszar, V.L., Marinho, M.M., Lürling, M., 2017. Chitosan as coagulant on cyanobacteria in lake restoration management may cause rapid cell lysis. *Water Res.* https://doi.org/10.1016/j.watres.2017.04.020

Müller, R., Stadelmann, P., 2004. Fish habitat requirements as the basis for rehabilitation of eutrophic lakes by oxygenation. *Fish. Manag. Ecol.* **11**, 251–260. https://doi.org/10.1111/j.1365-2400.2004.00393.x

Münch, M.A., van Kaam, R., As, K., Peiffer, S., Heerdt, G.T., Slomp, C.P., Behrends, T., 2024. Impact of iron addition on phosphorus dynamics in sediments of a shallow peat lake 10 years after treatment. *Water Res.* **248**, 120844. https://doi.org/10.1016/j.watres.2023.120844

Munger, Z.W., Carey, C.C., Gerling, A.B., Doubek, J.P., Hamre, K.D., McClure, R.P., Schreiber, M.E., 2019. Oxygenation and hydrologic controls on iron and manganese mass budgets in a drinking-water reservoir. *Lake Reserv. Manag.* **35**, 277–291. https://doi.org/10.1080/10402381.2018.1545811

Murphy, T.P., Hall, K.G., Northcote, T.G., 1988. Lime treatment of a hardwater lake to reduce eutrophication. *Lake Reserv. Manage.* **4**, 51–62. https://doi.org/10.1080/07438148809354813

Myers, B.J.E., Lynch, A.J., Bunnell, D.B., Chu, C., Falke, J.A., Kovach, R.P., Krabbenhoft, T.J., Kwak, T.J., Paukert, C.P., 2017. Global synthesis of the documented and projected effects of climate change on inland fishes. *Rev. Fish Biol. Fish.* **27**, 339–361. https://doi.org/10.1007/s11160-017-9476-z

Myrstener, E., Ninnes, S., Meyer-Jacob, C., Mighall, T., Bindler, R., 2021. Long-term development and trajectories of inferred lake-water organic carbon and pH in naturally acidic boreal lakes. *Limnol. Oceanogr.* **66**, 2408–2422. https://doi.org/10.1002/lno.11761

Nagul, E.A., McKelvie, I.D., Worsfold, P., Kolev, S.D., 2015. The molybdenum blue reaction for the determination of orthophosphate revisited: Opening the black box. *Anal. Chim. Acta* **890**, 60–82. https://doi.org/10.1016/j.aca.2015.07.030

Nalley, J.O., O'Donnell, D.R., Litchman, E., 2018. Temperature effects on growth rates and fatty acid content in freshwater algae and cyanobacteria. *Algal Res.* **35**, 500–507. https://doi.org/10.1016/j.algal.2018.09.018

Naselli-Flores, L., 2003. Man-made lakes in Mediterranean semi-arid climate: The strange case of Dr Deep Lake and Mr Shallow Lake. *Hydrobiologia.* **506–509**, 13–21. https://doi.org/10.1023/B:HYDR.0000008550.34409.06

Naymik, J., Larsen, C.A., Myers, R., Hoovestol, C., Gastelecutto, N., Bates, D., 2023. Long-term trends in inflowing chlorophyll *a* and nutrients and their relation to dissolved oxygen in a large western reservoir. *Lake Reserv. Manag.* 1–19. https://doi.org/10.1080/10402381.2022.2160395

Nelligan, C., Jeziorski, A., Rühland, K.M., Paterson, A.M., Meyer-Jacob, C., Smol, J.P., 2019a. A multi-basin comparison of historical water quality trends in Lake Manitou, Ontario, a provincially significant lake trout lake. *Lake Reserv. Manag.* 1–17. https://doi.org/10.1080/10402381.2019.1659889

Nelligan, C., Jeziorski, A., Rühland, K.M., Paterson, A.M., Smol, J.P., 2019b. Long-term trends in hypolimnetic volumes and dissolved oxygen concentrations in Boreal Shield lakes of south-central Ontario, Canada. *Can. J. Fish. Aquat. Sci.* 1–11. https://doi.org/10.1139/cjfas-2018-0278

Neumann, A., Blukacz-Richards, E.A., Saha, R., Alberto Arnillas, C., Arhonditsis, G.B., 2023. A Bayesian hierarchical spatially explicit modelling framework to examine phosphorus export between contrasting flow regimes. *J. Gt. Lakes Res.* **49**, 190–208. https://doi.org/10.1016/j.jglr.2022.10.003

Newell, S.E., Davis, T.W., Johengen, T.H., Gossiaux, D., Burtner, A., Palladino, D., McCarthy, M.J., 2019. Reduced forms of nitrogen are a driver of non-nitrogen-fixing harmful cyanobacterial blooms and toxicity in Lake Erie. *Harmful Algae.* **81**, 86–93. https://doi.org/10.1016/j.hal.2018.11.003

Ni, Z., Huang, D., Li, Y., Liu, X., Wang, S., 2022. Novel insights into molecular composition of organic phosphorus in lake sediments. *Water Res.* **214**, 118197. https://doi.org/10.1016/j.watres.2022.118197

Nielson, J.R., Henderson, S.M., 2022. Bottom boundary layer mixing processes across internal seiche cycles: Dominance of downslope flows. *Limnol. Oceanogr.* **67**, 1111–1125. https://doi.org/10.1002/lno.12060

Niemistö, J., Holmroos, H., Pekcan-Hekim, Z., Horppila, J., 2008. Interactions between sediment resuspension and sediment quality decrease the TN: TP ratio in a shallow lake. *Limnol. Oceanogr.* **53**, 2407–2415. https://doi.org/10.4319/lo.2008.53.6.2407

Niemistö, J., Köngäs, P., Härkönen, L., Horppila, J., 2016. Hypolimnetic aeration intensifies phosphorus recycling and increases organic material sedimentation in a stratifying lake: Effects through increased temperature and turbulence. *Boreal Environ. Res.* **21**, 571–587.

Niemistö, J., Tamminen, P., Ekholm, P., Horppila, J., 2012. Sediment resuspension: Rescue or downfall of a thermally stratified eutrophic lake? *Hydrobiologia.* **686**, 267–276. https://doi.org/10.1007/s10750-012-1021-4

Nishizaki, M., Ackerman, J.D., 2017. Mussels blow rings: Jet behavior affects local mixing. *Limnol. Oceanogr.* **62**, 125–136. https://doi.org/10.1002/lno.10380

Nogaro, G., et al., 2020. Alum treatment did not improve water quality in hypereutrophic Grand Lake St. Mary's, Ohio. In *Internal Phosphorus Loading: Causes, Case Studies, and Management* (Eds. Steinman, A.D., Spears, B.M.). J. Ross Publishing, Plantation, FL, pp. 147–168.

Nõges, P., Nõges, T., Laas, A., 2010. Climate-related changes of phytoplankton seasonality in large shallow Lake Võrtsjärv, Estonia. *Aquat. Ecosyst. Health Manag.* **13**, 154–163. https://doi.org/10.1080/14634981003788953

Nõges, T., Janatian, N., Laugaste, R., Nõges, P., 2020. Post-soviet changes in nitrogen and phosphorus stoichiometry in two large non-stratified lakes and the impact on phytoplankton. *Glob. Ecol. Conserv.* **24**, e01369. https://doi.org/10.1016/j.gecco.2020.e01369

Noonan, T.A., 1986. Water quality in Long Lake, Minnesota, following riplox sediment treatment. *Lake Reserv. Manag.* **2**, 131–137. https://doi.org/10.1080/07438148609354615

Noordhuis, R., Zuidam, B.G. van, Peeters, E.T.H.M., Geest, G.J. van, 2016. Further improvements in water quality of the Dutch Borderlakes: Two types of clear states at different nutrient levels. *Aquat. Ecol.* **50**, 521–539. https://doi.org/10.1007/s10452-015-9521-8

North, R.L., Guildford, S.J., Smith, R.E.H., Havens, S.M., Twiss, M.R., 2007. Evidence for phosphorus, nitrogen, and iron colimitation of phytoplankton communities in Lake Erie. *Limnol. Ocean.* **52**, 315–328. https://doi.org/10.4319/lo.2007.52.1.0315

North, R.P., North, R.L., Livingstone, D.M., Köster, O., Kipfer, R., 2014. Long-term changes in hypoxia and soluble reactive phosphorus in the hypolimnion of a large temperate lake: Consequences of a climate regime shift. *Glob. Change Biol.* **20**, 811–823. https://doi.org/10.1111/gcb.12371

Nürnberg, G.K., 1984a. *The Availability of Phosphorus from Anoxic Hypolimnia to Epilimnetic Plankton* (PhD Thesis). McGill University, Montreal, Quebec.

Nürnberg, G.K., 1984b. Iron and hydrogen sulfide interference in the analysis of soluble reactive phosphorus in anoxic waters. *Water Res.* **18**, 369–377. https://doi.org/10.1016/0043-1354(84)90114-3

Nürnberg, G.K., 1984c. The prediction of internal phosphorus load in lakes with anoxic hypolimnia. *Limnol. Oceanogr.* **29**, 111–124. https://doi.org/10.4319/lo.1984.29.1.0111

Nürnberg, G.K., 1985. Availability of phosphorus upwelling from iron-rich anoxic hypolimnia. *Arch. Hydrobiol.* **104**, 459–476.

Nürnberg, G.K., 1987. Hypolimnetic withdrawal as a lake restoration technique. *Soc. Civ. Eng. Am. J. Environ. Eng.* **113**, 1006–1017. https://doi.org/10.1061/(ASCE)0733-9372(1987)113:5(1006)

Nürnberg, G.K., 1988a. A simple model for predicting the date of fall turnover in thermally stratified lakes. *Limnol. Oceanogr.* **33**, 1190–1195. https://doi.org/10.4319/lo.1988.33.5.1190

Nürnberg, G.K., 1988b. Prediction of phosphorus release rates from total and reductant-soluble phosphorus in anoxic lake sediments. *Can. J. Fish. Aquat. Sci.* **45**, 453–462. https://doi.org/10.1139/f88-054

Nürnberg, G.K., 1991. Phosphorus from internal sources in the Laurentian Great Lakes, and the concept of threshold external load. *J. Gt. Lakes Res.* **17**, 132–140. https://doi.org/10.1016/S0380-1330(91)71348-8

Nürnberg, G.K., 1994. Phosphorus release from anoxic sediments: What we know and how we can deal with it. *Limnetica*. **10**, 1–4.

Nürnberg, G.K., 1995a. Quantifying anoxia in lakes. *Limnol. Oceanogr*. **40**, 1100–1111. https://doi.org/10.4319/lo.1995.40.6.1100

Nürnberg, G.K., 1995b. The anoxic factor, a quantitative measure of anoxia and fish species richness in Central Ontario Lakes. *Trans. Am. Fish. Soc.* **124**, 677–686. https://doi.org/10.1577/1548-8659(1995)124<0677:TAFAQM>2.3.CO;2

Nürnberg, G.K., 1996. Trophic state of clear and colored, soft- and hardwater lakes with special consideration of nutrients, anoxia, phytoplankton and fish. *Lake Reserv. Manag.* **12**, 432–447. https://doi.org/10.1080/07438149609354283

Nürnberg, G.K., 1997. Coping with water quality problems due to hypolimnetic anoxia in Central Ontario Lakes. *Water Qual. Res. J. Can.* **32**, 391–405. https://doi.org/10.2166/wqrj.1997.025

Nürnberg, G.K., 1998. Prediction of annual and seasonal phosphorus concentrations in stratified and polymictic lakes. *Limnol. Oceanogr*. **43**, 1544–1552. https://doi.org/10.4319/lo.1998.43.7.1544

Nürnberg, G.K., 1999. Determining trophic state in experimental lakes. *Limnol. Oceanogr*. **44**, 1176–1179. https://doi.org/10.4319/lo.1999.44.4.1176

Nürnberg, G.K., 2001. Eutrophication and trophic state – why does lake water (quality) differ from lake to lake? *LakeLine*. **21**(1), 29–33.

Nürnberg, G.K., 2002. Quantification of oxygen depletion in lakes and reservoirs with the hypoxic factor. *Lake Reserv. Manag.* **18**, 298–305. https://doi.org/10.1080/07438140209353936

Nürnberg, G.K., 2004. Quantified hypoxia and anoxia in lakes and reservoirs. *TheScientificWorld*. **4**, 42–54. https://doi.org/10.1100/tsw.2004.5

Nürnberg, G.K., 2005. Quantification of internal phosphorus loading in polymictic lakes. *Verh Intern. Ver. Limnol*. **29**, 623–626. https://doi.org/10.1080/03680770.2005.11902753

Nürnberg, G.K., 2007a. Low-Nitrate-Days (LND), a potential indicator of cyanobacteria blooms in a eutrophic hardwater reservoir. *Water Qual. Res. J. Can.* **42**(4), 269–283. https://doi.org/10.2166/wqrj.2007.029

Nürnberg, G.K., 2007b. Lake responses to long-term hypolimnetic withdrawal treatments. *Lake Reserv. Manag.* **23**, 388–409. https://doi.org/10.1080/07438140709354026

Nürnberg, G.K., 2009. Assessing internal phosphorus load – problems to be solved. *Lake Reserv. Manag.* **25**, 419–432. https://doi.org/10.1080/00357520903458848

Nürnberg, G.K., 2017. Attempted management of cyanobacteria by Phoslock (lanthanum-modified clay) in Canadian lakes, water quality results and predictions. *Lake Reserv. Manag.* **33**, 163–170. https://doi.org/10.1080/10402381.2016.1265618

Nürnberg, G.K., 2019. Quantification of anoxia and hypoxia in water bodies. In *Encyclopedia of Water: Science, Technology, and Society* (Ed. Maurice, P.A.), 2nd ed. John Wiley & Sons, Inc. https://doi.org/10.1002/9781119300762.wsts0081

Nürnberg, G.K., 2020a. Chapter 3: Internal phosphorus loading models: A critical review. In *Internal Phosphorus Loading: Causes, Case Studies, and Management* (Eds. Steinman, A.D., Spears, B.M.). J. Ross Publishing, Plantation, FL, pp. 45–62.

Nürnberg, G.K., 2020b. Hypolimnetic withdrawal as a lake restoration technique: Determination of feasibility and continued benefits. *Hydrobiologia*. **847**, 4487–4501. https://doi.org/10.1007/s10750-019-04094-z

Nürnberg, G.K., 2020c. Chapter 6: Observed and modelled internal phosphorus loads in stratified and polymictic basins of a mesotrophic lake. In *Internal Phosphorus Loading: Causes, Case Studies, and Management* (Eds. Steinman, A.D., Spears, B.M.). J. Ross Publishing, Plantation, FL, pp. 111–123.

Nürnberg, G.K., Dillon, P.J., 1993. Iron budgets in temperate lakes. *Can. J. Fish. Aquat. Sci.* **50**, 1728–1737. https://doi.org/doi/10.1139/f93-194

Nürnberg, G.K., Fischer, R., Paterson, A.M., 2018. Reduced phosphorus retention by anoxic bottom sediments after the remediation of an industrial acidified lake area: Indications from P, Al, and Fe sediment fractions. *Sci. Total Environ.* **626**, 412–422. https://doi.org/10.1016/j.scitotenv.2018.01.103

Nürnberg, G.K., Hartley, R., Davis, E., 1987. Hypolimnetic withdrawal in two North American lakes with anoxic phosphorus release from the sediment. *Water Res.* **21**, 923–928. https://doi.org/10.1016/S0043-1354(87)80009-X

Nürnberg, G.K., Howell, T., Palmer, M., 2019. Long-term impact of Central Basin hypoxia and internal phosphorus loading on north shore water quality in Lake Erie. *Inland Waters.* **9**, 362–373. https://doi.org/10.1080/20442041.2019.1568072

Nürnberg, G.K., LaZerte, B.D., 2001. Predicting lake water quality. In *Managing Lakes and Reservoirs* (Eds. Holdren, C., Jones, W., Taggart, J.). North American Lake Management Society, Terrene Institute in cooperation with Office Water Assessment Watershed Protection Division U.S. Environ. Prot. Agency, Madison, WI, pp. 139–163.

Nürnberg, G.K., LaZerte, B.D., 2004. Modeling the effect of development on internal phosphorus load in nutrient-poor lakes. *Water Resour. Res.* **40**, W01105. https://doi.org/10.1029/2003WR002410

Nürnberg, G.K., LaZerte, B.D., 2015. Influence of upwelling and onshore circulation of low oxygen water on the coastal water quality of the north shore of Lake Erie's Central Basin. *Report for the Ontario Ministry of the Environment and Climate Change*. Freshwater Research, Baysville, Ontario.

Nürnberg, G.K., LaZerte, B.D., 2016a. Trophic state decrease after lanthanum-modified bentonite (Phoslock) application to a hyper-eutrophic polymictic urban lake frequented by Canada geese (*Branta canadensis*). *Lake Reserv. Manag.* **32**, 74–88. https://doi.org/10.1080/10402381.2015.1133739

Nürnberg, G.K., LaZerte, B.D., 2016b. More than 20 years of estimated internal phosphorus loading in polymictic, eutrophic Lake Winnipeg, Manitoba. *J. Gt. Lakes Res.* **42**, 18–27. https://doi.org/10.1016/j.jglr.2015.11.003

Nürnberg, G.K., LaZerte, B.D., Loh, P.S., Molot, L.A., 2013a. Quantification of internal phosphorus load in large, partially polymictic and mesotrophic Lake Simcoe, Ontario. *J. Gt. Lakes Res.* **39**, 271–279. https://doi.org/10.1016/j.jglr.2013.03.017

Nürnberg, G.K., LaZerte, B.D., Olding, D.D., 2003. An artificially induced *Planktothrix rubescens* surface bloom in a small kettle lake in southern Ontario compared to blooms world-wide. *Lake Reserv. Manag.* **19**, 307–322. https://doi.org/10.1080/07438140309353941

Nürnberg, G.K., Molot, L.A., O'Connor, E., Jarjanazi, H., Winter, J.G., Young, J.D., 2013b. Evidence for internal phosphorus loading, hypoxia and effects on phytoplankton in partially polymictic Lake Simcoe, Ontario. *J. Gt. Lakes Res.* **39**, 259–270. https://doi.org/10.1016/j.jglr.2013.03.016

Nürnberg, G.K., Peters, R.H., 1984a. Biological availability of soluble reactive phosphorus in anoxic and oxic freshwater. *Can. J. Fish. Aquat. Sci.* **41**, 757–765. https://doi.org/10.1139/f84-088

Nürnberg, G.K., Peters, R.H., 1984b. The importance of internal phosphorus load to the eutrophication of lakes with anoxic hypolimnia. *Verh. Intern. Ver. Limnol.* **22**, 190–194. https://doi.org/10.1080/03680770.1983.11897287

Nürnberg, G.K., Shaw, M., 1999. Productivity of clear and humic lakes: Nutrients, phytoplankton, bacteria. *Hydrobiologia.* **382**, 97–112. https://doi.org/10.1023/A:1003445406964

Nürnberg, G.K., Shaw, M., Dillon, P.J., McQueen, D.J., 1986. Internal phosphorus load in an oligotrophic Precambrian Shield lake with an anoxic hypolimnion. *Can. J. Fish. Aquat. Sci.* **43**, 574–580. https://doi.org/10.1139/f86-068

Nürnberg, G.K., Tarvainen, M., Ventelä, A.-M., Sarvala, J., 2012. Internal phosphorus load estimation during biomanipulation in a large polymictic and mesotrophic lake. *Inland Waters.* **2**, 147–162. https://doi.org/10.5268/IW-2.3.469

Ohle, W., 1935. Organische Kolloide in ihrer Wirkung auf den Stoffhaushalt der Gewasser. *Naturwissenschaften.* **23**, 480–484.

Oliver, R.L., Hamilton, D.P., Brookes, J.D., Ganf, G.G., 2012. Physiology, blooms and prediction of planktonic cyanobacteria. In *Ecology of Cyanobacteria II*. Springer, Dordrecht, pp. 155–194.

Oliver, S.K., Collins, S.M., Soranno, P.A., Wagner, T., Stanley, E.H., Jones, J.R., Stow, C.A., Lottig, N.R., 2017. Unexpected stasis in a changing world: Lake nutrient and chlorophyll trends since 1990. *Glob. Change Biol.* **23**, 5455–5467.

O'Neil, J.M., Davis, T.W., Burford, M.A., Gobler, C.J., 2012. The rise of harmful cyanobacteria blooms: The potential roles of eutrophication and climate change. *Harmful Algae*. **14**, 313–334. https://doi.org/10.1016/j.hal.2011.10.027

Ontario Ministry of Environment, 2009. Phoslock toxicity testing with three sediment dwelling organisms (*Hyalella azteca, Hexagenia spp.* and *Chironomus dilutus*) and two water column dwelling organisms (Rainbow Trout and *Daphnia magna*). *Technical Memorandum*. Ontario Ministry of the Environment, Toronto.

O'Reilly, C.M., Sharma, S., Gray, D.K., Hampton, S.E., Read, J.S., Rowley, R.J., Schneider, P., Lenters, J.D., McIntyre, P.B., Kraemer, B.M., Weyhenmeyer, G.A., Straile, D., Dong, B., Adrian, R., Allan, M.G., Anneville, O., Arvola, L., Austin, J., Bailey, J.L., Baron, J.S., Brookes, J.D., de Eyto, E., Dokulil, M.T., Hamilton, D.P., Havens, K., Hetherington, A.L., Higgins, S.N., Hook, S., Izmest'eva, L.R., Joehnk, K.D., Kangur, K., Kasprzak, P., Kumagai, M., Kuusisto, E., Leshkevich, G., Livingstone, D.M., MacIntyre, S., May, L., Melack, J.M., Mueller-Navarra, D.C., Naumenko, M., Noges, P., Noges, T., North, R.P., Plisnier, P.-D., Rigosi, A., Rimmer, A., Rogora, M., Rudstam, L.G., Rusak, J.A., Salmaso, N., Samal, N.R., Schindler, D.E., Schladow, S.G., Schmid, M., Schmidt, S.R., Silow, E., Soylu, M.E., Teubner, K., Verburg, P., Voutilainen, A., Watkinson, A., Williamson, C.E., Zhang, G., 2015. Rapid and highly variable warming of lake surface waters around the globe. *Geophys. Res. Lett.* **2015**, GL066235. https://doi.org/10.1002/2015GL066235

Oreskes, N., 2004. Science and public policy: What's proof got to do with it? *Environ. Sci. Policy*. **7**, 369–383. https://doi.org/10.1016/j.envsci.2004.06.002

Orihel, D.M., Baulch, H.M., Casson, N.J., North, R.L., Parsons, C.T., Seckar, D.C.M., Venkiteswaran, J.J., 2017. Internal phosphorus loading in Canadian fresh waters: A critical review and data analysis. *Can. J. Fish. Aquat. Sci.* 1–25. https://doi.org/10.1139/cjfas-2016-0500

Ortiz-Cañavate, B.K., Wolinska, J., Agha, R., 2019. Fungicides at environmentally relevant concentrations can promote the proliferation of toxic bloom-forming cyanobacteria by inhibiting natural fungal parasite epidemics. *Chemosphere*. **229**, 18–21. https://doi.org/10.1016/j.chemosphere.2019.04.203

Osburn, F.S., Wagner, N.D., Taylor, R.B., Chambliss, C.K., Brooks, B.W., Scott, J.T., 2023. The effects of salinity and N: P on N-rich toxins by both an N-fixing and non-N-fixing cyanobacteria. *Limnol. Oceanogr. Lett.* **n/a**, 162–172. https://doi.org/10.1002/lol2.10234

Osgood, R.A., 1988. Lake mixis and internal phosphorus dynamics. *Arch. Hydrobiol.* **113**, 629–638.

Osgood, R.A., 2017. Inadequacy of best management practices for restoring eutrophic lakes in the United States: Guidance for policy and practice. *Inland Waters*. **7**, 401–407. https://doi.org/10.1080/20442041.2017.1368881

Ostrofsky, M.L., Marbach, R.M., 2019. Predicting internal phosphorus loading in stratified lakes. *Aquat. Sci.* **81**. https://doi.org/10.1007/s00027-018-0618-8

Ostrovsky, I., Yacobi, Y.Z., Walline, P., Kalikhman, I., 1996. Seiche-induced mixing: Its impact on lake productivity. *Limnol. Ocean.* **41**, 323–332. https://doi.org/10.4319/lo.1996.41.2.0323

Özkan, K., Jeppesen, E., Davidson, T.A., Bjerring, R., Johansson, L.S., Søndergaard, M., Lauridsen, T.L., Svenning, J.-C., 2016. Long-term trends and temporal synchrony in plankton richness, diversity and biomass driven by re-oligotrophication and climate across 17 Danish lakes. *Water*. **8**, 427. https://doi.org/10.3390/w8100427

Padisák, J., 1998. Sudden and gradual responses of phytoplankton to global climate change: Case studies from two large, shallow lakes (Balaton, Hungary; Neusiedlersee Austria/Hungary). In 'In: Management of lakes and reservoirs during global change. NATO advanced research workshop on management of lakes and reservoirs during global climate change, 1995. November 11–15., Prága, Czech Republic. In *Management of Lakes and Reservoirs during Global Change. Proceedings of the NATO Advanced Research Workshop, Prague* (Eds. George, D.G., Jones, J.G., Puncochar, P., Reynolds, C.S.). Kluwer, Dordrecht, pp. 111–125.

Padisák, J., Crossetti, L.O., Naselli-Flores, L., 2009. Use and misuse in the application of the phytoplankton functional classification: A critical review with updates. *Hydrobiologia*. **621**, 1–19. https://doi.org/10.1007/s10750-008-9645-0

Paerl, H.W., 2016. Impacts of climate change on cyanobacteria in aquatic environments. In *Climate Change and Microbial Ecology: Current Research and Future Trends*. Justus Liebig University, Giessen, Germany, Caister Academic Press, pp. 5–22.

Paerl, H.W., 2017. Controlling cyanobacterial harmful blooms in freshwater ecosystems. *Microb. Biotechnol.* **10**, 1106–1110. https://doi.org/10.1111/1751-7915.12725

Paerl, H.W., 2018. Mitigating toxic planktonic cyanobacterial blooms in aquatic ecosystems facing increasing anthropogenic and climatic pressures. *Toxins*. **10**, 76. https://doi.org/10.3390/toxins10020076

Paerl, H.W., Havens, K.E., Xu, H., Zhu, G., McCarthy, M.J., Newell, S.E., Scott, J.T., Hall, N.S., Otten, T.G., Qin, B., 2020. Mitigating eutrophication and toxic cyanobacterial blooms in large lakes: The evolution of a dual nutrient (N and P) reduction paradigm. *Hydrobiologia*. **847**, 4359–4375. https://doi.org/10.1007/s10750-019-04087-y

Paerl, H.W., Huisman, J., 2008. Blooms like it hot. *Science*. **320**, 57–58. https://doi.org/10.1126/science.1155398

Paerl, H.W., Huisman, J., 2009. Climate change: A catalyst for global expansion of harmful cyanobacterial blooms. *Environ. Microbiol. Rep.* **1**, 27–37. https://doi.org/10.1111/j.1758-2229.2008.00004.x

Paerl, H.W., Otten, T.G., 2013. Harmful cyanobacterial blooms: Causes, consequences, and controls. *Microb. Ecol.* **65**, 995–1010. https://doi.org/10.1007/s00248-012-0159-y

Paerl, H.W., Otten, T.G., Kudela, R., 2018. Mitigating the expansion of harmful algal blooms across the freshwater-to-marine continuum. *Environ. Sci. Technol.* **52**, 5519–5529. https://doi.org/10.1021/acs.est.7b05950

Paerl, H.W., Paul, V.J., 2012. Climate change: Links to global expansion of harmful cyanobacteria. *Water Res.* **46**, 1349–1363. https://doi.org/10.1016/j.watres.2011.08.002

Palmer, M.E., Yan, N.D., Somers, K.M., 2014. Climate change drives coherent trends in physics and oxygen content in North American lakes. *Clim. Change*. **124**, 285–299.

Paludan, C., Jensen, H.S., 1995. Sequential extraction of phosphorus in freshwater wetland and lake sediment, significance of humic acids. *Wetlands*. **15**, 365–373. https://doi.org/10.1007/BF03160891

Panksep, K., Tamm, M., Mantzouki, E., Rantala-Ylinen, A., Laugaste, R., Sivonen, K., Tammeorg, O., Kisand, V., 2020. Using microcystin gene copies to determine potentially-toxic blooms, example from a shallow eutrophic Lake Peipsi. *Toxins*. 23. https://doi.org/10.3390/toxins12040211

Paterson, A.M., Rühland, K.M., Anstey, C.V., Smol, J.P., 2017. Climate as a driver of increasing algal production in Lake of the Woods, Ontario, Canada. *Lake Reserv. Manag.* **33**, 403–414. https://doi.org/10.1080/10402381.2017.1379574

Paterson, M.J., Podemski, C.L., Wesson, L.J., Dupuis, A.P., 2011. The effects of an experimental freshwater cage aquaculture operation on Mysis diluviana. *J. Plankton. Res.* **33**, 25–36. https://doi.org/10.1093/plankt/fbq096

Patiño, R., Christensen, V.G., Graham, J.L., Rogosch, J.S., Rosen, B.H., 2023. Toxic algae in inland waters of the conterminous United States – a review and synthesis. *Water*. **15**, 2808. https://doi.org/10.3390/w15152808

Patrick, R., Crum, B., Coles, J., 1969. Temperature and manganese as determining factors in the presence of diatom or blue-green algal floras in streams. *Proc. Natl. Acad. Sci.* **64**, 472–478. https://doi.org/10.1073/pnas.64.2.47

Penn, M.R., Auer, M.T., Doerr, S.M., Driscoll, C.T., Brooks, C.M., Effler, S.W., 2000. Seasonality in phosphorus release rates from the sediments of a hypereutrophic lake under a matrix of pH and redox conditions. *Can. J. Fish. Aquat. Sci.* **57**, 1033–1041. https://doi.org/10.1139/f00-035

Perakis, S.S., Welch, E.B., Jacoby, J.M., 1996. Sediment-to-water blue-green algal recruitment in response to alum and environmental factors. *Hydrobiologia*. **318**, 165–177. https://doi.org/10.1007/BF00016678

Pereira, A.C., Mulligan, C.N., 2023. Practices for eutrophic shallow lake water remediation and restoration: A critical literature review. *Water*. **15**, 2270. https://doi.org/10.3390/w15122270

Perga, M.-E., Minaudo, C., Doda, T., Arthaud, F., Beria, H., Chmiel, H.E., Escoffier, N., Lambert, T., Napolleoni, R., Obrador, B., Perolo, P., Rüegg, J., Ulloa, H., Bouffard, D., 2023. Near-bed stratification controls bottom hypoxia in ice-covered alpine lakes. *Limnol. Oceanogr.* **68**, 1232–1246. https://doi.org/10.1002/lno.12341

Peters, R.H., 1986. The role of prediction in limnology. *Limnol. Oceanogr*. **31**, 1143–1159. https://doi.org/10.4319/lo.1986.31.5.1143

Pettersson, K., Herlitz, E., Istvanovics, V., 1993. The role of Gloetrichia echinulata in the transfer of phosphorus from sediments to water in lake Erken. *Hydrobiologia*. **253**, 123–129. https://doi.org/10.1007/978-94-011-1598-8_15

Petty, E.L., Obrecht, D.V., North, R.L., 2020. Filling in the flyover zone: High phosphorus in midwestern (USA) reservoirs results in high phytoplankton biomass but not high primary productivity. *Front. Environ. Sci*. **8**. https://doi.org/10.3389/fenvs.2020.00111

Picardo, M., Filatova, D., Nuñez, O., Farré, M., 2019. Recent advances in the detection of natural toxins in freshwater environments. *TrAC Trends Anal. Chem*. **112**, 75–86. https://doi.org/10.1016/j.trac.2018.12.017

Piccolroaz, S., Zhu, S., Ladwig, R., Carrea, L., Oliver, S., Piotrowski, A.P., Ptak, M., Shinohara, R., Sojka, M., Woolway, R.I., Zhu, D.Z., 2024. Lake water temperature modeling in an era of climate change: Data sources, models, and future prospects. *Rev. Geophys*. **62**, e2023RG000816. https://doi.org/10.1029/2023RG000816

Pick, F.R., 2016. Blooming algae: A Canadian perspective on the rise of toxic cyanobacteria. *Can. J. Fish. Aquat. Sci*. **73**, 1149–1158. https://doi.org/10.1139/cjfas-2015-0470

Pick, F.R., Lean, D.R.S., 1987. The role of macronutrients (C, N, P) in controlling cyanobacterial dominence in temperate lakes. *N Z J Mar. Freshw. Res*. **21**, 425–434. https://doi.org/10.1080/00288330.1987.9516238

Pilla, R.M., Williamson, C.E., 2022. Earlier ice breakup induces changepoint responses in duration and variability of spring mixing and summer stratification in dimictic lakes. *Limnol. Oceanogr*. **67**, S173–S183. https://doi.org/10.1002/lno.11888

Pilla, R.M., Williamson, C.E., Adamovich, B.V., Adrian, R., Anneville, O., Chandra, S., Colom-Montero, W., Devlin, S.P., Dix, M.A., Dokulil, M.T., 2020. Deeper waters are changing less consistently than surface waters in a global analysis of 102 lakes. *Sci. Rep*. **10**, 1–15. https://doi.org/10.1038/s41598-020-76873-x

Pilla, R.M., Williamson, C.E., Zhang, J., Smyth, R.L., Lenters, J.D., Brentrup, J.A., Knoll, L.B., Fisher, T.J., 2018. Browning-related decreases in water transparency lead to long-term increases in surface water temperature and thermal stratification in two small lakes. *J. Geophys. Res. Biogeosci*. **123**, 1651–1665. https://doi.org/10.1029/2017JG004321

Pilla, R.M., et al., 2021. Global data set of long-term data descriptor summertime vertical temperature profiles in 153 lakes. *Sci. Data*. 13. https://doi.org/doi.org/10.1038/s41597-021-00983-y

Pilon, S., Zastepa, A., Taranu, Z.E., Gregory-Eaves, I., Racine, M., Blais, J.M., Poulain, A.J., Pick, F.R., 2019. Contrasting histories of microcystin-producing cyanobacteria in two temperate lakes as inferred from quantitative sediment DNA analyses. *Lake Reserv. Manag*. **35**, 102–117. https://doi.org/10.1080/10402381.2018.1549625

Poirier-Larabie, S., Hudon, C., Richard, H.-P.P., Gagnon, C., 2020. Cyanotoxin release from the benthic, mat-forming cyanobacterium Microseira (Lyngbya) wollei in the St. Lawrence River, Canada. *Env. Sci. Pollut. Res*. **11**. https://doi.org/10.1007/s11356-020-09290-2

Poléo, A.B.S., 1995. Aluminium polymerization – a mechanism of acute toxicity of aqueous aluminium to fish. *Aquat. Toxicol*. **31**, 347–356. https://doi.org/10.1016/0166-445X(94)00083-3

Popper, K.R., 1959. *The Logic of Scientific Discove*. Harper Books, New York.

Preece, E.P., Moore, B.C., Skinner, M.M., Child, A., Dent, S., 2019. A review of the biological and chemical effects of hypolimnetic oxygenation. *Lake Reserv. Manag*. **35**, 229–246. https://doi.org/10.1080/10402381.2019.1580325

Prepas, E.E., Babin, J., Murphy, T.P., Chambers, P.A., Sandland, G., Ghadouani, A., Serediak, M., 2001. Long-term effects of successsive Ca(OH)2 and CaCO3 treatments on the water quality of two eutrophic hardwater lakes. *Freshw. Biol*. **46**, 1089–1103. https://doi.org/10.1046/j.1365-2427.2001.00792.x

Psenner, R., Bostrom, B., Dinka, M, Pettersson, K., Pucsko, R., Sager, M., 1988. Fractionation of phosphorus in suspended matter and sediment. *Ergeb. Limnol*. **30**, 98–110.

Psenner, R., Pucsko, R., 1988. Phosphorus fractionation: Advantages and limits of the method for the study of sediment P origins and interactions. *Arch. Hydrobiol.* **30**, 43–59.

Qin, B., Paerl, H.W., Brookes, J.D., Liu, J., Jeppesen, E., Zhu, G., Zhang, Y., Xu, H., Shi, K., Deng, J., 2019. Why Lake Taihu continues to be plagued with cyanobacterial blooms through 10 years (2007–2017) efforts. *Sci. Bull.* **64**, 354–356. https://doi.org/10.1016/j.scib.2019.02.008

Qin, B., Zhang, Y., Zhu, G., Gong, Z., Deng, J., Hamilton, D.P., Gao, G., Shi, K., Zhou, J., Shao, K., Zhu, M., Zhou, Y., Tang, X., Li, L., 2020a. Are nitrogen-to-phosphorus ratios of Chinese lakes actually increasing? *Proc. Natl. Acad. Sci.* **117**, 21000–21002. https://doi.org/10.1073/pnas.2013445117

Qin, B., Zhou, J., Elser, J.J., Gardner, W.S., Deng, J., Brookes, J.D., 2020b. Water depth underpins the relative roles and fates of nitrogen and phosphorus in lakes. *Environ. Sci. Technol.* **54**, 3191–3198. https://doi.org/10.1021/acs.est.9b05858

Quinlan, R., Filazzola, A., Mahdiyan, O., Shuvo, A., Blagrave, K., Ewins, C., Moslenko, L., Gray, D.K., O'Reilly, C.M., Sharma, S., 2021. Relationships of total phosphorus and chlorophyll in lakes worldwide. *Limnol. Oceanogr.* **66**, 392–404. https://doi.org/10.1002/lno.11611

Rabalais, N.N., Turner, R.E., 2019. Gulf of Mexico hypoxia: Past, present, and future. *Limnol. Oceanogr. Bull.* https://doi.org/10.1002/lob.10351

Raberg, J.H., Harning, D.J., Crump, S.E., de Wet, G., Blumm, A., Kopf, S., Geirsdóttir, Á., Miller, G.H., Sepúlveda, J., 2021. Revised fractional abundances and warm-season temperatures substantially improve brGDGT calibrations in lake sediments. *Biogeosciences.* **18**, 3579–3603. https://doi.org/10.5194/bg-18-3579-2021

Radbourne, A.D., Elliott, J.A., Maberly, S.C., Ryves, D.B., Anderson, N.J., 2019. The impacts of changing nutrient load and climate on a deep, eutrophic, monomictic lake. *Freshw. Biol.* https://doi.org/10.1111/fwb.13293

Rahel, F.J., Olden, J.D., 2008. Assessing the effects of climate change on aquatic invasive species. *Conserv. Biol.* **22**, 521–533. https://doi.org/0.1111/j.1523-1739.2008.00950.x

Randall, M.C., Carling, G.T., Dastrup, D.B., Miller, T., Nelson, S.T., Rey, K.A., Hansen, N.C., Bickmore, B.R., Aanderud, Z.T., 2019. Sediment potentially controls in-lake phosphorus cycling and harmful cyanobacteria in shallow, eutrophic Utah Lake. *PLoS ONE.* **14**, e0212238. https://doi.org/10.1371/journal.pone.0212238

Ranjbar, M.H., Hamilton, D.P., Etemad-Shahidi, A., Helfer, F., 2021. Individual-based modelling of cyanobacteria blooms: Physical and physiological processes. *Sci. Total Environ.* **792**, 148418. https://doi.org/10.1016/j.scitotenv.2021.148418

Rantala, A., Rajaniemi-Wacklin, P., Lyra, C., Lepisto, L., Rintala, J., Mankiewicz-Boczek, J., Sivonen, K., 2006. Detection of microcystin-producing cyanobacteria in finnish lakes with genus-specific microcystin synthetase gene E (mcyE) PCR and associations with environmental factors. *Appl. Environ. Microbiol.* **72**, 6101–6110. https://doi.org/10.1128/AEM.01058-06

Rastetter, E.B., Ohman, M.D., Elliott, K.J., Rehage, J.S., Rivera-Monroy, V.H., Boucek, R.E., Castañeda-Moya, E., Danielson, T.M., Gough, L., Groffman, P.M., Jackson, C.R., Miniat, C.F., Shaver, G.R., 2021. Time lags: Insights from the U.S. long term ecological research network. *Ecosphere* **12**. https://doi.org/10.1002/ecs2.3431

Rastogi, R.P., Madamwar, D., Incharoensakdi, A., 2015. Bloom dynamics of cyanobacteria and their toxins: Environmental health impacts and mitigation strategies. *Front. Microbiol.* **6**. https://doi.org/10.3389/fmicb.2015.01254

Rastogi, R.P., Sinha, R.P., Incharoensakdi, A., 2014. The cyanotoxin-microcystins: Current overview. *Rev. Environ. Sci. Biotechnol.* **13**, 215–249. https://doi.org/10.1007/s11157-014-9334-6

Reddy, K.R., Newman, S., Osborne, T.Z., White, J.R., Fitz, H.C., 2011. Phosphorus cycling in the greater Everglades ecosystem: Legacy phosphorus implications for management and restoration. *Crit. Rev. Environm. Sci. Technol.* **41**, 149–186. https://doi.org/10.1080/10643389.2010.530932

Redfield, A.C., 1958. The biological control of chemical factors in the environment. *Am. Sci.* **46**, 205–221.

Reichwaldt, E.S., Ghadouani, A., 2012. Effects of rainfall patterns on toxic cyanobacterial blooms in a changing climate: Between simplistic scenarios and complex dynamics. *Water. Res.* **46**, 1372–1393. https://doi.org/10.1016/j.watres.2011.11.052

Reid, A.J., Carlson, A.K., Creed, I.F., Eliason, E.J., Gell, P.A., Johnson, P.T.J., Kidd, K.A., MacCormack, T.J., Olden, J.D., Ormerod, S.J., Smol, J.P., Taylor, W.W., Tockner, K., Vermaire, J.C., Dudgeon, D., Cooke, S.J., 2019. Emerging threats and persistent conservation challenges for freshwater biodiversity. *Biol. Rev.* **94**, 849–873. https://doi.org/10.1111/brv.12480

Reinl, K.L., Brookes, J.D., Carey, C.C., Harris, T.D., Ibelings, B.W., Morales-Williams, A.M., De Senerpont Domis, L.N., Atkins, K.S., Isles, P.D.F., Mesman, J.P., North, R.L., Rudstam, L.G., Stelzer, J.A.A., Venkiteswaran, J.J., Yokota, K., Zhan, Q., 2021. Cyanobacterial blooms in oligotrophic lakes: Shifting the high-nutrient paradigm. *Freshw. Biol.* **n/a**. https://doi.org/10.1111/fwb.13791

Reinl, K.L., Harris, T.D., North, R.L., Almela, P., Berger, S.A., Bizic, M., Burnet, S.H., Grossart, H.-P., Ibelings, B.W., Jakobsson, E., Knoll, L.B., Lafrancois, B.M., McElarney, Y., Morales-Williams, A.M., Obertegger, U., Ogashawara, I., Paule-Mercado, M.C., Peierls, B.L., Rusak, J.A., Sarkar, S., Sharma, S., Trout-Haney, J.V., Urrutia-Cordero, P., Venkiteswaran, J.J., Wain, D.J., Warner, K., Weyhenmeyer, G.A., Yokota, K., 2023. Blooms also like it cold. *Limnol. Oceanogr. Lett.* **n/a**. https://doi.org/10.1002/lol2.10316

Reiss, R.S., Lemmin, U., Barry, D.A., 2022. Wind-induced hypolimnetic upwelling between the multi-depth basins of Lake Geneva during winter: An overlooked deepwater renewal mechanism? *J. Geophys. Res. Oceans.* **127**. https://doi.org/10.1029/2021JC018023

Reitzel, K., Ahlgren, J., DeBrabandere, H., Waldebäck, M., Gogoll, A., Tranvik, L., Rydin, E., 2007. Degradation rates of organic phosphorus in lake sediment. *Biogeochem.* **82**, 15–28. https://doi.org/10.1007/s10533-006-9049-z

Reitzel, K., Jensen, H.S., Egemose, S., 2013. pH dependent dissolution of sediment aluminum in six Danish lakes treated with aluminum. *Water Res.* **47**, 1409–1420. https://doi.org/10.1016/j.watres.2012.12.004

Ren, L., Ding, K., Hu, Z., Wang, H., Qi, N., Xu, W., 2022. Processes and mechanisms of phosphorus mobility among sediment, water, and cyanobacteria under hydrodynamic conditions. *Environ. Sci. Pollut. Res.* **29**, 9354–9368. https://doi.org/10.1007/s11356-021-16255-6

Rengefors, K., Gustafsson, S., Ståhl-Delbanco, A., 2004. Factors regulating the recruitment of cyanobacterial and eukaryotic phytoplankton from littoral and profundal sediments. *Aquat. Microb. Ecol.* **36**, 213–226. https://doi.org/10.3354/ame036213

Reynolds, C.S., 1984. Phytoplankton periodicity: The interactions of form, function and environmental variability. *Freshw. Biol.* **14**, 111–142. https://doi.org/10.1111/j.1365-2427.1984.tb00027.x

Reynolds, C.S., 1992. Eutrophication and management of planktonic algae: what Vollenweider couldn't tell us. In *Eutrophication: Research and Application to Water Supply* (Eds. Sutcliffe, D.W., Jones, J.G.). The Freshwater Biological Association, Ambleside, Cumbria, Ambleside, pp. 4–29.

Reynolds, C.S., Elliott, J.A., Frassl, M.A., 2014. Predictive utility of trait-separated phytoplankton groups: A robust approach to modeling population dynamics. *J. Gt. Lakes Res.* **40**, 143–150. https://doi.org/10.1016/j.jglr.2014.02.005

Reynolds, C.S., Huszar, V., Kruk, C., Naselli-Flores, L., Melo, S., 2002. Towards a functional classification of the freshwater phytoplankton. *J. Plankton Res.* **24**, 417–428. https://doi.org/10.1093/plankt/24.5.417

Richardson, D.C., Melles, S.J., Pilla, R.M., Hetherington, A.L., Knoll, L.B., Williamson, C.E., Kraemer, B.M., Jackson, J.R., Long, E.C., Moore, K., Rudstam, L.G., Rusak, J.A., Saros, J.E., Sharma, S., Strock, K.E., Weathers, K.C., Wigdahl-Perry, C.R., 2017. Transparency, geomorphology and mixing regime explain variability in trends in lake temperature and stratification across northeastern North America (1975–2014). *Water.* **9**, 442. https://doi.org/10.3390/w9060442

Richardson, J., Miller, C., Maberly, S.C., Taylor, P., Globevnik, L., Hunter, P., Jeppesen, E., Mischke, U., Moe, S.J., Pasztaleniec, A., Søndergaard, M., Carvalho, L., 2018. Effects of multiple stressors on cyanobacteria abundance vary with lake type. *Glob. Change Biol.* **24**, 5044–5055. https://doi.org/10.1111/gcb.14396

Rigosi, A., Carey, C.C., Ibelings, B.W., Brookes, J.D., 2014. The interaction between climate warming and eutrophication to promote cyanobacteria is dependent on trophic state and varies among taxa. *Limnol. Ocean.* **59**, 99–114. https://doi.org/99-114

Rinta-Kanto, J.M., Konopko, E.A., DeBruyn, J.M., Bourbonniere, R.A., Boyer, G.L., Wilhelm, S.W., 2009. Lake Erie Microcystis: Relationship between microcystin production, dynamics of genotypes and environmental parameters in a large lake. *Harmful Algae.* **8**, 665–673. https://doi.org/10.1016/j.hal.2008.12.004

Ripl, W., 1976. Biochemical oxidation of polluted lake sediment with nitrate – a new lake restoration method. *Ambio*. **5**, 132–135.

Rippey, B., McElarney, Y., Thompson, J., Allen, M., Gallagher, M., Douglas, R., 2022. Recovery targets and timescales for Lough Neagh and other lakes. *Water Res*. **222**, 118858. https://doi.org/10.1016/j.watres.2022.118858

Robb, M., Greenop, B., Goss, Z., Douglas, G., Adeney, J., 2003. Application of PhoslockTM, an innovative phosphorus binding clay, to two Western Australian waterways: Preliminary findings. *Hydrobiologia* **494**, 237–243. https://doi.org/10.1007/978-94-017-3366-3_32

Robson, B.J., 2014. When do aquatic systems models provide useful predictions, what is changing, and what is next? *Environ. Model. Softw*. **61**, 287–296. https://doi.org/10.1016/j.envsoft.2014.01.009

Rocha, M. de J.D., Lima Neto, I.E., 2022. Internal phosphorus loading and its driving factors in the dry period of Brazilian semiarid reservoirs. *J. Environ. Manage*. **312**, 114983. https://doi.org/10.1016/j.jenvman.2022.114983

Rohwer, R.R., Hale, R.J., Vander Zanden, M.J., Miller, T.R., McMahon, K.D., 2023. Species invasions shift microbial phenology in a two-decade freshwater time series. *Proc. Natl. Acad. Sci*. **120**, e2211796120. https://doi.org/10.1073/pnas.2211796120

Rohwer, R.R., Ladwig, R., Hanson, P.C., Walsh, J.R., Vander Zanden, M.J., Dugan, H.A., 2024. Increased anoxia following species invasion of a eutrophic lake. *Limnol. Oceanogr. Lett*. **9**, 33–42. https://doi.org/10.1002/lol2.10364

Rolighed, J., Jeppesen, E., Søndergaard, M., Bjerring, R., Janse, J., Mooij, W., Trolle, D., 2016. Climate change will make recovery from eutrophication more difficult in shallow Danish Lake Søbygaard. *Water*. **8**, 459. https://doi.org/10.3390/w8100459

Rönicke, H., Frassl, M.A., Rinke, K., Tittel, J., Beyer, M., Kormann, B., Gohr, F., Schultze, M., 2021. Suppression of bloom-forming colonial cyanobacteria by phosphate precipitation: A 30 years case study in Lake Barleber (Germany). *Ecol. Eng*. **162**, 106171. https://doi.org/10.1016/j.ecoleng.2021.106171

Rose, K.C., Winslow, L.A., Read, J.S., Hansen, G.J.A., 2016. Climate-induced warming of lakes can be either amplified or suppressed by trends in water clarity. *Limnol. Oceanogr. Lett*. **1**, 44–53. https://doi.org/10.1002/lol2.10027

Ross, G., Haghseresht, F., Cloete, T.E., 2008. The effect of pH and anoxia on the performance of Phoslock®, a phosphorus binding clay. *Harmful Algae*. **7**, 545–550. https://doi.org/10.1016/j.hal.2007.12.007

Rothe, M., Frederichs, T., Eder, M., Kleeberg, A., Hupfer, M., 2014. Evidence for vivianite formation and its contribution to long-term phosphorus retention in a recent lake sediment: A novel analytical approach. *Biogeosciences*. **11**, 5169–5180. https://doi.org/10.5194/bg-11-5169-2014

Rousso, B.Z., Bertone, E., Stewart, R., Hamilton, D.P., 2020. A systematic literature review of forecasting and predictive models for cyanobacteria blooms in freshwater lakes. *Water Res*. **182**, 115959. https://doi.org/10.1016/j.watres.2020.115959

Rowe, M.D., Anderson, E.J., Wang, J., Vanderploeg, H.A., 2015. Modeling the effect of invasive quagga mussels on the spring phytoplankton bloom in Lake Michigan. *J. Gt. Lakes Res*. **41**, 49–65. https://doi.org/10.1016/j.jglr.2014.12.018

Rühland, K.M., Evans, M., Smol, J.P., 2023. Arctic warming drives striking twenty-first century ecosystem shifts in Great Slave Lake (Subarctic Canada), North America's deepest lake. *Proc. R. Soc. B Biol. Sci*. **290**, 20231252. https://doi.org/10.1098/rspb.2023.1252

Rühland, K.M., Paterson, A.M., Smol, J.P., 2015. Lake diatom responses to warming: Reviewing the evidence. *J. Paleolimnol*. **54**, 1–35. https://doi.org/10.1007/s10933-015-9837-3

Rydin, E., 2000. Potentially mobile phosphorus in Lake Erken sediment. *Water Res*. **34**, 2037–2042. https://doi.org/10.1016/S0043-1354(99)00375-9

Saar, K., Nõges, P., Søndergaard, M., Jensen, M., Jørgensen, C., Reitzel, K., Jeppesen, E., Lauridsen, T.L., Jensen, H.S., 2022. The impact of climate change and eutrophication on phosphorus forms in sediment: Results from a long-term lake mesocosm experiment. *Sci. Total Environ*. 153751. https://doi.org/10.1016/j.scitotenv.2022.153751

Sakamoto, M., 1971. Chemical factors involved in the control of phytoplankton production in the experimental lakes area, northwestern Ontario. *J. Fish. Res. Board Can.* **28**, 203–213. https://doi.org/10.1139/f71-032

Sarnelle, O., Morrison, J., Kaul, R., Horst, G., Wandell, H., Bednarz, R., 2010. Citizen monitoring: Testing hypotheses about the interactive influences of eutrophication and mussel invasion on a cyanobacterial toxin in lakes. *Water Res.* **44**, 141–150. https://doi.org/10.1016/j.watres.2009.09.014

Sarnelle, O., White, J.D., Horst, G.P., Hamilton, S.K., 2012. Phosphorus addition reverses the positive effect of zebra mussels (Dreissena polymorpha) on the toxic cyanobacterium, Microcystis aeruginosa. *Water Res.* **46**, 3471–3478. https://doi.org/10.1016/j.watres.2012.03.050

Saros, J.E., Stone, J.R., Pederson, G.T., Slemmons, K.E.H., Spanbauer, T., Schliep, A., Cahl, D., Williamson, C.E., Engstrom, D.R., 2012. Climate-induced changes in lake ecosystem structure inferred from coupled neo- and paleoecological approaches. *Ecology.* **93**, 2155–2164. https://doi.org/10.1890/11-2218.1

Sarvala, J., Helminen, H., 2023. Impacts of chemical precipitation of phosphorus with polyaluminum chloride in two eutrophic lakes in southwest Finland. *Inland Waters.* **13**, 412–427. https://doi.org/10.1080/20442041.2023.2266177

Sas, H., 1989. *Lake Restoration by Reduction of Nutrient Loading: Expectations, Experiences, Extrapolations.* Hans Richarz Publikations-Service, Sankt Augustin, Germany.

Savenko, V.S., Savenko, A.V., 2022. The main features of phosphorus transport in world rivers. *Water* **14**, 16. https://doi.org/10.3390/w14010016

Scarlett, K.R., Kim, S., Lovin, L.M., Chatterjee, S., Scott, J.T., Brooks, B.W., 2020. Global scanning of cylindrospermopsin: Critical review and analysis of aquatic occurrence, bioaccumulation, toxicity and health hazards. *Sci. Total Environ.* **738**, 139807. https://doi.org/10.1016/j.scitotenv.2020.139807

Scheffer, M., Bascompte, J., Brock, W.A., Brovkin, V., Carpenter, S.R., Dakos, V., Held, H., Van Nes, E.H., Rietkerk, M., Sugihara, G., 2009. Early-warning signals for critical transitions. *Nature.* **461**, 53–59. https://doi.org/10.1038/nature08227

Scheffer, M., Hosper, S.H., Meijer, M.-L., Moss, B., Jeppesen, E., 1993. Alternative equilibria in shallow lakes. *Trends Ecol. Evol.* **8**, 275–279. https://doi.org/10.1016/0169-5347(93)90254-M

Scheffer, M., Portielje, R., Zambrano, L., 2003. Fish facilitate wave resuspension of sediment. *Limnol. Oceanogr.* **48**, 1920–1926. https://doi.org/10.4319/lo.2003.48.5.1920

Schelske, C.L., 2009. Eutrophication: Focus on phosphorus. *Science.* **324**, 722–722. https://doi.org/10.1126/science.324_722

Schindler, D.W., 1974. Eutrophication and recovery in experimental lakes: Implications for lake management. *Science.* **184**, 897–899. https://doi.org/10.1126/science.184.4139.897

Schindler, D.W., Hecky, R.E., 2009. Eutrophication: More nitrogen data needed. *Science.* **324**, 721–722. https://doi.org/10.1126/science.324_721b

Schindler, D.W., Hecky, R.E., Findlay, D.L., Stainton, M.P., Parker, B.R., Paterson, M.J., Beaty, K.G., Lyng, M., Kasian, S.E.M., 2008. Eutrophication of lakes cannot be controlled by reducing nitrogen input: Results of a 37-year whole-ecosystem experiment. *Proc. Nat. Acad. Sci.* **105**, 11254–11258. https://doi.org/10.1073/pnas.0805108105

Schmidt, D.F., Grise, K.M., Pace, M.L., 2019. High-frequency climate oscillations drive ice-off variability for Northern Hemisphere lakes and rivers. *Clim. Change.* **152**, 517–532.

Scholz, S.N., Esterhuizen-Londt, M., Pflugmacher, S., 2017. Rise of toxic cyanobacterial blooms in temperate freshwater lakes: Causes, correlations and possible countermeasures. *Toxicol. Environ. Chem.* **99**, 543–577. https://doi.org/10.1080/02772248.2016.1269332

Schroth, A.W., Giles, C.D., Isles, P.D.F., Xu, Y., Perzan, Z., Druschel, G.K., 2015. Dynamic coupling of iron, manganese, and phosphorus behavior in water and sediment of shallow ice-covered eutrophic lakes. *Environ. Sci. Technol.* **49**, 9758–9767. https://doi.org/10.1021/acs.est.5b02057

Schütz, J., Rydin, E., Huser, B.J., 2017. A newly developed injection method for aluminum treatment in eutrophic lakes: Effects on water quality and phosphorus binding efficiency. *Lake Reserv. Manag.* **33**, 152–162. https://doi.org/10.1080/10402381.2017.1318418

Schwefel, R., MacIntyre, S., Cortés, A., Sadro, S., 2023. Oxygen depletion and sediment respiration in ice-covered arctic lakes. *Limnol. Oceanogr.* **68**, 1470–1489. https://doi.org/10.1002/lno.12357

Scott, J.T., McCarthy, M.J., Paerl, H.W., 2019. Nitrogen transformations differentially affect nutrient-limited primary production in lakes of varying trophic state. *Limnol. Oceanogr. Lett.* https://doi.org/10.1002/lol2.10109

Senar, O.E., Creed, I.F., Trick, C.G., 2021. Lake browning may fuel phytoplankton biomass and trigger shifts in phytoplankton communities in temperate lakes. *Aquat. Sci.* **83**, 21. https://doi.org/10.1007/s00027-021-00780-0

Sereda, J.M., Hudson, J.J., McLoughlin, P.D., 2008. General empirical models for predicting the release of nutrients by fish, with a comparison between detritivores and non-detritivores. *Freshw. Biol.* **53**, 2133–2144. https://doi.org/10.1111/j.1365-2427.2008.02029.x

Sharma, S., Blagrave, K., Filazzola, A., Imrit, M.A., Franssen, H.-J.H., 2021. Forecasting the permanent loss of lake ice in the northern hemisphere within the 21st century. *Geophys. Res. Lett.* **48**, e2020GL091108. https://doi.org/10.1029/2020GL091108

Sharma, S., Blagrave, K., Magnuson, J.J., O'Reilly, C.M., Oliver, S., Batt, R.D., Magee, M.R., Straile, D., Weyhenmeyer, G.A., Winslow, L., Woolway, R.I., 2019. Widespread loss of lake ice around the Northern Hemisphere in a warming world. *Nat. Clim. Change.* **9**, 227–231. https://doi.org/10.1038/s41558-018-0393-5

Sharpley, A., Jarvie, H.P., Buda, A., May, L., Spears, B., Kleinman, P., 2013. Phosphorus legacy: Overcoming the effects of past management practices to mitigate future water quality impairment. *J. Environ. Qual.* **42**, 1308. https://doi.org/10.2134/jeq2013.03.0098

Shatwell, T., Köhler, J., 2019. Decreased nitrogen loading controls summer cyanobacterial blooms without promoting nitrogen-fixing taxa: Long-term response of a shallow lake. *Limnol. Oceanogr.* **64**, S166–S178. https://doi.org/10.1002/lno.11002

Shen, C., Liao, Q., Bootsma, H.A., 2020. Modelling the influence of invasive mussels on phosphorus cycling in Lake Michigan. *Ecol. Model.* **416**, 108920. https://doi.org/10.1016/j.ecolmodel.2019.108920

Shen, J., Du, J., Lucas, L.V., 2022. Simple relationships between residence time and annual nutrient retention, export, and loading for estuaries. *Limnol. Oceanogr.* **67**, 918–933. https://doi.org/10.1002/lno.12045

Shinohara, R., Matsuzaki, S.S., Watanabe, M., Nakagawa, M., Yoshida, H., Kohzu, A., 2023. Heat waves can cause hypoxia in shallow lakes. *Geophys. Res. Lett.* **50**, e2023GL102967. https://doi.org/10.1029/2023GL102967

Shuvo, A., O'Reilly, C.M., Blagrave, K., Ewins, C., Filazzola, A., Gray, D., Mahdiyan, O., Moslenko, L., Quinlan, R., Sharma, S., 2021. Total phosphorus and climate are equally important predictors of water quality in lakes. *Aquat. Sci.* **83**, 16. https://doi.org/10.1007/s00027-021-00776-w

Silvonen, S., Niemistö, J., Myyryläinen, J., Kinnunen, O., Huotari, S., Nurminen, L., Horppila, J., Jilbert, T., 2022. Extracting phosphorus and other elements from lake water: Chemical processes in a hypolimnetic withdrawal and treatment system. *Water Res.* 118507. https://doi.org/10.1016/j.watres.2022.118507

Silvonen, S., Nurminen, L., Horppila, J., Niemistö, J., Jilbert, T., 2023. Closed-circuit hypolimnetic withdrawal and treatment: Impact of effluent discharge on epilimnetic P and N concentrations. *Limnology.* https://doi.org/10.1007/s10201-023-00732-7

Simpson, Z.P., McDowell, R.W., Condron, L.M., McDaniel, M.D., Jarvie, H.P., Abell, J.M., 2021. Sediment phosphorus buffering in streams at baseflow: A meta-analysis. *J. Environ. Qual.* **50**, 287–311. https://doi.org/10.1002/jeq2.20202

Singh, N.K., Van Meter, K.J., Basu, N.B., 2023. Widespread increases in soluble phosphorus concentrations in streams across the transboundary Great Lakes Basin. *Nat. Geosci.* **16**, 893–900. https://doi.org/10.1038/s41561-023-01257-5

Sinha, E., Michalak, A.M., Balaji, V., 2017. Eutrophication will increase during the 21st century as a result of precipitation changes. *Science.* **357**, 405–408. https://doi.org/10.1126/science.aan2409

Sivarajah, B., Simmatis, B., Favot, E.J., Palmer, M.J., Smol, J.P., 2021. Eutrophication and climatic changes lead to unprecedented cyanobacterial blooms in a Canadian sub-Arctic landscape. *Harmful Algae.* **105**, 102036. https://doi.org/10.1016/j.hal.2021.102036

Smayda, T.J., 1997. What is a bloom? A commentary. *Limnol. Oceanogr.* **42**, 1132–1136. https://doi.org/10.4319/lo.1997.42.5_part_2.1132

Smith, L., Watzin, M.C., Druschel, G., 2011. Relating sediment phosphorus mobility to seasonal and diel redox fluctuations at the sediment-water interface in a eutrophic freshwater lake. *Limnol Ocean.* **56**, 2251–2264. https://doi.org/10.4319/lo.2011.56.6.2251

Smith, R.B., Bass, B., Sawyer, D., Depew, D., Watson, S.B., 2019. Estimating the economic costs of algal blooms in the Canadian Lake Erie Basin. *Harmful Algae.* **87**, 101624. https://doi.org/10.1016/j.hal.2019.101624

Smith, V.H., 1983. Low nitrogen to phosphorus ratios favor dominance by blue-green algae in lake phytoplankton. *Science.* **221**, 669–671. https://doi.org/10.1126/science.221.4611.669

Smith, V.H., 1985. Predictive models for the biomass of blue-green algae in lakes. *JAWRA J. Am. Water Resour. Assoc.* **21**, 433–439. https://doi.org/10.1111/j.1752-1688.1985.tb00153.x

Smith, V.H., Schindler, D.W., 2009. Eutrophication science: Where do we go from here? *Trends Ecol. Evol.* **24**, 201–207. https://doi.org/10.1016/j.tree.2008.11.009

Smits, A.P., MacIntyre, S., Sadro, S., 2020. Snowpack determines relative importance of climate factors driving summer lake warming. *Limnol. Oceanogr. Lett.* **5**, 271–279. https://doi.org/10.1002/lol2.10147

Smol, J.P., 2008. *Pollution of Lakes and Rivers – A Paleoenvironmental Perspective*, 2nd ed. Blackwell Publishing, Oxford.

Smol, J.P., Douglas, M.S., 2007. From controversy to consensus: Making the case for recent climate change in the Arctic using lake sediments. *Front. Ecol. Environ.* **5**, 466–474. https://doi.org/10.1890/060162

Smolders, A.J.P., Lamers, L.P.M., Lucassen, E.C.H.E.T., Velde, G. van der, Roelofs, J.G.M., 2006. Internal eutrophication: How it works and what to do about it – a review. *Chem Ecol.* **22**, 93–111. https://doi.org/10.1080/02757540600579730

Solovchenko, A., Gorelova, O., Karpova, O., Selyakh, I., Semenova, L., Chivkunova, O., Baulina, O., Vinogradova, E., Pugacheva, T., Scherbakov, P., Vasilieva, S., Lukyanov, A., Lobakova, E., 2020. Phosphorus feast and famine in cyanobacteria: Is luxury uptake of the nutrient just a consequence of acclimation to its shortage? *Cells.* **9**, 1933. https://doi.org/10.3390/cells9091933

Søndergaard, M., Jensen, J.P., Jeppesen, E., 2003. Role of sediment and internal loading of phosphorus in shallow lakes. *Hydrobiologia.* **506–509**, 135–145. https://doi.org/10.1023/B:HYDR.0000008611.12704.dd

Søndergaard, M., Jeppesen, E., 2020. Chapter 4: Understanding the drivers on internal phosphorus loading in lakes. In *Internal Phosphorus Loading: Causes, Case Studies, and Management* (Eds. Steinman, A.D., Spears, B.M.). J. Ross Publishing, Plantation, FL, pp. 63–76.

Søndergaard, M., Lauridsen, T.L., Johansson, L.S., Jeppesen, E., 2017. Nitrogen or phosphorus limitation in lakes and its impact on phytoplankton biomass and submerged macrophyte cover. *Hydrobiologia.* **795**, 35–48. https://doi.org/10.1007/s10750-017-3110-x

Søndergaard, M., Liboriussen, L., Pedersen, A.R., Jeppesen, E., 2008. Lake restoration by fish removal: Short- and long-term effects in 36 Danish lakes. *Ecosystems.* **11**, 1291–1305. https://doi.org/10.1007/s10021-008-9193-5

Søndergaard, M., Nielsen, A., Johansson, L.S., Davidson, T.A., 2023a. Temporarily summer-stratified lakes are common: Profile data from 436 lakes in lowland Denmark. *Inland Waters.* **13**, 153–166. https://doi.org/10.1080/20442041.2023.2203060

Søndergaard, M., Nielsen, A., Skov, C., Baktoft, H., Reitzel, K., Kragh, T., Davidson, T.A., 2023b. Temporarily and frequently occurring summer stratification and its effects on nutrient dynamics, greenhouse gas emission and fish habitat use: Case study from Lake Ormstrup (Denmark). *Hydrobiologia.* **850**, 65–79. https://doi.org/10.1007/s10750-022-05039-9

Soo, R.M., Hemp, J., Parks, D.H., Fischer, W.W., Hugenholtz, P., 2017. On the origins of oxygenic photosynthesis and aerobic respiration in Cyanobacteria. *Science.* **355**, 1436–1440. https://doi.org/10.1126/science.aal3794

Sorichetti, R.J., Creed, I.F., Trick, C.G., 2016. Iron and iron-binding ligands as cofactors that limit cyanobacterial biomass across a lake trophic gradient. *Freshw. Biol.* **61**, 146–157. https://doi.org/10.1111/fwb.12689/full

Sosiak, A., 2022. Assessment of the hypolimnetic withdrawal system at Pine Lake, Alberta. *Lake Reserv. Manag.* **38**, 47–66. https://doi.org/10.1080/10402381.2021.2001609

Spears, B.M., Carvalho, L., Futter, M.N., May, L., Thackeray, S.J., Adrian, R., Angeler, D.G., Burthe, S.J., Davidson, T.A., Daunt, F., Gsell, A.S., Hessen, D.O., Moorhouse, H., Huser, B., Ives, S.C., Janssen, A.B.G., Mackay, E.B., Søndergaard, M., Jeppesen, E., 2016a. Ecological instability in lakes: A predictable condition? *Environ. Sci. Technol.* **50**, 3285–3286. https://doi.org/10.1021/acs.est.6b00865

Spears, B.M., Hamilton, D.P., Pan, Y., Zhaosheng, C., May, L., 2022. Lake management: Is prevention better than cure? *Inland Waters*. **12**, 173–186. https://doi.org/10.1080/20442041.2021.1895646

Spears, B.M., Lürling, M., Yasseri, S., Castro-Castellon, A.T., Gibbs, M., Meis, S., McDonald, C., McIntosh, J., Sleep, D., Van Oosterhout, F., 2013a. Lake responses following lanthanum-modified bentonite clay (Phoslock®) application: An analysis of water column lanthanum data from 16 case study lakes. *Water Res.* **47**, 5930–5942. https://doi.org/10.1016/j.watres.2013.07.016

Spears, B.M., Mackay, E.B., Yasseri, S., Gunn, I.D.M., Waters, K.E., Andrews, C., Cole, S., De Ville, M., Kelly, A., Meis, S., Moore, A.L., Nürnberg, G.K., van Oosterhout, F., Pitt, J.-A., Madgwick, G., Woods, H.J., Lürling, M., 2016b. A meta-analysis of water quality and aquatic macrophyte responses in 18 lakes treated with lanthanum modified bentonite (Phoslock®). *Water Res.* **97**, 111–121. https://doi.org/10.1016/j.watres.2015.08.020

Spears, B.M., Meis, S., Anderson, A., Kellou, M., 2013b. Comparison of phosphorus (P) removal properties of materials proposed for the control of sediment P release in UK lakes. *Sci. Tot. Environm.* **442**, 103–110. https://doi.org/10.1016/j.scitotenv.2012.09.066

Spiese, C.E., Bowling, M.N., Moeller, S.E.M., 2023. Is glyphosate an underlying cause of increased dissolved reactive phosphorus loading in the Western Lake Erie basin? *J. Gt. Lakes Res.* **49**, 631–639. https://doi.org/10.1016/j.jglr.2023.03.009

Spoof, L., Catherine, A., 2016. Appendix 3: Tables of microcystins and nodularins. In *Handbook of Cyanobacterial Monitoring and Cyanotoxin Analysis*. John Wiley & Sons, Ltd, pp. 526–537. https://doi.org/10.1002/9781119068761.app3

Stackpoole, S.M., Stets, E.G., Sprague, L.A., 2019. Variable impacts of contemporary versus legacy agricultural phosphorus on US river water quality. *Proc. Natl. Acad. Sci.* 201903226. https://doi.org/10.1073/pnas.1903226116

Stainsby, E.A., Winter, J.G., Jarjanazi, H., Paterson, A.M., Evans, D.O., Young, J.D., 2011. Changes in the thermal stability of Lake Simcoe from 1980 to 2008. *J. Gt. Lakes Res.* **37**, 55–62.

Steinman, A.D., Chu, X., Ogdahl, M., 2009. Spatial and temporal variability of internal and external phosphorus loads in Mona Lake, Michigan. *Aquat. Ecol.* **43**, 1–18. https://doi.org/10.1007/s10452-007-9147-6

Steinman, A.D., Spears, B.M., 2020a. Chapter 1: What is internal phosphorus loading and why does it occur? In *Internal Phosphorus Loading: Causes, Case Studies, and Management* (Eds. Steinman, A.D., Spears, B.M.). J. Ross Publishing, Plantation, FL, pp. 3–13.

Steinman, A.D., Spears, B.M., 2020b. *Internal Phosphorus Loading: Causes, Case Studies, and Management*. J. Ross Publishing, Plantation, FL.

Sterner, R.W., 2008. On the phosphorus limitation paradigm for lakes. *Int. Rev. Hydrobiol.* **93**, 433–445. https://doi.org/10.1002/iroh.200811068

Sterner, R.W., Andersen, T., Elser, J.J., Hessen, D.O., Hood, J.M., McCauley, E., Urabe, J., 2008. Scale-dependent carbon: Nitrogen : Phosphorus seston stoichiometry in marine and freshwaters. *Limnol. Ocean.* **53**, 1169–1180. https://doi.org/10.4319/lo.2008.53.3.1169

Sterner, R.W., Reinl, K.L., Lafrancois, B.M., Brovold, S., Miller, T.R., 2020. A first assessment of cyanobacterial blooms in oligotrophic Lake Superior. *Limnol. Oceanogr.* **65**, 2984–2998. https://doi.org/10.1002/lno.11569

Stetler, J.T., Knoll, L.B., Driscoll, C.T., Rose, K.C., 2021. Lake browning generates a spatiotemporal mismatch between dissolved organic carbon and limiting nutrients. *Limnol. Oceanogr. Lett.* **6**, 182–192. https://doi.org/10.1002/lol2.10194

Stockwell, J.D., Doubek, J.P., Adrian, R., Anneville, O., Carey, C.C., Carvalho, L., Domis, L.N.D.S., Dur, G., Frassl, M.A., Grossart, H.-P., Ibelings, B.W., Lajeunesse, M.J., Lewandowska, A.M., Llames, M.E.,

Matsuzaki, S.-I.S., Nodine, E.R., Nõges, P., Patil, V.P., Pomati, F., Rinke, K., Rudstam, L.G., Rusak, J.A., Salmaso, N., Seltmann, C.T., Straile, D., Thackeray, S.J., Thiery, W., Urrutia-Cordero, P., Venail, P., Verburg, P., Woolway, R.I., Zohary, T., Andersen, M.R., Bhattacharya, R., Hejzlar, J., Janatian, N., Kpodonu, A.T.N.K., Williamson, T.J., Wilson, H.L., 2020. Storm impacts on phytoplankton community dynamics in lakes. *Glob. Change Biol.* **26**, 2756–2784. https://doi.org/10.1111/gcb.15033

Straskraba, M., 1994. Ecotechnological models for reservoir water quality management. *Ecol. Model.* **74**, 1–38. https://doi.org/10.1016/0304-3800(94)90108-2

Straskraba, M., 1998. Limnological differences between deep valley reservoirs and deep lakes. *Int. Rev. Gesamt. Hydrobiol.* **83**, 1–12.

Stumpf, R.P., Johnson, L.T., Wynne, T.T., Baker, D.B., 2016. Forecasting annual cyanobacterial bloom biomass to inform management decisions in Lake Erie. *J. Gt. Lakes Res.* **42**, 1174–1183. https://doi.org/10.1016/j.jglr.2016.08.006

Stumpf, R.P., Wynne, T.T., Baker, D.B., Fahnenstiel, G.L., 2012. Interannual variability of cyanobacterial blooms in Lake Erie. *PLoS ONE.* **7**, e42444. https://doi.org/10.1371/journal.pone.0042444

Su, H., Chen, J., Li, Y., Rao, Q., Luo, C., Deng, X., Shen, H., Li, R., Chen, J., Sun, Y., Pan, J., Ma, S., Feng, Y., Wang, H., Fang, J., Xie, P., 2023. Carp stocking and climate change are potentially more important factors than nutrient enrichment driving water quality deterioration in subtropical freshwater lakes in China. *Limnol. Oceanogr.* **68**, S131–S143. https://doi.org/10.1002/lno.12280

Su, L., Zhong, C., Gan, L., He, X., Yu, J., Zhang, X., Liu, Z., 2021. Effects of lanthanum modified bentonite and polyaluminium chloride on the environmental variables in the water and sediment phosphorus form in Lake Yanglan, China. *Water.* **13**, 1947. https://doi.org/10.3390/w13141947

Sukenik, A., Quesada, A., Salmaso, N., 2015. Global expansion of toxic and non-toxic cyanobacteria: Effect on ecosystem functioning. *Biodivers. Conserv.* **24**, 889–908. https://doi.org/10.1007/s10531-015-0905-9

Summers, E.J., Ryder, J.L., 2023. A critical review of operational strategies for the management of harmful algal blooms (HABs) in Inland reservoirs. *J. Environ. Manage.* **330**, 117141. https://doi.org/10.1016/j.jenvman.2022.117141

Summers, J.C., Kurek, J., Kirk, J.L., Muir, D.C., Wang, X., Wiklund, J.A., Cooke, C.A., Evans, M.S., Smol, J.P., 2016. Recent warming, rather than industrial emissions of bioavailable nutrients, is the dominant driver of lake primary production shifts across the Athabasca Oil Sands Region. *PLoS ONE.* **11**, e0153987.

Sun, C., Zhong, J., Pan, G., Mortimer, R.J.G., Yu, J., Wen, S., Zhang, L., Yin, H., Fan, C., 2023a. Controlling internal nitrogen and phosphorus loading using Ca-poor soil capping in shallow eutrophic lakes: Long-term effects and mechanisms. *Water Res.* **233**, 119797. https://doi.org/10.1016/j.watres.2023.119797

Sun, L., Cai, Y., Yang, W., Yi, Y., Yang, Z., 2019. Climatic variations within the dry valleys in southwestern China and the influences of artificial reservoirs. *Clim. Change.* **155**, 111–125. https://doi.org/10.1007/s10584-019-02457-y

Sun, Y.-F., Guo, Y., Xu, C., Liu, Y., Zhao, X., Liu, Q., Jeppesen, E., Wang, H., Xie, P., 2023b. Will "air eutrophication" increase the risk of ecological threat to public health? *Environ. Sci. Technol.* acs.est.3c01368. https://doi.org/10.1021/acs.est.3c01368

Svirčev, Z., Lalić, D., Bojadžija Savić, G., Tokodi, N., Drobac Backović, D., Chen, L., Meriluoto, J., Codd, G.A., 2019. Global geographical and historical overview of cyanotoxin distribution and cyanobacterial poisonings. *Arch. Toxicol.* **93**, 2429–2481. https://doi.org/10.1007/s00204-019-02524-4

Swann, M.M., Cortes, A., Forrest, A.L., Framsted, N., Sadro, S., Schladow, S.G., De Palma-Dow, A., 2024. Internal phosphorus loading alters nutrient limitation and contributes to cyanobacterial blooms in a polymictic lake. *Aquat. Sci.* **86**, 46. https://doi.org/10.1007/s00027-024-01045-2

Swarbrick, V.J., Quiñones-Rivera, Z.J., Leavitt, P.R., 2020. Seasonal variation in effects of urea and phosphorus on phytoplankton abundance and community composition in a hypereutrophic hardwater lake. *Freshw. Biol.* **65**, 1765–1781. https://doi.org/10.1111/fwb.13580

Swarbrick, V.J., Simpson, G.L., Glibert, P.M., Leavitt, P.R., 2018. Differential stimulation and suppression of phytoplankton growth by ammonium enrichment in eutrophic hardwater lakes over 16 years. *Limnol. Oceanogr.* https://doi.org/10.1002/lno.11093

Tammeorg, O., Chorus, I., Spears, B., Nõges, P., Nürnberg, G.K., Tammeorg, P., Søndergaard, M., Jeppesen, E., Paerl, H., Huser, B., Horppila, J., Jilbert, T., Budzyńska, A., Dondajewska-Pielka, R., Gołdyn, R., Haasler, S., Hellsten, S., Härkönen, L.H., Kiani, M., Kozak, A., Kotamäki, N., Kowalczewska-Madura, K., Newell, S., Nurminen, L., Nõges, T., Reitzel, K., Rosińska, J., Ruuhijärvi, J., Silvonen, S., Skov, C., Važić, T., Ventelä, A., Waajen, G., Lürling, M., 2024a. Sustainable lake restoration: From challenges to solutions. *WIREs Water*. **11**, e1689. https://doi.org/10.1002/wat2.1689

Tammeorg, O., Haldna, M., Nõges, P., Appleby, P., Möls, T., Niemistö, J., Tammeorg, P., Horppila, J., 2018. Factors behind the variability of phosphorus accumulation in Finnish lakes. *J. Soils Sediments*. https://doi.org/10.1007/s11368-018-1973-8

Tammeorg, O., Horppila, J., Tammeorg, P., Haldna, M., Niemistö, J., 2016. Internal phosphorus loading across a cascade of three eutrophic basins: A synthesis of short- and long-term studies. *Sci. Total Environ.* **572**, 943–954. https://doi.org/10.1016/j.scitotenv.2016.07.224

Tammeorg, O., Möls, T., Niemistö, J., Holmroos, H., Horppila, J., 2017. The actual role of oxygen deficit in the linkage of the water quality and benthic phosphorus release: Potential implications for lake restoration. *Sci. Total Environ.* **599–600**, 732–738. https://doi.org/10.1016/j.scitotenv.2017.04.244

Tammeorg, O., Niemistö, J., Möls, T., Laugaste, R., Panksep, K., Kangur, K., 2013. Wind-induced sediment resuspension as a potential factor sustaining eutrophication in large and shallow Lake Peipsi. *Aquat. Sci.* **75**, 559–570. https://doi.org/10.1007/s00027-013-0300-0

Tammeorg, O., Nürnberg, G., Horppila, J., Haldna, M., Niemistö, J., 2020a. Redox-related release of phosphorus from sediments in large and shallow Lake Peipsi: Evidence from sediment studies and long-term monitoring data. *J. Gt. Lakes Res.* https://doi.org/10.1007/s00027-020-00724-0

Tammeorg, O., Nürnberg, G., Niemistö, J., Haldna, M., Horppila, J., 2020b. Internal phosphorus loading due to sediment anoxia in shallow areas: Implications for lake aeration treatments. *Aquat. Sci.* **82**. https://doi.org/10.1007/s00027-020-00724-0

Tammeorg, O., Nürnberg, G., Nõges, P., Niemistö, J., 2022b. The role of humic substances in sediment phosphorus release in northern lakes. *Sci. Total Environ.* **833**, 155257. https://doi.org/10.1016/j.scitotenv.2022.155257

Tammeorg, O., Nürnberg, G., Tõnno, I., Kisand, A., Tuvikene, L., Nõges, T., Nõges, P., 2022a. Sediment phosphorus mobility in Võrtsjärv, a large shallow lake: Insights from phosphorus sorption experiments and long-term monitoring. *Sci. Total Environ.* **829**, 154572. https://doi.org/10.1016/j.scitotenv.2022.154572

Tammeorg, O., Nürnberg, G., Tõnno, I., Toom, L., Nõges, P., 2024b. Spatio-temporal variations in sediment phosphorus dynamics in a large shallow lake: Mechanisms and impacts of redox-related internal phosphorus loading. *Sci. Total Environ.* **907**, 168044. https://doi.org/10.1016/j.scitotenv.2023.168044

Tanentzap, A.J., Morabito, G., Volta, P., Rogora, M., Yan, N.D., Manca, M., 2020. Climate warming restructures an aquatic food web over 28 years. *Glob. Change Biol.* **26**, 6852–6866. https://doi.org/10.1111/gcb.15347

Taranu, Z.E., Gregory-Eaves, I., Leavitt, P.R., Bunting, L., Buchaca, T., Catalan, J., Domaizon, I., Guilizzoni, P., Lami, A., McGowan, S., Moorhouse, H., Morabito, G., Pick, F.R., Stevenson, M.A., Thompson, P.L., Vinebrooke, R.D., 2015. Acceleration of cyanobacterial dominance in north temperate-subarctic lakes during the Anthropocene. *Ecol. Lett.* **18**, 375–384. https://doi.org/10.1111/ele.12420

Taranu, Z.E., Zurawell, R.W., Pick, F., Gregory-Eaves, I., 2012. Predicting cyanobacterial dynamics in the face of global change: The importance of scale and environmental context. *Glob. Change Biol.* **18**, 3477–3490. https://doi.org/10.1111/gcb.12015

Tassone, S.J., Besterman, A.F., Buelo, C.D., Ha, D.T., Walter, J.A., Pace, M.L., 2023. Increasing heatwave frequency in streams and rivers of the United States. *Limnol. Oceanogr. Lett.* **8**, 295–304. https://doi.org/10.1002/lol2.10284

Tassone, S.J., Besterman, A.F., Buelo, C.D., Walter, J.A., Pace, M.L., 2022. Co-occurrence of aquatic heatwaves with atmospheric heatwaves, low dissolved oxygen, and low pH events in estuarine ecosystems. *Estuaries Coasts*. **45**, 707–720. https://doi.org/10.1007/s12237-021-01009-x

Tay, C.J., Mohd, M.H., Teh, S.Y., Koh, H.L., 2022. Internal phosphorus recycling promotes rich and complex dynamics in an algae-phosphorus model: Implications for eutrophication management. *J. Theor. Biol.* **532**, 110913. https://doi.org/10.1016/j.jtbi.2021.110913

Tee, H.S., Waite, D., Payne, L., Middleditch, M., Wood, S., Handley, K.M., 2020. Tools for successful proliferation: Diverse strategies of nutrient acquisition by a benthic cyanobacterium. *ISME J.* **14**, 2164–2178. https://doi.org/10.1038/s41396-020-0676-5

Tellier, J.M., Kalejs, N.I., Leonhardt, B.S., Cannon, D., Höök, T.O., Collingsworth, P.D., 2021. Widespread prevalence of hypoxia and the classification of hypoxic conditions in the Laurentian Great Lakes. *J. Gt. Lakes Res.* S0380133021002458. https://doi.org/10.1016/j.jglr.2021.11.004

Thayne, M.W., Kraemer, B.M., Mesman, J.P., Ibelings, B.W., Adrian, R., 2022. Antecedent lake conditions shape resistance and resilience of a shallow lake ecosystem following extreme wind storms. *Limnol. Oceanogr.* **67**, S101–S120. https://doi.org/10.1002/lno.11859

Thomas, M.K., Litchman, E., 2016. Effects of temperature and nitrogen availability on the growth of invasive and native cyanobacteria. *Hydrobiologia.* **763**, 357–369. https://doi.org/10.1007/s10750-015-2390-2

Tillmanns, A.R., Wilson, A.E., Pick, F.R., Sarnelle, O., 2008. Meta-analysis of cyanobacterial effects on zooplankton population growth rate: Species-specific responses. *Fundam. Appl. Limnol.* **171**, 285–295. https://doi.org/10.1127/1863-9135/2008/0171-0285

Trevino-Garrison, I., DeMent, J., Ahmed, F.S., Haines-Lieber, P., Langer, T., Ménager, H., Neff, J., Van der Merwe, D., Carney, E., 2015. Human illnesses and animal deaths associated with freshwater harmful algal blooms – Kansas. *Toxins.* **7**, 353–366. https://doi.org/10.3390/toxins7020353

Trisos, C.H., Merow, C., Pigot, A.L., 2020. The projected timing of abrupt ecological disruption from climate change. *Nature.* **580**, 496–501. https://doi.org/10.1038/s41586-020-2189-9

Tu, L., Gilli, A., Lotter, A.F., Vogel, H., Moyle, M., Boyle, J.F., Grosjean, M., 2021. The nexus among long-term changes in lake primary productivity, deep-water anoxia, and internal phosphorus loading, explored through analysis of a 15,000-year varved sediment record. *Glob. Planet. Change.* **207**, 103643. https://doi.org/10.1016/j.gloplacha.2021.103643

Tu, L., Zander, P., Szidat, S., Lloren, R., Grosjean, M., 2020. The influences of historic lake trophy and mixing regime changes on long-term phosphorus fraction retention in sediments of deep eutrophic lakes: A case study from Lake Burgäschi, Switzerland. *Biogeosciences.* **17**, 2715–2729. https://doi.org/10.5194/bg-17-2715-2020

Uk, S., Yang, H., Vouchlay, T., Ty, S., Sokly, S., Sophal, T., Chantha, O., Chihiro, Y., 2022. Dynamics of phosphorus fractions and bioavailability in a large shallow tropical lake characterized by monotonal flood pulse in Southeast Asia. *J. Gt. Lakes Res.* **48**, 944–960. https://doi.org/10.1016/j.jglr.2022.04.005

Urban, M.C., 2015. Accelerating extinction risk from climate change. *Science.* **348**, 571–573. https://doi.org/10.1126/science.aaa4984

Vanderploeg, H.A., Liebig, J.R., Carmichael, W.W., Agy, M.A., Johengen, T.H., Fahnenstiel, G.L., Nalepa, T.F., 2001. Zebra mussel (*Dreissena polymorpha*) selective filtration promoted toxic *Microcystis* blooms in Saginaw Bay (Lake Huron) and Lake Erie. *Can. J. Fish. Aquat. Sci.* **58**, 1208–1221. https://doi.org/10.1139/f01-066

Van de Waal, D.B., Gsell, A.S., Harris, T., Paerl, H.W., de Senerpont Domis, L.N., Huisman, J., 2024. Hot summers raise public awareness of toxic cyanobacterial blooms. *Water Res.* **249**, 120817. https://doi.org/10.1016/j.watres.2023.120817

Van de Waal, D.B., Smith, V.H., Declerck, S.A.J., Stam, E.C.M., Elser, J.J., 2014. Stoichiometric regulation of phytoplankton toxins. *Ecol. Lett.* **17**, 736–742. https://doi.org/10.1111/ele.12280

Van de Waal, D.B., Verspagen, J.M.H., Lürling, M., Van Donk, E., Visser, P.M., Huisman, J., 2009. The ecological stoichiometry of toxins produced by harmful cyanobacteria: An experimental test of the carbon-nutrient balance hypothesis. *Ecol. Lett.* **12**, 1326–1335. https://doi.org/10.1111/j.1461-0248.2009.01383.x

van Nes, E., Rip, W., Scheffer, M., 2007. A theory for cyclic shifts between alternative states in shallow lakes. *Ecosyst.* **10**, 17–28. https://doi.org/10.1007/s10021-006-0176-0

van Oosterhout, F., Yasseri, S., Noyma, N., Huszar, V., Manzi Marinho, M., Mucci, M., Waajen, G., Lürling, M., 2022. Assessing the long-term efficacy of internal loading management to control eutrophication in Lake Rauwbraken. *Inland Waters.* **12**, 61–77. https://doi.org/10.1080/20442041.2021.1969189

Vautard, R., Cattiaux, J., Yiou, P., Thépaut, J.-N., Ciais, P., 2010. Northern Hemisphere atmospheric stilling partly attributed to an increase in surface roughness. *Nat. Geosci.* **3**, 756–761. https://doi.org/10.1038/ngeo979

Verspagen, J.M.H., Visser, P.M., Huisman, J., 2006. Aggregation with clay causes sedimentation of the buoyant cyanobacteria Microcystis spp. *Aquat. Microb. Ecol.* **44**, 165–174. https://doi.org/10.3354/ame044165

Visser, P.M., Ibelings, B.W., Bormans, M., Huisman, J., 2016. Artificial mixing to control cyanobacterial blooms: A review. *Aquat. Ecol.* **50**, 423–441. https://doi.org/10.1007/s10452-015-9537-0

Vollenweider, R.A., 1968. *Water Management Research (OECD Technical Report No. OECD/DAS/CSI/68.27)*. Organisation for Economic Co-operation and Development, Paris.

Vollenweider, R.A., 1976. Advances in defining critical loading levels for phosphorus in lake eutrophication. *Mem. Ist. Ital. Idrobiol.* **33**, 53–83.

Vollenweider, V.R., 1969. Possibilities and limits of elementary models concerning the budget of substances in lakes (German: Möglichkeiten und Grenzen elementarer Modelle der Stoffbilanz von Seen). *Arch. Hydrobiol.* **66**, 1–36.

Vörösmarty, C.J., McIntyre, P.B., Gessner, M.O., Dudgeon, D., Prusevich, A., Green, P., Glidden, S., Bunn, S.E., Sullivan, C.A., Liermann, C.R., Davies, P.M., 2010. Global threats to human water security and river biodiversity. *Nature.* **467**, 555–561. https://doi.org/10.1038/nature09440

Vrede, T., Tranvik, L., 2006. Iron constraints on planktonic primary production in oligotrophic lakes. *Ecosyst. N. Y. Print.* **9**, 1094–1105. https://doi.org/10.1007/s10021-006-0167-1

Vuorio, K., Järvinen, M., Kotamäki, N., 2020. Phosphorus thresholds for bloom-forming cyanobacterial taxa in boreal lakes. *Hydrobiologia.* **847**, 4389–4400. https://doi.org/10.1007/s10750-019-04161-5

Waajen, G., van Oosterhout, F., Douglas, G., Lürling, M., 2016. Management of eutrophication in Lake De Kuil (The Netherlands) using combined flocculant – Lanthanum modified bentonite treatment. *Water Res.* **97**, 83–95. https://doi.org/10.1016/j.watres.2015.11.034

Wagner, C., Adrian, R., 2009. Cyanobacteria dominance: Quantifying the effects of climate change. *Limnol. Oceanogr.* **54**, 2460–2468.

Wagner, K., 2015. *Oxygenation and Circulation to Aid Water Supply Reservoir Management (No. 4222c)*. Water Research Foundation, Alexandria, VA.

Walsby, A.E., 1994. Gas vesicles. *Microb. Rev.* **58**, 94–144. https://doi.org/10.1128/mr.58.1.94-144.1994

Walsby, A.E., Schanz, F., Schmid, M., 2006. The Burgundy-blood phenomenon: A model of buoyancy change explains autumnal waterblooms by *Planktothrix rubescens* in Lake Zürich. *New Phytol.* **169**, 109–122. https://doi.org/10.1111/j.1469-8137.2005.01567.x

Wang, D., 2012. Redox chemistry of molybdenum in natural waters and its involvement in biological evolution. *Front. Microbiol.* **3**. https://doi.org/10.3389/fmicb.2012.00427

Wang, H., Li, Q., Xu, J., 2023a. Climate warming does not override eutrophication, but facilitates nutrient release from sediment and motivates eutrophic process. *Microorganisms.* **11**, 910. https://doi.org/10.3390/microorganisms11040910

Wang, J., Chen, J., Yu, P., Yang, X., Zhang, L., Geng, Z., He, K., 2020. Oxygenation and synchronous control of nitrogen and phosphorus release at the sediment-water interface using oxygen nano-bubble modified material. *Sci. Total Environ.* **725**, 138258. https://doi.org/10.1016/j.scitotenv.2020.138258

Wang, J., Chen, Q., Huang, S., Wang, Z., Li, D., 2023b. Cyanobacterial organic matter (COM) positive feedback aggravates lake eutrophication by changing the phosphorus release characteristics of sediments. *Sci. Total Environ.* **892**, 164540. https://doi.org/10.1016/j.scitotenv.2023.164540

Wang, J., Wang, S., Jin, X., Zhu, S., Wu, F., 2008. Ammonium release characteristics of the sediments from the shallow lakes in the middle and lower reaches of Yangtze River region, China. *Environ. Geol.* **55**, 37–45. https://doi.org/10.1007/s00254-007-0962-9

Wang, X., Qin, B., Gao, G., Paerl, H.W., 2010. Nutrient enrichment and selective predation by zooplankton promote Microcystis (Cyanobacteria) bloom formation. *J. Plankton. Res.* **32**, 457–470. https://doi.org/10.1093/plankt/fbp143

Ward, N.K., Steele, B.G., Weathers, K.C., Cottingham, K.L., Ewing, H.A., Hanson, P.C., Carey, C.C., 2020. Differential responses of maximum versus median chlorophyll-a to air temperature and nutrient loads in an oligotrophic lake over 31 years. *Water Resour. Res.* **56**, e2020WR027296. https://doi.org/10.1029/2020WR027296

Waters, S., Atalah, J., Thompson, L., Thomson-Laing, G., Pearman, J.K., Puddick, J., Howarth, J.D., Reyes, L., Vandergoes, M.J., Wood, S.A., 2023. It's all in the mud – the use of sediment geochemistry to estimate contemporary water quality in lakes. *Appl. Geochem.* **153**, 105667. https://doi.org/10.1016/j.apgeochem.2023.105667

Waters, S., Hamilton, D., Pan, G., Michener, S., Ogilvie, S., 2022. Oxygen nanobubbles for lake restoration – where are we at? A review of a new-generation approach to managing lake eutrophication. *Water*. **14**, 1989. https://doi.org/10.3390/w14131989

Waters, S., Verburg, P., Schallenberg, M., Kelly, D., 2021. Sedimentary phosphorus in contrasting, shallow New Zealand lakes and its effect on water quality. *N. Z. J. Mar. Freshw. Res.* **55**, 592–611. https://doi.org/10.1080/00288330.2020.1848884

Watson, S.B., McCauley, E., Downing, J.A., 1992. Sigmoid relationship between phosphorus, algal biomass, and algal community structure. *Can. J. Fish. Aquat. Sci.* **49**, 2605–2610. https://doi.org/10.1139/f92-288

Watson, S.B., McCauley, E., Downing, J.A., 1997. Patterns in phytoplankton taxonomic composition across temperate lakes of differing nutrient status. *Limnol. Ocean.* **42**, 487–495. https://doi.org/10.4319/lo.1997.42.3.0487

Watson, S.B., Whitton, B.A., Higgins, S.N., Paerl, H.W., Brooks, B.W., Wehr, J.D., 2015. Harmful algal blooms. In *Freshwater Algae of North America*. Academic Press, Cambridge, MA, pp. 873–920.

Wauer, G., Gonsiorczyk, T., Kretschmer, K., Casper, P., Koschel, R., 2005. Sediment treatment with a nitrate-storing compound to reduce phosphorus release. *Water Res.* **39**, 494–500. https://doi.org/10.1016/j.watres.2004.10.017

Wauer, G., Teien, H.-C., 2010. Risk of acute toxicity for fish during aluminium application to hardwater lakes. *Sci. Total Environ.* **408**, 4020–4025. https://doi.org/10.1016/j.scitotenv.2010.05.033

Weenink, E.F., Luimstra, V.M., Schuurmans, J.M., Van Herk, M.J., Visser, P.M., Matthijs, H.C.P., 2015. Combatting cyanobacteria with hydrogen peroxide: A laboratory study on the consequences for phytoplankton community and diversity. *Front. Microbiol.* **6**. https://doi.org/10.3389/fmicb.2015.00714

Wehr, J., Janse van Vuuren, S., 2024. Chapter 17: Algae and cyanobacteria communities. In *Wetzel's Limnology* (Eds. Jones, I.D., Smol, J.P.), 4th ed. Academic Press, San Diego, pp. 463–510. https://doi.org/10.1016/B978-0-12-822701-5.00017-3

Welch, E.B., Jacoby, J.M., 2009. Phosphorus reduction by dilution and shift in fish species in Moses Lake, WA. *Lake Reserv. Manage.* **25**, 276–283. https://doi.org/10.1080/07438140903083906

Wells, M.G., Troy, C.D., 2022. Surface mixed layers in Lakes. In *Encyclopedia of Inland Waters*. Elsevier, pp. 546–561. https://doi.org/10.1016/B978-0-12-819166-8.00126-2

Wetzel, R.G., 1975. *Limnology*. W B Saunders, Toronto, Canada.

Wetzel, R.G., 2001. *Limnology – Lake and River Ecosystems*, 3rd ed. Academic Press, New York.

WHO – World Health Organization, 1998. *Guidelines for Drinking-Water Quality: Second Edition, Addendum to Volume 2, Health Criteria and Other Supporting Information*. World Health Organization, Geneva.

WHO – World Health Organization, 2020. *Cyanobacterial Toxins: Background Document for Development of WHO Guidelines for Drinking-Water Quality and Guidelines for Safe Recreational Water Environments*. https://www.who.int/water_sanitation_health/water-quality/guidelines/chemicals/en/

WHO – World Health Organization, 2021. *Guidelines on Recreational Water Quality. Volume 1, Coastal and Fresh Waters*, Licence: CC BY-NC-SA 3.0 IGO. ed. World Health Organization, Geneva, Switzerland.

WHO – World Health Organization, 2022. *Guidelines for Drinking-Water Quality*, 4th ed. incorporating the first and second addenda. ed. World Health Organization, Geneva.

Wilhelm, S., 1995. Ecology of iron-limited cyanobacteria: A review of physiological responses and implications for aquatic systems. *Aquat. Microb. Ecol.* **9**, 295–303. https://doi.org/10.3354/ame009295

Wilhelm, S., Adrian, R., 2008. Impact of summer warming on the thermal characteristics of a polymictic lake and consequences for oxygen, nutrients and phytoplankton. *Freshw. Biol.* **53**, 226–237. https://doi.org/10.1111/j.1365-2427.2007.01887.x

Wilkinson, G.M., Walter, J.A., Buelo, C.D., Pace, M.L., 2022. No evidence of widespread algal bloom intensification in hundreds of lakes. *Front. Ecol. Environ.* **20**, 16–21. https://doi.org/10.1002/fee.2421

Wilkinson, G.M., Walter, J.A., Fleck, R., Pace, M.L., 2020. Beyond the trends: The need to understand multiannual dynamics in aquatic ecosystems. *Limnol. Oceanogr. Lett.* **5**, 281–286. https://doi.org/10.1002/lol2.10153

Willis, A., Woodhouse, J.N., 2020. Defining cyanobacterial species: Diversity and description through genomics. *Crit. Rev. Plant Sci.* **39**, 101–124. https://doi.org/10.1080/07352689.2020.1763541

Winslow, L.A., Read, J.S., Hansen, G.J.A., Hanson, P.C., 2015. Small lakes show muted climate change signal in deepwater temperatures. *Geophys. Res. Lett.* **42**, 355–361. https://doi.org/10.1002/2014GL062325

Winslow, L.A., Read, J.S., Hansen, G.J.A., Rose, K.C., Robertson, D.M., 2017. Seasonality of change: Summer warming rates do not fully represent effects of climate change on lake temperatures. *Limnol. Oceanogr.* **62**, 2168–2178. https://doi.org/10.1002/lno.10557

Winter, J.G., DeSellas, A.M., Fletcher, R., Heintsch, L., Morley, A., Nakamoto, L., Utsumi, K., 2011. Algal blooms in Ontario, Canada: Increases in reports since 1994. *Lake Reserv. Manage.* **27**, 107–114. https://doi.org/10.1080/07438141.2011.557765

WMO, 2023. *WMO [World Meteorological Organization] Provisional State of the Global Climate 2023.* World Meteorological Organization, Geneva, Switzerland.

Wood, S.A., Borges, H., Puddick, J., Biessy, L., Atalah, J., Hawes, I., Dietrich, D.R., Hamilton, D.P., 2017. Contrasting cyanobacterial communities and microcystin concentrations in summers with extreme weather events: Insights into potential effects of climate change. *Hydrobiologia.* **785**, 71–89. https://doi.org/10.1007/s10750-016-2904-6

Woolway, R.I., Denfeld, B., Tan, Z., Jansen, J., Weyhenmeyer, G.A., Fuente, S.L., 2022a. Winter inverse lake stratification under historic and future climate change. *Limnol. Oceanogr. Lett.* **7**, 302–311. https://doi.org/10.1002/lol2.10231

Woolway, R.I., Jennings, E., Shatwell, T., Golub, M., Pierson, D.C., Maberly, S.C., 2021a. Lake heatwaves under climate change. *Nature.* **589**, 402–407. https://doi.org/10.1038/s41586-020-03119-1

Woolway, R.I., Merchant, C.J., 2019. Worldwide alteration of lake mixing regimes in response to climate change. *Nat. Geosci.* **12**, 271–276. https://doi.org/10.1038/s41561-019-0322-x

Woolway, R.I., Merchant, C.J., Van Den Hoek, J., Azorin-Molina, C., Nõges, P., Laas, A., Mackay, E.B., Jones, I.D., 2019. Northern Hemisphere atmospheric stilling accelerates lake thermal responses to a warming world. *Geophys. Res. Lett.* **46**, 11983–11992. https://doi.org/10.1029/2019GL082752

Woolway, R.I., Sharma, S., Smol, J.P., 2022b. Lakes in hot water: The impacts of a changing climate on aquatic ecosystems. *BioScience.* **72**, 1050–1061. https://doi.org/10.1093/biosci/biac052

Woolway, R.I., Sharma, S., Weyhenmeyer, G.A., Debolskiy, A., Golub, M., Mercado-Bettín, D., Perroud, M., Stepanenko, V., Tan, Z., Grant, L., Ladwig, R., Mesman, J., Moore, T.N., Shatwell, T., Vanderkelen, I., Austin, J.A., DeGasperi, C.L., Dokulil, M., La Fuente, S., Mackay, E.B., Schladow, S.G., Watanabe, S., Marcé, R., Pierson, D.C., Thiery, W., Jennings, E., 2021b. Phenological shifts in lake stratification under climate change. *Nat. Commun.* **12**, 2318. https://doi.org/10.1038/s41467-021-22657-4

Wu, Z., Liu, Y., Liang, Z., Wu, S., Guo, H., 2017. Internal cycling, not external loading, decides the nutrient limitation in eutrophic lake: A dynamic model with temporal Bayesian hierarchical inference. *Water Res.* **116**, 231–240. https://doi.org/10.1016/j.watres.2017.03.039

Wunderling, N., von der Heydt, A., Aksenov, Y., Barker, S., Bastiaansen, R., Brovkin, V., Brunetti, M., Couplet, V., Kleinen, T., Lear, C.H., Lohmann, J., Roman-Cuesta, R.M., Sinet, S., Swingedouw, D., Winkelmann, R., Anand, P., Barichivich, J., Bathiany, S., Baudena, M., Bruun, J.T., Chiessi, C.M., Coxall, H.K., Docquier, D., Donges, J.F., Falkena, S.K.J., Klose, A.K., Obura, D., Rocha, J., Rynders, S., Steinert, N.J., Willeit, M., 2023. Climate tipping point interactions and cascades: A review. *EGUsphere* 1–45. https://doi.org/10.5194/egusphere-2023-1576

Wurtsbaugh, W.A., Paerl, H.W., Dodds, W.K., 2019. Nutrients, eutrophication and harmful algal blooms along the freshwater to marine continuum. *WIREsWiley Interdiscip. Rev. Water.* **7**, e1373. https://doi.org/10.1002/wat2.1373

Xiao, M., Willis, A., Burford, M.A., 2017. Differences in cyanobacterial strain responses to light and temperature reflect species plasticity. *Harmful Algae*. **62**, 84–93. https://doi.org/10.1016/j.hal.2016.12.008

Xie, L., Su, X., Xu, H., 2020. Chapter 11: Phosphorus dynamics and its relationship with cyanobacterial blooms in Lake Taihu, China. In *Internal Phosphorus Loading: Causes, Case Studies, and Management* (Eds. Steinman, A.D., Spears, B.M.). J. Ross Publishing, Plantation, FL, pp. 211–221.

Xie, L., Xie, P., Li, S., Tang, H., Liu, H., 2003. The low TN:TP ratio, a cause or a result of Microcystis blooms? *Water Res*. **37**, 2073–2080. https://doi.org/10.1016/S0043-1354(02)00532-8

Xu, H., McCarthy, M.J., Paerl, H.W., Brookes, J.D., Zhu, G., Hall, N.S., Qin, B., Zhang, Y., Zhu, M., Hampel, J.J., Newell, S.E., Gardner, W.S., 2021. Contributions of external nutrient loading and internal cycling to cyanobacterial bloom dynamics in Lake Taihu, China: Implications for nutrient management. *Limnol. Oceanogr*. **66**, 1492–1509. https://doi.org/10.1002/lno.11700

Xu, Z., Gao, G., Tu, B., Qiao, H., Ge, H., Wu, H., 2019. Physiological response of the toxic and non-toxic strains of a bloom-forming cyanobacterium Microcystis aeruginosa to changing ultraviolet radiation regimes. *Hydrobiologia*. **833**, 143–156. https://doi.org/10.1007/s10750-019-3896-9

Yan, M., Wang, D., Qu, J., He, W., Chow, C.W.K., 2007. Relative importance of hydrolyzed Al(III) species (Ala, Alb, and Alc) during coagulation with polyaluminum chloride: A case study with the typical micro-polluted source waters. *J. Colloid. Interface Sci*. **316**, 482–489. https://doi.org/10.1016/j.jcis.2007.08.036

Yan, N.D., Paterson, A.M., Somers, K.M., Scheider, W.A., 2008. An introduction to the Dorset special issue: Transforming understanding of factors that regulate aquatic ecosystems on the southern Canadian Shield. *Can. J. Fish. Aquat. Sci*. **65**, 781–785. https://doi.org/10.1139/f08-077

Yankova, Y., Neuenschwander, S., Köster, O., Posch, T., 2017. Abrupt stop of deep water turnover with lake warming: Drastic consequences for algal primary producers. *Sci. Rep.* **7**. https://doi.org/www.nature.com/scientificreports/

Yao, F., Livneh, B., Rajagopalan, B., Wang, J., Crétaux, J.-F., Wada, Y., Berge-Nguyen, M., 2023. Satellites reveal widespread decline in global lake water storage. *Science*. **380**, 743–749. https://doi.org/10.1126/science.abo2812

Yao, Y., Wang, P., Wang, C., Hou, J., Miao, L., Yuan, Y., Wang, T., Liu, C., 2016. Assessment of mobilization of labile phosphorus and iron across sediment-water interface in a shallow lake (Hongze) based on in situ high-resolution measurement. *Environ. Pollut*. **219**, 873–882. https://doi.org/10.1016/j.envpol.2016.08.054

Yasarer, L.M.W., Sturm, B.S.M., 2016. Potential impacts of climate change on reservoir services and management approaches. *Lake Reserv. Manag*. **32**, 13–26. https://doi.org/10.1080/10402381.2015.1107665

Yasseri, S., Epe, T.S., 2016. Analysis of the La:P ratio in lake sediments – vertical and spatial distribution assessed by a multiple-core survey. *Water Res*. **97**, 96–100. https://doi.org/10.1016/j.watres.2015.07.037

Yin, H., Douglas, G.B., Cai, Y., Liu, C., Copetti, D., 2018. Remediation of internal phosphorus loads with modified clays, influence of fluvial suspended particulate matter and response of the benthic macroinvertebrate community. *Sci. Total Environ*. **610–611**, 101–110. https://doi.org/10.1016/j.scitotenv.2017.07.243

Yin, H., Yin, P., Yang, Z., 2023. Seasonal sediment phosphorus release across sediment-water interface and its potential role in supporting algal blooms in a large shallow eutrophic Lake (Lake Taihu, China). *Sci. Total Environ*. **896**, 165252. https://doi.org/10.1016/j.scitotenv.2023.165252

Yindong, T., Xiwen, X., Miao, Q., Jingjing, S., Yiyan, Z., Wei, Z., Mengzhu, W., Xuejun, W., Yang, Z., 2021. Lake warming intensifies the seasonal pattern of internal nutrient cycling in the eutrophic lake and potential impacts on algal blooms. *Water Res*. **188**, 116570. https://doi.org/10.1016/j.watres.2020.116570

Yuan, L.L., Pollard, A.I., Pather, S., Oliver, J.L., D'Anglada, L., 2014. Managing microcystin: Identifying national-scale thresholds for total nitrogen and chlorophyll *a*. *Freshw. Biol*. **59**, 1970–1981. https://doi.org/10.1111/fwb.12400

Zamparas, M.G., Gianni, A., Stathi, P., Deligiannakis, Y., Zacharias, I., 2012. Removal of phosphate from natural waters using innovative modified bentonites. *Appl. Clay Sci*. **62**, 101–106. https://doi.org/10.1016/j.clay.2012.04.020

Zamparas, M.G., Kyriakopoulos, G.L. (Eds.), 2021. *Chemical Lake Restoration: Technologies, Innovations and Economic Perspectives*. Springer International Publishing, Cham. https://doi.org/10.1007/978-3-030-76380-0

Zamparas, M.G., Zacharias, I., 2014. Restoration of eutrophic freshwater by managing internal nutrient loads. A review. *Sci. Total Environ.* **496**, 551–562. https://doi.org/10.1016/j.scitotenv.2014.07.076

Zastepa, A., Comte, J., Crevecoeur, S., 2023a. Prevalence and ecological features of deep chlorophyll layers in Lake of the Woods, a complex hydrological system with strong trophic, physical, and chemical gradients. *J. Gt. Lakes Res.* **49**, 122–133. https://doi.org/10.1016/j.jglr.2022.09.007

Zastepa, A., Westrick, J.A., Liang, A., Birbeck, J.A., Furr, E., Watson, L.C., Stockdill, J.L., Ramakrishna, B.S., Crevecoeur, S., 2023b. Broad screening of toxic and bioactive metabolites in cyanobacterial and harmful algal blooms in Lake of the Woods (Canada and USA), 2016–2019. *J. Gt. Lakes Res.* **49**, 134–146. https://doi.org/10.1016/j.jglr.2022.12.006

Zhan, Q., Stratmann, C.N., van der Geest, H.G., Veraart, A.J., Brenzinger, K., Lürling, M., de Senerpont Domis, L.N., 2021. Effectiveness of phosphorus control under extreme heatwaves: Implications for sediment nutrient releases and greenhouse gas emissions. *Biogeochemistry*. **156**, 421–436. https://doi.org/10.1007/s10533-021-00854-z

Zhan, Q., Teurlincx, S., van Herpen, F., Raman, N.V., Lürling, M., Waajen, G., de Senerpont Domis, L.N., 2022. Towards climate-robust water quality management: Testing the efficacy of different eutrophication control measures during a heatwave in an urban canal. *Sci. Total Environ.* **828**, 154421. https://doi.org/10.1016/j.scitotenv.2022.154421

Zhan, Y., Wu, X., Lin, J., Zhang, Z., Zhao, Y., Yu, Y., Wang, Y., 2019. Combined use of calcium nitrate addition and anion exchange resin capping to control sedimentary phosphorus release and its nitrate-nitrogen releasing risk. *Sci. Total Environ.* **689**, 203–214. https://doi.org/10.1016/j.scitotenv.2019.06.406

Zhang, M., Yang, Z., Yu, Y., Shi, X., 2020. Interannual and seasonal shift between Microcystis and Dolichospermum: A 7-year investigation in Lake Chaohu, China. *Water*. **12**, 1978. https://doi.org/10.3390/w12071978

Zhang, T., He, J., Luo, X., 2017. Effect of Fe and EDTA on freshwater cyanobacteria bloom formation. *Water*. **9**, 326. https://doi.org/10.3390/w9050326

Zhang, Z., Wang, Z., Xie, Q., Wu, D., 2024. Inactivation of phosphorus in a highly eutrophic pond using Zeofixer® to eliminate the free-floating aquatic plant (Spirodela polyrhiza). *Ecol. Eng.* **199**, 107171. https://doi.org/10.1016/j.ecoleng.2023.107171

Zhao, K., Wang, L., You, Q., Zhang, J., Pang, W., Wang, Q., 2022. Impact of cyanobacterial bloom intensity on plankton ecosystem functioning measured by eukaryotic phytoplankton and zooplankton indicators. *Ecol. Indic.* **140**, 109028. https://doi.org/10.1016/j.ecolind.2022.109028

Zhao, L., Zhu, R., Zhou, Q., Jeppesen, E., Yang, K., 2023. Trophic status and lake depth play important roles in determining the nutrient-chlorophyll a relationship: Evidence from thousands of lakes globally. *Water Res.* 120182. https://doi.org/10.1016/j.watres.2023.120182

Zhao, S., Hermans, M., Niemistö, J., Jilbert, T., 2024. Elevated internal phosphorus loading from shallow areas of eutrophic boreal lakes: Insights from porewater geochemistry. *Sci. Total Environ.* **907**, 167950. https://doi.org/10.1016/j.scitotenv.2023.167950

Zhong, Y., Notaro, M., Vavrus, S.J., 2018. Spatially variable warming of the Laurentian Great Lakes: An interaction of bathymetry and climate. *Clim. Dyn.* https://doi.org/10.1007/s00382-018-4481-z

Zhong, Y., Notaro, M., Vavrus, S.J., Foster, M.J., 2016. Recent accelerated warming of the Laurentian Great Lakes: Physical drivers. *Limnol. Oceanogr.* **61**, 1762–1786. https://doi.org/10.1002/lno.10331

Zhou, Y., Michalak, A.M., Beletsky, D., Rao, Y.R., Richards, R.P., 2015. Record-breaking Lake Erie hypoxia during 2012 drought. *Environ. Sci. Technol.* **49**, 800–807. https://doi.org/10.1021/es503981n

Zia, A., Schroth, A.W., Hecht, J.S., Isles, P., Clemins, P.J., Turnbull, S., Bitterman, P., Tsai, Y., Mohammed, I.N., Bucini, G., Doran, E.M.B., Koliba, C., Bomblies, A., Beckage, B., Winter, J., Adair, E.C., Rizzo, D.M., Gibson, W., Pinder, G., 2022. Climate change-legacy phosphorus synergy hinders lake response to aggressive water policy targets. *Earths Future*. **10**, e2021EF002234. https://doi.org/10.1029/2021EF002234

Zinnert, J.C., Nippert, J.B., Rudgers, J.A., Pennings, S.C., González, G., Alber, M., Baer, S.G., Blair, J.M., Burd, A., Collins, S.L., Craft, C., Di Iorio, D., Dodds, W.K., Groffman, P.M., Herbert, E., Hladik, C., Li, F., Litvak, M.E., Newsome, S., O'Donnell, J., Pockman, W.T., Schalles, J., Young, D.R., 2021. State changes: Insights from the U.S. long term ecological research network. *Ecosphere*. **12**. https://doi.org/10.1002/ecs2.3433

Zou, W., Zhu, G., Cai, Y., Xu, H., Zhu, M., Gong, Z., Zhang, Y., Qin, B., 2020. Quantifying the dependence of cyanobacterial growth to nutrient for the eutrophication management of temperate-subtropical shallow lakes. *Water Res.* **177**, 115806. https://doi.org/10.1016/j.watres.2020.115806

Zurawell, R.W., Chen, H., Burke, J.M., Prepas, E.E., 2005. Hepatotoxic cyanobacteria: A review of the biological importance of microcystins in freshwater environments. *J. Toxicol. Environ. Health Part B.* **8**, 1–37. https://doi.org/10.1080/10937400590889412

Index